T0122897

Learn Git in a Month of Lunches

Learn Git in a Month of Lunches

RICK UMALI

MANNING
SHELTER ISLAND

For online information and ordering of this and other Manning books, please visit www.manning.com. The publisher offers discounts on this book when ordered in quantity. For more information, please contact

Special Sales Department
Manning Publications Co.
20 Baldwin Road
PO Box 761
Shelter Island, NY 11964
Email: orders@manning.com

♾ Recognizing the importance of preserving what has been written, it is Manning's policy to have the books we publish printed on acid-free paper, and we exert our best efforts to that end. Recognizing also our responsibility to conserve the resources of our planet, Manning books are printed on paper that is at least 15 percent recycled and processed without elemental chlorine.

Manning Publications Co.
20 Baldwin Road
PO Box 761
Shelter Island, NY 11964

Development editor: Helen Sturgis
Technical development editor: Jonathan Thoms
Copyeditor: Sharon Wilkey
Proofreader: Corbin Collins
Technical proofreader: Karsten Strøbaek
Typesetter: Marija Tudor
Cover designer: Leslie Haimes

ISBN: 9781617292415
Printed in the United States of America

brief contents

contents

preface

A few years ago, while watching an instructional video about how to play the guitar, I heard a great expression. The instructor was demonstrating a complicated strumming pattern. At the correct speed, it looked incredibly fast. He said he'd try to slow it down so we could see what was happening but acknowledged that it would be hard to do so. He said it would be like trying to fall down slowly.

I liked that: *fall down slowly.*

When Git is demonstrated, it can seem incredibly fast. The fact that it is often demonstrated on the command line adds even more mystery to what exactly is happening with all those commands.

This book is my attempt to slow things down so that you can see and think about every single step that is happening when you interact with Git. I am taking this approach because, when I have presented Git at local user groups, people want to know what each command is doing. In a presentation, it's hard to cater to each question, but in a book, there is room to explore the details.

Another guitarist, this time on YouTube, gave the sage advice that before you play musical pieces at their correct tempo, you should learn them much more slowly. When you take things slowly, your fingers learn how to properly move, and only after you build confidence can you play a piece faster and faster.

I liked that as well: *learn things slowly.*

The tutorial in the Git documentation covers details in short paragraphs, but this book will take the opposite approach. We'll devote a whole chapter to what the tutorial covers in a single sentence. As you slowly build confidence with one command, you will find yourself using it faster the next time.

I have spent much of my career in customer-facing roles such as technical support, onsite consulting, and training. I have learned from speaking with people that "slowing down" goes a long way toward making people comfortable with the nontrivial, technical details. This book is a result of that approach.

about this book

This book is aimed at coding professionals who are beginners to either source control or Git. Anyone who types code into files (whether it is a computer program, a CSS file, or an HTML file) can benefit from learning Git to keep track of their work. This book covers beginner to intermediate level topics.

Using version control is one of the characteristics of a professional developer, but this book does not assume you know another version control system. In fact, this book will explain Git entirely in Git's terminology, so I won't mention any other version control system.

I hope you are comfortable on the command line, but do not worry if you are not. I will go over each command, and over time you will become more comfortable with this skill. In addition, the book will show various GUIs that make Git easier to use.

If you are a complete beginner, you should read the chapters in order. Each chapter is designed to be read during your lunch hour. In that hour, you should be able to read the text and go through each of the "Try it now" exercises for a particular chapter. The exercises at the end of each chapter will help reinforce the points of the chapter, but they can be done at your leisure.

Making time for the "Try it now" exercises is key. Performing these exercises will help you understand Git better. In each chapter, you will work on a repository that is completely your own. Each chapter creates an environment or a situation that is safe to experiment in. In certain chapters, you can re-create the repository to a known working state using code that is available from the book's website (http://www.manning.com/umali).

Each chapter is full of illustrations and screen shots. The diagrams will give you a mental model for how Git organizes your code and its history. Take time to think about these drawings, and how they apply to the exercises. Finally, the end of each chapter will list the Git commands covered in that chapter, where appropriate.

Chapters are grouped into basic topics, intermediate topics, advanced topics, branching and merging, collaborating, and the Git ecosystem. The first chapter provides a thorough outline of these groupings. Some chapters build on earlier chapters, so keep that in mind if you are past the beginner level and you decide you want to skip around.

Basic topics, covered in chapters 1 through 6, include setting up a repository, adding and committing files to the repository, and inquiring about the status and history of this repository. If you're a solo developer, these chapters will probably be sufficient for 80% of your needs.

Intermediate topics (chapters 7 and 8) focus on the Git staging area and accessing different parts of your repository. Advanced topics (chapters 15 and 16) reveal interesting ways to query your repository history, as well as how to manipulate your history.

Branching and merging, addressed in chapters 9 and 10, will break down how to use one of Git's foremost features: fast branching. After reading these chapters, you won't think twice about creating a branch to experiment with something in your code base. The collaborating chapters (chapters 11 through 14) will explain how to collaborate correctly using Git's collaboration commands. If you've ever been confused about pushing and pulling, these four chapters will clear things up!

There are three chapters (chapters 17, 18, and 19) on Git's ecosystem: third-party tools, user interfaces, and GitHub. Git has gained more users thanks to user interfaces and tools that make Git easier, but it is GitHub, the Git repository hosting company, that helped increase its adoption by the open-source community at large.

Finally, there is a chapter (chapter 20) on customizing Git for your environment. It appears last because I believe that learning a new system in its default or stock settings is the best way to get started. That said, skimming this chapter early in your reading may be helpful.

Source code downloads

As mentioned before, the publisher's website for the book contains code that will help you re-create a working environment for the text, as well as files and scripts necessary for the various exercises. This code (in the form of scripts) is available for download as a zip file from www.manning.com/books/learn-git-in-a-month-of-lunches.

Author Online

Support for this book will come from Author Online, a web-based forum at https://forums.manning.com/forums/learn-git-in-a-month-of-lunches. A link to the forum is also available from the publisher's website at www.manning.com/books/learn-git-in-a-month-of-lunches.

Readers are encouraged to post questions and feedback there. I plan to monitor the forum and to chime in as necessary.

About the author

Rick Umali is a senior-level technology professional who lives and works in greater Boston, Massachusetts.

He has worked for high-tech companies his entire career and enjoys speaking to beginner audiences about technology.

His experience ranges from enterprise software (search, e-commerce) to web development. He has spent time in customer-facing roles (training, support, and consulting) and programming in a range of languages (Java, PHP, Ruby).

acknowledgments

I had a lot of help throughout the writing and publishing process. First and foremost, I want to thank Helen Stergius at Manning, who gave me a lot of support during the development of the manuscript. She was always positive and cheerful, and she helped me get the book to the finish line. Other editors who helped along the way include Supriya Savkoor, Susie Pitzken, and Sean Dennis.

Jonathan Thoms and Karsten Strøbaek provided a detailed technical reading and proofing during the manuscript phase. They can take the credit when the code and examples work well.

Kevin Sullivan and his team took my drawings and made them much clearer. Sharon Wilkey copyedited the manuscript and made the words gleam. Mary Piergies, with Janet Vail, oversaw the production of the book through layout and printing.

Aleksandar Dragosavljevic managed the peer reviews for this book. Each reviewer weighed in on issues big and small, and in doing so, helped improve the book. Their names: Art Bergquist, Boris Vasile, Changgeng Li, Ernesto Cardenas Cangahuala, Harinath Mallepally, Kathleen Estrada, Keith Webster, Luciano Favaro, Michel Graciano, Miguel Biraud, Mohsen Mostafa Jokar, Nacho Ormeño, Nitin Gode, Patrick Dennis, Ralph Leibmann, Richard Butler, Scott King, Stuart Ellis, and Travis Nelson.

Ozren Harlovic, review editor, was the first person at Manning who noticed my interest in writing a book. Michael Stephens, Manning's associate publisher, got me to sign on the dotted line.

The initial idea for writing the book came from my involvement with the Boston PHP Meetup, the largest education-focused Meetup in New England. Organizers Matt Murphy, Bobby Cahill, and Gene Babon were a source of great encouragement.

My neighbor, Peter Loshin, provided early inspiration. Emails from Mike McQuaid, Ben Melançon, and Larry Ullman kept me going.

My wife Jenn and my daughter Mia were always good-natured about my book writing. I am grateful for their patience and love.

1 Before you begin

You may have heard of Git, the wondrous new software that puts the fun back into the laborious work of version control. You might even have browsed the many software offerings on GitHub, the popular social coding website. (You might even be confused about Git and GitHub!) Maybe you're using one of those other version control systems, which are now considered old-school. Maybe you've been working without version control (gasp!), because you think it's only for programmers (it's not). Maybe you've become curious about how to contribute to open source software, but Git has always been a roadblock. However you got to this book, I'm glad you're here exploring Git!

As more and more corporate IT shops begin to embrace open source software, more and more IT developers and administrators will encounter Git. Git has become the de facto source-code control system for open source developers. Tinkering and modifying open source software to suit your needs is one of the benefits of open source, but you'll want to use the safety net of source-code control, and Git is that safety net.

As you use Git, you'll see that it encourages an attitude of being careful about changes. *Commit often* is a mantra you'll hear often in the Git community (as well as *continuous integration* and *continuous deployment camps*), and for good reason. Version control is the most important thing you can practice as a developer, and Git makes it easy.

1.1 What makes Git so special?

Git is a distributed version control system (DVCS). This means that you don't need to run a Git server to get all its benefits. You don't even need a network to run Git's commands.

Figure 1.1 The "single point of contention" that developers have to deal with: the version control system that houses the repository

Earlier version control systems put code in a castle, but developers have to be given access to read and write to a *repository*, which is where the code is stored. This repository, also called a *repo*, often exists on some other machine, which requires a network. As developers are added, the increased load to this one server decreases reliability. Certain version control commands such as branching and tagging become specialized and often require special access. Finally, the version control server must be up and running in order for developers to do work. This uptime requirement requires oversight and adds to the cost of running a version control system. All this makes the server a single point of failure, as shown in figure 1.1.

Git inverts this by giving each developer a version control repository, as shown in figure 1.2. Each repository runs entirely on the developer's local machine. Each developer can access any part of a project's history, compare versions, make branches, and perform any other operation that would normally require special permissions or network access with a server. This liberating scenario is an idea that takes some getting used to. Instead of making requests to the specialists who run your version control system, you can perform any and all of these specialized tasks—but this means that you have to learn more about these tasks.

Figure 1.2 Distributed version control systems are liberating because each developer has a copy of the entire repository.

This idea of giving every developer a repository makes Git a *distributed* version control system. Being distributed allows every developer to have the same capabilities as everyone else. Large open source projects like Drupal and Linux have thousands of developers in many locations, some with sparse Internet connectivity. With Git, all of these developers can make their contributions with the same ease as the project leader.

This may sound like a free-for-all but, because Git doesn't require a central repository, many projects have self-organized in ideal ways. Some projects have small development teams with a single project leader who can manage all the commits that might be made to a repository, but many projects have multiple people that help with commits.

1.2 Is this book for you?

Version control is akin to basic hygiene, and everyone needs hygiene. If you produce or modify files on a computer, this book might be for you. This book is geared toward people who are coding professionals: software engineers, developers, programmers, web developers, system administrators, and quality assurance and testing people.

Git is available on the three major platforms: Windows, Mac, and Unix/Linux. It was born and bred on the command line, so you'll have a lot of command-line work. Ideally, you're someone who embraces this concept, or at least is open to putting down the mouse and typing in a lot of commands.

If you have directories or files that look like the following listing, you're a candidate for this book!

Listing 1.1 Are you a candidate for learning Git with this book?

```
C:\buildtools>dir
 Volume in drive C is GNU
 Volume Serial Number is 5101-E64D

 Directory of C:\buildtools

03/15/2014  08:22 PM    <DIR>          .
03/15/2014  08:22 PM    <DIR>          ..
03/01/2014  08:22 AM            11,843 filefixup-01.bat
03/03/2014  08:52 AM            11,943 filefixup-02.bat
03/08/2014  11:22 AM            12,050 filefixup-03.bat
03/10/2014  02:22 PM            12,352 filefixup-04.bat
03/15/2014  03:21 PM            11,878 filefixup.bat
               5 File(s)          60,066 bytes
               2 Dir(s)  467,705,196,544 bytes free
```

Finally, this book is geared to Git beginners. The book demonstrates Git tasks step-by-step in tutorial fashion. Each chapter has a set of tasks to try, and they start out slowly but eventually build up speed.

1.3 How to use this book

This book is designed to be read one chapter each day. Each chapter should take only 40 minutes to read, and if you take an hour for lunch, you'll have 20 minutes left to do the practice assignments contained in each chapter. If you can type and eat lunch at the same time (I've seen people who do this), you might be able to squeeze in a chapter in a half hour!

MAIN CHAPTERS

Chapters 2 through 20 contain the main content. This gives you roughly one month (four five-day business weeks) of lunch learning. You don't need to rush through this content. It's helpful to let the content marinate in your brain before you start the next day's chapter. Each chapter also has several TRY IT NOW sections.

HANDS-ON LABS

Most of the main-content chapters include a short lab for you to complete. You'll need to install Git on your local machine to serve as your lab environment. The labs consist of tasks that enable you to practice your new knowledge. In some tasks, you'll repeat the commands you encountered in the chapter. In other tasks, you'll experiment with commands you just learned. Some of the labs require you to dig around and find answers.

All of the tasks and answers are rooted in the material of the current chapter or previous chapters. Sample answers are available on the book's website, but persist before giving in. That's the best way to learn!

FURTHER EXPLORATION

Git is deep. I can't plumb all the depths in this one book, but I can point you to some resources and say, "Go in that direction." One of the goals of this book is to teach you how to teach yourself about Git.

ABOVE AND BEYOND

Figuring out how Git implements its commands has been a source of great learning for me. The Above and Beyond sections share some of that learning, but they're not necessary for the labs or for the subject matter discussed in the chapter. If you're the type of person who likes to know why something works the way it does, these sections can provide some additional insight, but feel free to skip these or bookmark them for a later time.

1.4 Installing Git

This book will teach you Git by having you run Git commands in its native command-line environment. Installing Git is as straightforward as installing any other software package on your platform. I'd argue that it's easier because there's no server to start up. Git installs a directory of commands and documentation, and then places the `git` command in your `PATH`. (For Windows, it also installs a command line that you must use.) Keep repeating to yourself that there's no server installation at all, and the only thing to watch out for is your `PATH`.

Remember, this installation will be used for your labs, so be sure to complete this step.

UNIX/LINUX

On Unix/Linux, use the package manager of your distribution to install both the Git and Git GUI packages. This is covered in plenty of places on the web, but the definitive guide is http://git-scm.com/download/linux. You'll type into the command-line window.

MAC

To keep things simple, install the DMG package at http://git-scm.com/download/mac. This installation includes the Git GUI application as well as the Git commands. You'll type commands into the Mac Terminal client (a.k.a. the command-line window).

WINDOWS

In Windows, install Git from http://msysgit.github.io/. This package contains the Git GUI application as well as the Git commands. You'll type commands into the Git BASH window (a.k.a. a command-line window).

Git requires a specific command-line environment, so you can't run Git commands in Windows Command (CMD) or Windows PowerShell. The installation provides the correct command-line environment for Git (namely, Git BASH).

1.5 Your learning path

The first three chapters introduce version control concepts and how to become oriented to your Git installation.

The basics are covered in chapters 4 through 6. The Git commands covered in these chapters are arguably the ones you'll be using almost half of the time. Any developer who works with a version control system needs to know how to add files to a repository and how to inspect its history. You'll also learn how to make a Git repository. These are the basics, and if you're using Git on your own, these are probably the only chapters you need to learn.

Chapters 7 and 8 present intermediate commands. You'll learn about the Git staging area. You'll also learn other ways to inspect a repository's history, beyond the basics. These techniques become important if you work as a contributor to repositories that you don't own or create.

Chapters 9 and 10 cover branching and merging. Branching is a key feature of Git that requires you to merge more often.

Chapters 11 through 14 discuss how to collaborate with others: how to get changes from other people and how to submit your own changes.

The rest of the book touches on advanced topics (such as the `git rebase` command), the Git ecosystem (a few third-party tools are examined), and the `git config` command.

Table 1.1 outlines this learning path.

Table 1.1 Your learning path

Chapter 1: Before you begin Chapter 2: An overview of Git and version control Chapter 3: Getting oriented with Git	An Introduction to Key Concepts If you're unfamiliar with version control in general, this is where to start. You'll start with key version control concepts and then take a whirlwind tour of Git.
Chapter 4: Making and using a Git repository Chapter 5: Using Git with a GUI Chapter 6: Tracking and updating files with Git	The Basics This covers the Git commands you'll end up using 70% of the time: `init`, `add`, `commit`, `log`, `status`, `diff`.
Chapter 7: Committing parts of changes Chapter 8: The time machine that is Git	Intermediate Topics You'll study the staging area (manipulated by `git add`) and learn how to manipulate the Git time machine with `git checkout`.
Chapter 9: Taking a fork in the road Chapter 10: Merging branches	Branching and Merging This section covers `git branch` and `merge`. You'll learn how to use Git's ability to perform lightning-fast branching and then how to merge your branches back to your mainline.
Chapter 11: Cloning Chapter 12: Collaborating with remotes Chapter 13: Pushing your changes Chapter 14: Keeping in sync	Collaborating This four-chapter block covers the basics of collaboration using `git clone`, `remote`, `push`, `pull`, and `fetch`.
Chapter 15: Software archaeology Chapter 16: Understanding git rebase	Advanced Topics You'll explore the ability to deeply probe your history (via `git log`) and change your history (via `git rebase`).
Chapter 17: Workflows and branching conventions Chapter 18: Working with GitHub Chapter 19: Third-party tools and Git	Git Ecosystem Git's flexible architecture has spawned a wide ecosystem of workflows (git-flow and GitHub Flow), hosting (we'll focus on GitHub), and third-party tools (we'll focus on the Atlassian SourceTree and Eclipse GUIs for Git).
Chapter 20: Sharpening your Git	Staying Sharp This chapter covers `git config` and how to keep your Git skills sharp.

1.6 Online resources

The book's website (www.manning.com/umali) is a place I hope you'll visit as you progress through this book. Several supplementary resources for this book are available on that site, including these:

- Companion videos for some chapters
- Answers for each end-of-chapter lab
- Additional articles
- Consolidated up-to-date links to all the resources mentioned in this book

I'll be watching and responding to reader requests and feedback made on that site. You'll find the book's forum there, where I hope to interact with people.

1.7 Being immediately effective

The goal of this book is for you to immediately become effective with Git. *Being effective* means being able to quickly translate a version control task ("I want to make a tag") to the Git command (`git tag`). Git makes this easy because its verbs (`tag`, `branch`, `commit`, and so forth) directly map to version control tasks.

Being effective also means learning how to map Git commands to their equivalents in the Git GUIs (where appropriate). Even though Git is primarily accessed via the command line, some important aspects of Git are easier to do via a GUI. You'll learn this too.

Provided you have installed a local copy of Git, this book will show you how to safely operate Git in sandbox environments. Practicing the Git commands will increase your familiarity and confidence with the system.

You have no lab for this chapter, other than installing Git, which shouldn't take too much time. See you tomorrow, when you'll look at an overview of Git and take a quick tour.

2 An overview of Git and version control

Welcome back to lunch! Today you're going to learn about high-level version control concepts and take a quick tour of Git. Some of this may be familiar, so please skim as necessary, but be sure to pay attention to the Git features section and the tour! If you're completely new to version control, this chapter gives you enough background to follow the rest of this book. For people new to Git, the whirlwind tour gives you a fast introduction to Git's features.

2.1 Version control concepts

Version control is an essential practice for computer programmers. It's the act of keeping track of changes you make to a file or set of files. By extension, version control is an essential practice for organizations: it's the act of keeping track of versions of software.

For programmers, developing software isn't easy; you sometimes have to try multiple things before you come up with an acceptable solution. Versions are these multiple things. Keeping track of what you've tried is a good discipline that version control helps with (not only that, but it preserves working copies in case something goes awry in the current version)!

For organizations, maintaining software that's out in the field is made easier by knowing its version. If a bug is found, the organization needs to know which version contains that bug in order to properly fix it.

2.1.1 Version control for the software developer

If you're a software developer (and I'm guessing you are), figure 2.1 may look familiar.

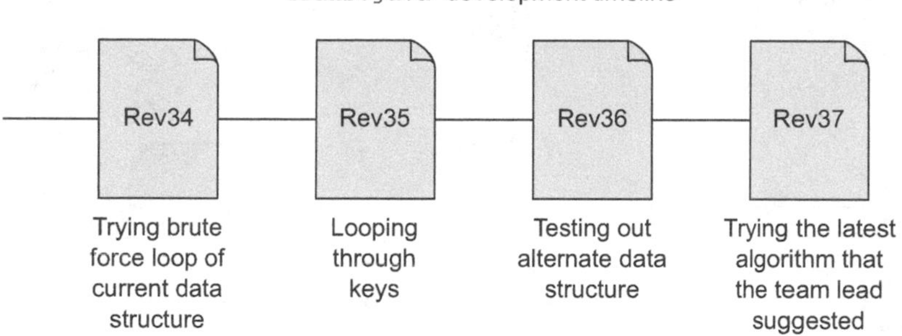

Figure 2.1 Version control for the individual developer

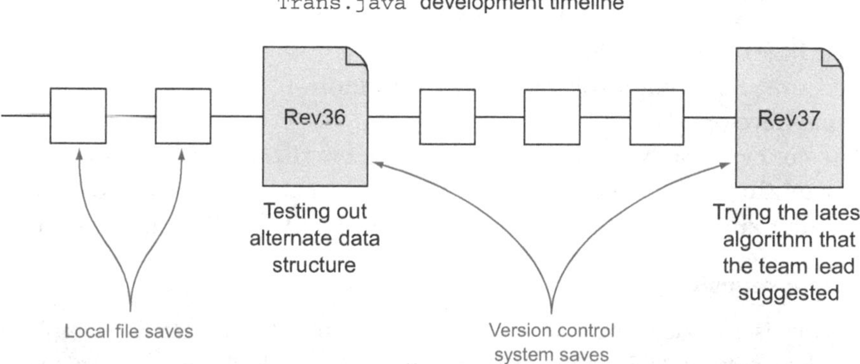

Figure 2.2 Saving to disk happens more often than saving to the version control system.

Figure 2.1 presents a timeline for the Java source-code file Trans.java. You'll notice that each box has a revision number, and let's assume further that there were 33 earlier revisions (boxes) of this file. Each box represents a logical step you made in the development of this particular file (a version of the file). As a developer, you're keeping track of these detailed versions. These versions are tracked over time, and as the changes are strung together, a timeline, or history, is formed. You may want to revisit certain versions in this timeline; version control helps with that.

You might save the file to your hard drive often, but for each of these versions, you've made a change that represents a complete thought. At this point, you save the file not just to the hard drive, but also to the version control system, as shown in figure 2.2.

2.1.2 Version control for the organization

For an organization, version control is more like figure 2.3.

Instead of the detailed steps that an individual would be concerned with, the organization is concerned with the big picture. For an organization that releases software, a version of software usually consists of many files. Version control must be able to

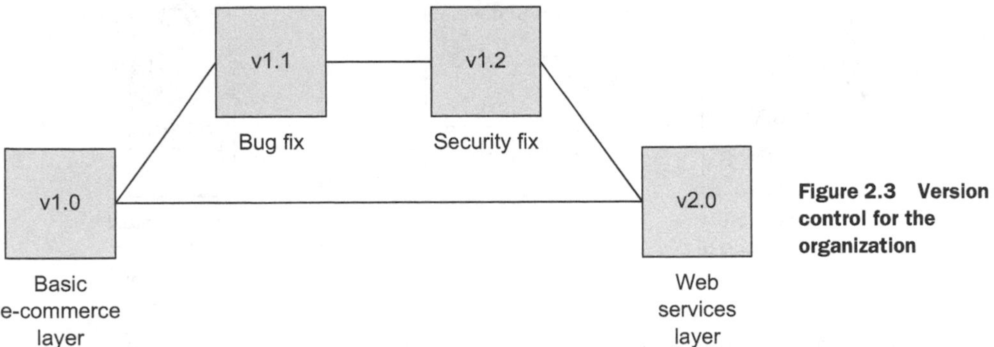

Figure 2.3 Version control for the organization

answer questions such as, what files make up version 1.0? What files make up the bug fix that's on top of version 1.1? What version contains the security fix?

Just like the single file that we're tracking in figure 2.1, the software system is tracked over time; but a software system has more files, typically coming from more than one developer.

The concerns of the version control system are the same for the individual and for the organization, despite the difference in scope. Each party wants to be able to go to a particular known version and be able to view the files within that version.

2.1.3 What is a repository?

A *repository* is a storage area for your files. Conveniently, in version control, this is typically a directory or folder that contains all the files for whatever project you're working on. There may even be subdirectories (subfolders) inside your repository, depending on how your project is organized.

If your directory looks like the following listing, you're practicing a manual form of version control.

Listing 2.1 Manual version control (annotated command-line session)

```
C:\buildtools>dir
 Volume in drive C is GNU
 Volume Serial Number is 5101-E64D

 Directory of C:\buildtools

03/15/2014  08:22 PM    <DIR>          .
03/15/2014  08:22 PM    <DIR>          ..
03/01/2014  08:22 AM            11,843 filefixup-01.bat
03/03/2014  08:52 AM            11,943 filefixup-02.bat
03/08/2014  11:22 AM            12,050 filefixup-03.bat
03/10/2014  02:22 PM            12,352 filefixup-04.bat
03/15/2014  03:21 PM            11,878 filefixup.bat
               5 File(s)         60,066 bytes
               2 Dir(s)  467,705,196,544 bytes free
```

Get a directory listing of the files in the current directory (on Windows)

Every time you make some kind of change to a file, you append a number to its name.

This manual system may make sense to you, but after a month or two, will you remember what has changed between filefixup-03.bat and filefixup-04.bat? Was it an important change?

If you work on multiple files, all in the same directory, another manual form of version control is making a copy of the entire directory. The following listing shows a terminal session on my server.

Listing 2.2 Manual version control (annotated command-line session)

```
% pwd                                                  <- Get the name of the
/home/rumali/RickUmaliVanityWebsite                       current directory
% ls                                                   <- Get a listing of the files in this directory
README.txt            make_new_index.pl         process_sports_feed.pl
bio.tmpl              make_ramblings_tmpl.sh    process_tech_feed.pl
blog_start.tmpl       make_rick_index.sh        processfeed.pl
contact.tmpl          make_sports_tmpl.sh       rick-yui.tmpl
footer.tmpl           make_tech_tmpl.sh         sports_start.tmpl
getfeed.pl            pictures_start.tmpl       tech_start.tmpl
make_flickr_tmpl.sh   process_flickr_feed.pl
% cd ..                                                <- Change directory to the parent directory
% ls                                                   <- Get a listing of the files in this directory
RickUmaliVanityWebsite         RickUmaliVanityWebsite.v01
RickUmaliVanityWebsite.v02     RickUmaliVanityWebsite.v03
RickUmaliVanityWebsite.v04     RickUmaliVanityWebsite.v05
% cp -r RickUmaliVanityWebsite RickUmaliVanityWebsite.v06
                                                       <- Create a copy of the website directory recursively,
                                                          using the cp command with -r switch (note that
                                                          we give it a new name, an incremented number)
```

The listing is annotated, so don't worry if this is new to you.

On my server, I first examine the list of files in the directory containing my vanity website. Then I go up to the parent directory and make a copy of this directory. This is yet another form of manual version control on a so-called repository.

2.1.4 What is a commit?

A *commit* is a saved change.

Listing 2.1 made a new version of a file by appending a number to the end of the filename. For example, the file filefixup-03.bat represents the third version of that Windows batch file.

In figure 2.1, I made a new version of the Trans.java file at every unique change. How do you know when to make a new version? Whenever you've made a change that's worth saving. In version control, *to commit* (the verb) is to save a change you made to your file (or files) back to the repository.

In listing 2.2, you saw that the website had five previous versions, prior to making a sixth version. What could those other versions be? Maybe version 1 is the initial set of

HTML pages, and version 2 is when I introduced CSS files. Version 3 might be the version where I added an image. Clearly, the variations are endless. Versions are saved changes that are worth saving.

Keeping track of changes is important because it allows you to go back to an earlier change. Remember that repository with the utility batch script (shown in the following listing)?

Listing 2.3 List of batch scripts in that manual repository

```
03/01/2014  08:22 AM              11,843 filefixup-01.bat
03/03/2014  08:52 AM              11,943 filefixup-02.bat
03/08/2014  11:22 AM              12,050 filefixup-03.bat
03/10/2014  02:22 PM              12,352 filefixup-04.bat
```

You have multiple versions (multiple saved copies) of a Windows batch file. Let's imagine that you share filefixup-03.bat with a colleague. If for some reason the BAT file didn't work for that person, you could now offer an earlier version, filefix-02.bat. This is an important capability! Having versions, and knowing what is in these versions, is a key first step in being a professional software developer.

2.1.5 What is a branch?

Branches are other paths, or lines, of development.

Imagine that you're making a website for a client, and it consists of an HTML file and a CSS file. You work on these two files in a default repository. This default repository is also conveniently associated with a default branch, which is often called the *main branch,* or the *trunk.* When you commit your changes to the repository, your changes go into this default branch, as in figure 2.4.

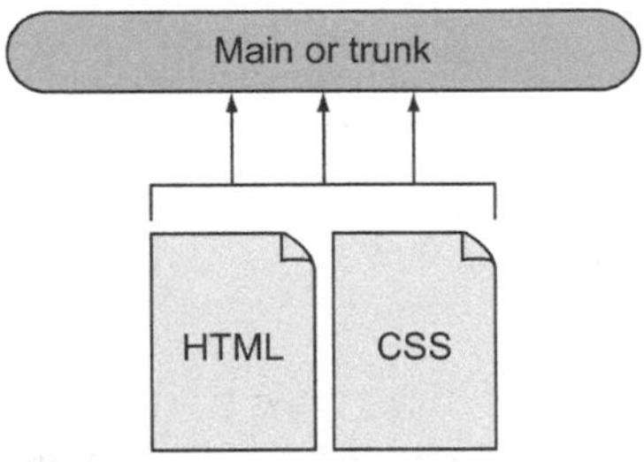

Figure 2.4 Saving files to a main, or trunk, branch

In the figure, the arrows represent commits to the branch. The HTML and CSS files are committed three times. Let's now suppose your client is comfortable with how these two files look, and asks you to install them on their web server. Because you're smart and you use version control, you make a copy of the repository, calling it Version 1. This is the creation of a new branch! In figure 2.5, the files you had on the main branch are copied to the Version 1 branch.

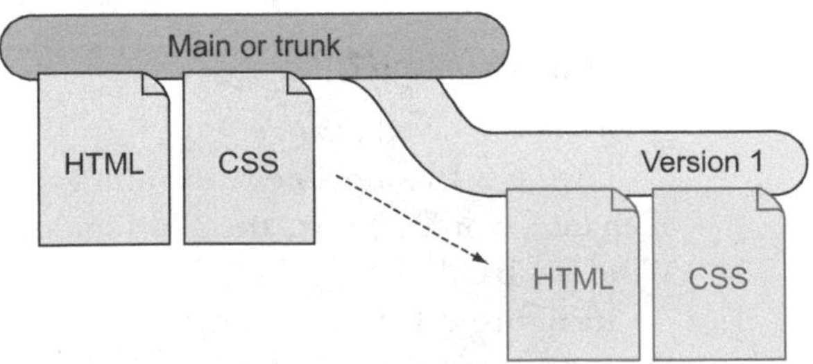

Figure 2.5 Making Version 1 in the repository

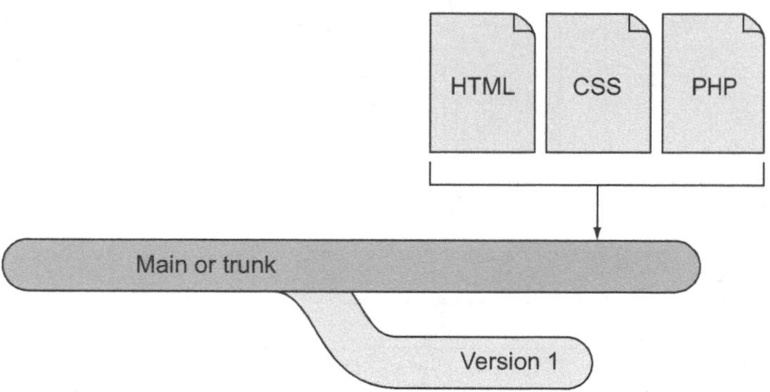

Figure 2.6 New commits happen after Version 1, as indicated by the arrow. The trunk grows bigger.

After you install the files on the web server, you go back to work. You do this work on the default branch. Let's suppose you've added a PHP file to your website to make it more dynamic. This causes you to update your HTML and CSS files. You now commit these changes to the repository. This commit is new work that was done after Version 1 (see figure 2.6).

Your client calls you up and wants you to make a change to the existing files on the web server. You might be bold enough to make changes directly on the server; this temptation is more common than you think! You might be able to make these changes on your main branch, and then copy the files from that main branch up to the website, but then you'll inadvertently install that PHP file. What if that PHP file isn't finished? It's more likely that the HTML and CSS changes you made are such that you can't make the fix in the main branch and then copy them up to the server. In this situation, usually too much has changed between commits.

You need a way to go back to the Version 1 branch and make the changes to that branch. You need the ability to go back to an earlier version of your entire repository. If you had this capability, you could make the changes in Version 1, upload those changes to the web server, and then resume your work back on the main branch.

Figure 2.7 shows all the commits that were made to this repository. Note the additional changes to the Version 1 branch. Another way to visualize the commits and branches is to use an ASCII timeline, as in the following listing.

Listing 2.4 Visualizing a branch in ASCII

```
          D---F   Version 1
         /
A---B---C---E     Main
```

It might be more apparent from this ASCII diagram that you have two paths of development: Main and Version 1.

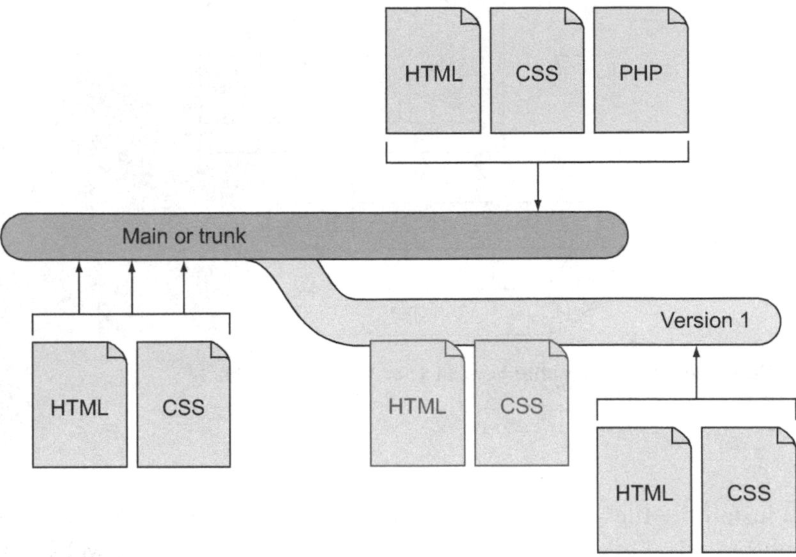

Figure 2.7 All of your commits and branches

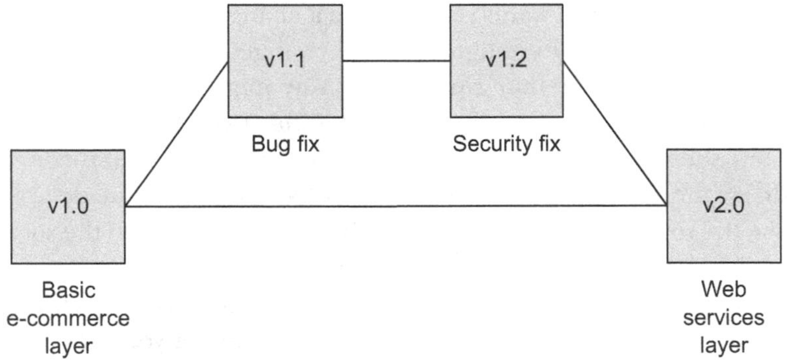

Figure 2.8 The organization's view of version control

Released code often gets updated as part of regular maintenance. Consider figure 2.8.

We made changes to the v1.0 code to add a bug fix (v1.1) and a security fix (v1.2). Some versions might not even get published!

2.2 Git's key features

Let's talk about Git, the software that puts the fun back into the drudgery of version control. The previous section presented the key concepts of version control. Every version control system today implements these concepts: versioning, auditing (via commit messages), and branching.

You'll now look at three of Git's key features, which distinguish it from other version control systems: distributed repositories, fast branching, and the staging area. (The inventor of Git, Linus Torvalds, placed a premium on these features for his project, the Linux kernel.)

2.2.1 Distributed repositories

I touched briefly on distributed repositories in the previous chapter, but let's dig a little deeper. Every time you make a commit of your file, it is stored in a repository. Earlier, I said that a repository is like a folder that contains all the files for a specific project. Where do you think that repository or folder exists? Do you think of a centralized location, perhaps a server, that's clearly labeled *repository*? If so, it's not surprising. Many of the common version control systems have a centralized server that houses the repository. Commits send your changed files up to this server. If you want to work on the file, you check it out of the repository, an operation that tells the repository you're now manipulating the file.

Git is the opposite of this. Git doesn't require a central server to be installed anywhere. With Git, every developer is given a copy of the repository, as shown in figure 2.9. All version control operations are done locally.

A brag that you may hear about Git is that you can commit changes to a repository even while you're flying in an airplane. This phrase didn't hit home until I remembered an evening I had flown from Boston to Minneapolis. On the plane, I realized that I couldn't connect to my version control system, so I'd have to wait until I landed before committing the work I was doing on the flight.

This happened to me sometime between 2006 and 2008. Nowadays, you can find Wi-Fi on airplanes, but at what cost? Also, what might the performance be on an airplane's Wi-Fi? With Git, you don't need to worry about cost or network performance; you can do everything to the repository because the repository is entirely local to you.

Being distributed allows source code to be extensively shared. There's no hassle of setting up permissions. With Git, you copy a repository you're interested in, and you can immediately make commits (on your copy).

Being distributed does require conventions and workflows, but you and your collaborators can define

Figure 2.9 Distributed version control systems are liberating because every developer has a copy of the entire repository.

these. (Conventions and workflows help eliminate issues when multiple people work on the same files and then share them with others.) All large projects need organization and conventions, and projects that use Git are no exception, but because Git is decentralized, each developer has full control of a local copy.

Backups for your repositories come for free when you use Git. Because everyone working on a project has the repository, you don't have to worry about losing the entire repository. A backup (current up to the last time you shared your changes with others) can be obtained by copying someone else's repository. People can still store the repository in a common location, but no one person's repository is more important than anyone else's.

2.2.2 Fast branching

As you saw in section 2.1.5, branching is the ability to make a copy of the repository, so you can work on that copy. In that section, you made a branch called Version 1 because you released some website code to a web server. Later, you used that branch to fix the website, all the while keeping your current work isolated from the fixes.

In large software companies, whole departments (often named *release engineering*) were tasked with managing branches. This is natural because releasing software almost always involves separating the code to be released from the developers who want to work on that code. It's for this reason that the ability to branch was often limited to restricted personnel.

Figure 2.10 represents the code in a repository at a software company that has released two versions of its product.

All the code that the company develops is stored in a main, or trunk, repository.

When a version of the software is ready to be released, a copy is made of the repository at that point in time, and given a name (for example, Version 1.0). This snapshot of the code base is made in case that version needs to be updated in the future (for example, as Version 1.1).

In earlier version control systems, making a branch to release software was painstaking and intricate. Making a branch usually involved copying files to a separate

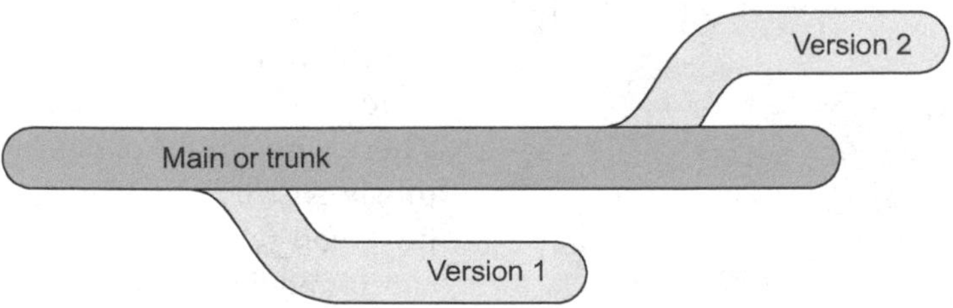

Figure 2.10 Software is developed on code that resides in the main (or trunk) part of the repository. As versions are ready to be released, a branch is made. This branch is a copy, or a snapshot, of all the code that represents that version.

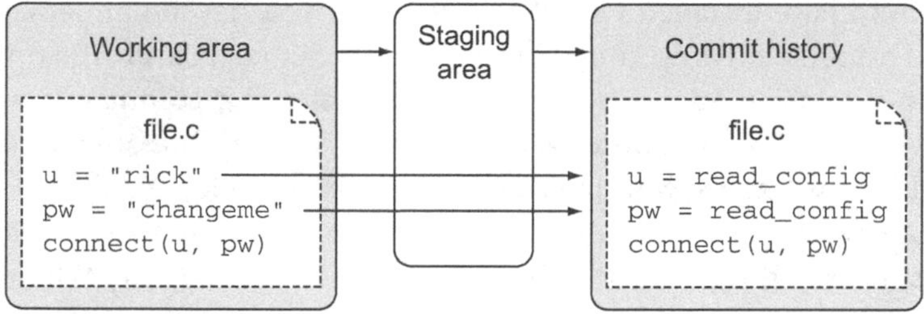

Figure 2.11 The staging area

location. Once this was ready, developers would then check out this branch, at which point they had two copies of the code: their local main code, and a copy of a particular version. If this sounds arduous, it is.

Git, on the other hand, makes branching as easy as changing a directory, and just as fast too. In most OSs, to switch to a directory via the command line, you type `cd some_directory`. In Git, to switch to a branch, you type `git checkout some_branch`. There's no copying of files and no interaction with a remote server to make this special copy. (Underneath the covers, Git updates a pointer to the specific commit of the new path, or line, of development.)

The speed with which you can create a new branch and begin developing on it introduces a new model for doing work: if you want to try a new idea in your code, create your own branch. Because you're working in a local repository, no other developer is disturbed by this new code stream. Your work is safe and isolated. In fact, working in such feature branches (as these are known) is the hallmark of a competent Git user!

2.2.3 The staging area

The *staging area* is an advanced architectural feature of Git that allows you, as the developer, to pick and choose exactly which parts of your changes should be committed. As you do work in a working copy of your files, you might make multiple changes in the course of your development. Most of these changes will ultimately be committed to the repository, but some shouldn't go in the repo. Debugging statements, sensitive information such as passwords, and other items for the developer's eyes can be cleanly removed only by using the staging area.

In figure 2.11, I've hardcoded my username/password into a file named file.c. Clearly, this isn't something to commit into the repository. I could commit a sanitized version of file.c, but what if I've made other changes to file.c that should be committed? Using the Git staging area allows me to commit just the parts of file.c that I want.

2.3 A quick tour of Git

Everything we've discussed and more is covered in great length in the upcoming chapters, but, for now, let's do something at the computer. Let's take a whirlwind tour of Git!

By now, you should have installed Git. Please do this if you haven't already. To properly understand a tool like Git, you need to try its commands. This is the only way to get comfortable with Git's way of doing things. Initially, the commands may seem strange, especially if you've used other version control systems, but after you get Git's commands under your fingers, you'll become more comfortable.

To begin your initial exploration of Git, it's best to obtain an existing repository. You'll take a fast tour of this work in both the GUI and command line. Whenever possible, this book provides instruction on using Git with either the GUI or the command-line mechanisms.

2.3.1 Using the GUI to tour a Git repository

Earlier in this chapter, you learned that repositories are distributed. To get a repository onto your computer, you have to clone it. In Git, *cloning* is the act of copying a repository from a remote location to a local directory on your machine.

Cloning a repository is the initial step in collaborating on source code, or any set of versioned files in Git. Someone has made a repository full of work, and in order for you to contribute changes to it, you must first clone it. Learning how to clone repositories enables you to begin collaborating with others who use Git.

The first thing you need in order to clone a repository is the URL for the existing repository. Visit https://github.com/rickumali/RickUmaliVanityWebsite. On this page, down the right-hand side, you'll see a text box labeled HTTPS Clone URL, as in figure 2.12. Use the browser to select this URL.

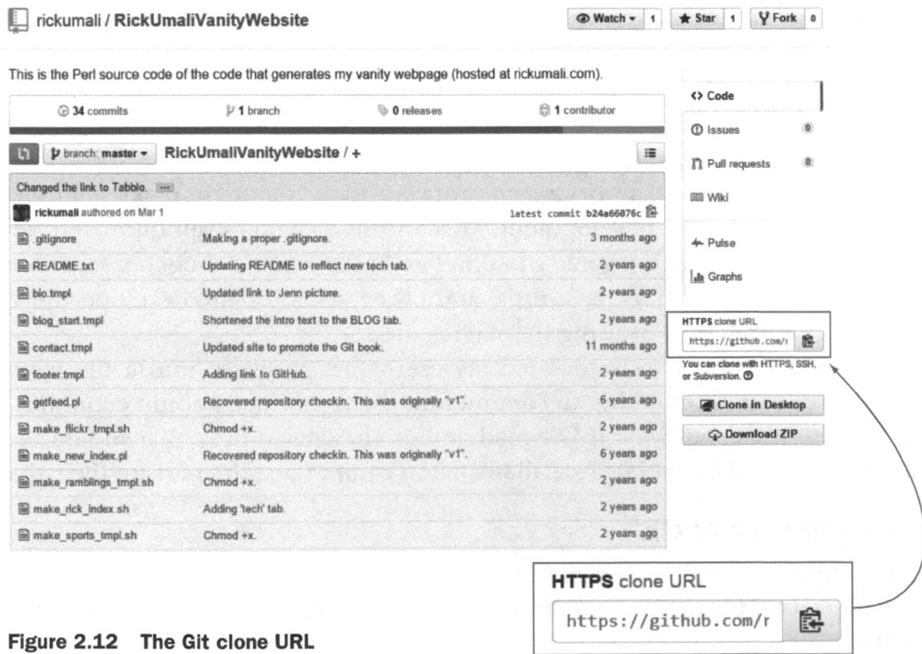

Figure 2.12 The Git clone URL

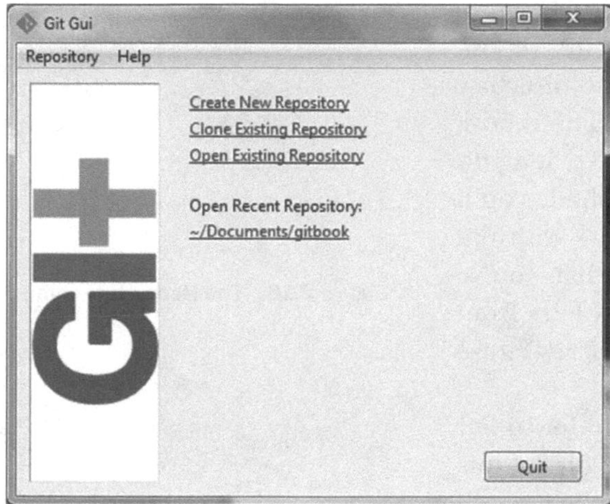

Figure 2.13 The initial screen for the Git GUI

Once you have the URL, open the Git GUI tool and select Clone Existing Repository, as in figure 2.13. Depending on which GUI you've selected for Git, there may be a different mechanism to access the clone feature.

> **WINDOWS NOTES** The Git GUI is usually available from the Start menu. Alternatively, you can navigate your directories by using the standard Windows Explorer, and then right-click to choose Git GUI Here from the context menu. In some cases, you may need to type `git gui` at the command line.

> **MAC NOTES** The Mac may have difficulty bringing up the Git GUI, depending on which Git you've installed. You may have to manually start the Git GUI by typing `git gui` at the Terminal. Please visit this book's forum for other tips on Git GUI startup issues.

In the dialog box, after pressing Clone Existing Repository, type the source URL `https://github.com/rickumali/RickUmaliVanityWebsite`, as shown in figure 2.14. This URL is known as a *Git URL*. Note that the Target Directory must be a directory that doesn't exist yet (Git will create it for you).

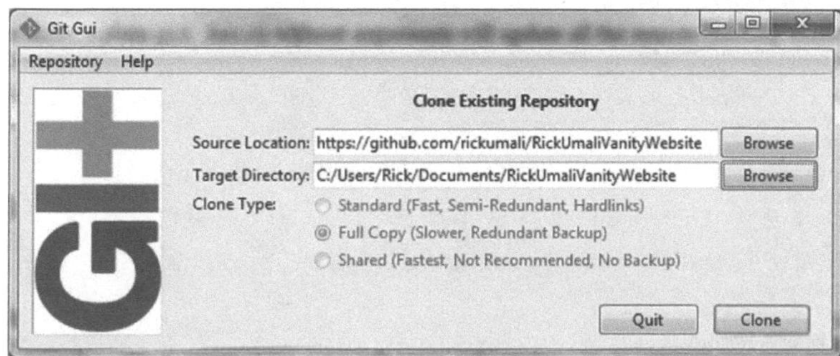

Figure 2.14 The Git clone prompt

After you click Clone, the Git GUI automatically downloads the entire repository to your machine, into the directory specified by Target Directory. This repository isn't big, so it won't take long to download. Once you're finished, you'll be at a default window to work with the repository. To see exactly what you've copied, select Browse master's Files from the Repository menu, shown in figure 2.15.

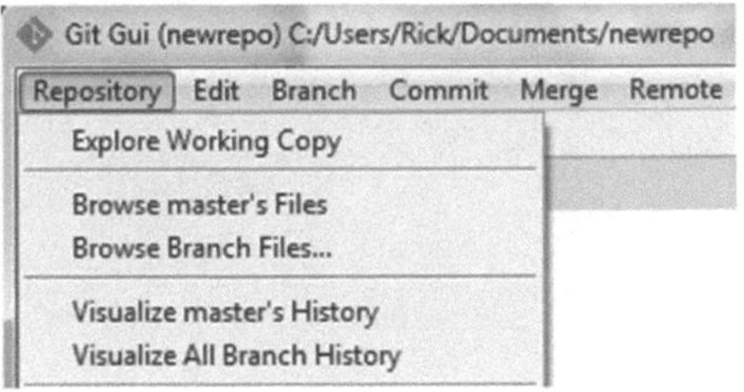

Figure 2.15 The Repository menu

You should see a small window titled File Browser that shows all the files of this repository, as in figure 2.16.

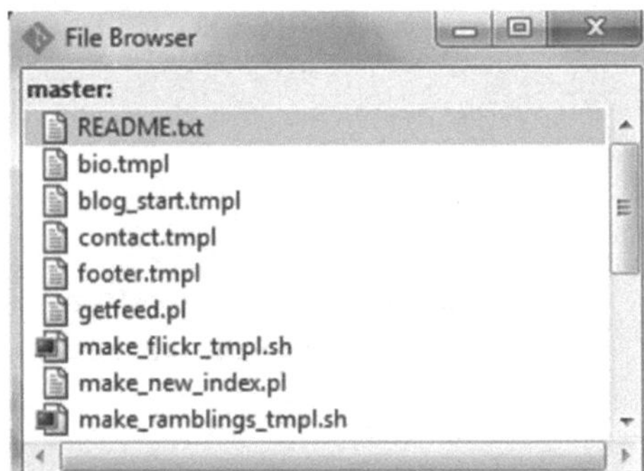

Figure 2.16 Viewing all the files of a repository

Double-clicking any file brings up that file in a specialized window showing the history for each line. For example, double-click the file README.txt, and you'll see the window shown in figure 2.17. This window shows the file, and for every line in the file, it shows the specific date and reason for bringing that line into the file. Hover over the lines, and a tooltip will show these details. This view is known as Git's *blame output*.

In summary, this repository now exists as a directory on your own machine. The directory is the Target Directory you specified in the prompt in figure 2.14. You can view

```
File Viewer
Commit:    master
 |    |      10 20  10   *     *      *   ./make_flickr_tmpl.sh
7308 7308    11 25  10   *     *      *   ./make_tech_tmpl.sh
436c 436c    12 30  10   *     *      *   ./make_rick_index.sh
 RU   RU     13
 |    |      14 ---------------------------------------------------
 |    |      15
 |    |      16 A high-level view of which Perl scripts (.pl) get calle
 |    |      17 (.tmpl) are used / generated:
 |    |      18
 |    |      19 make_sports_tmpl.sh -> getfeed.pl -> process_sports_fee
 |    |      20
 |    |      21 make_ramblings_tmpl.sh -> getfeed.pl -> processfeed.pl
 |    |      22
7308 7308    23 make_tech_tmpl.sh -> getfeed.pl -> process_tech_feed.pl
 RU   RU     24
```

Figure 2.17 Git blame output

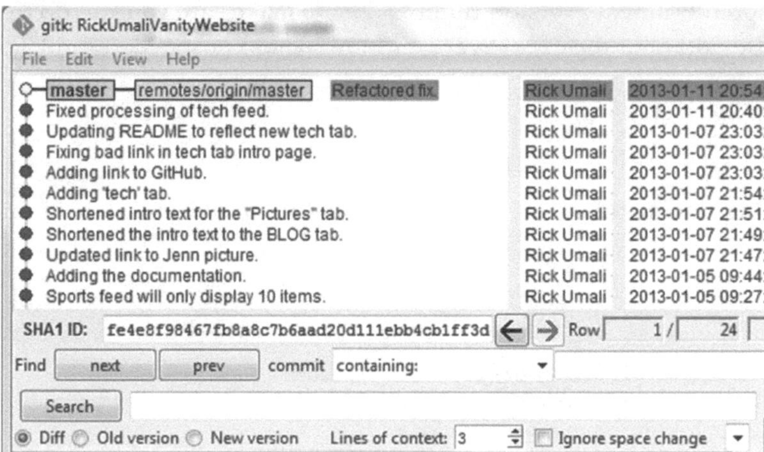

Figure 2.18 A view of this repository's history

its contents in whatever file-browsing tool you like (Mac Finder, Windows Explorer). The files appear as normal files, and you can update them whenever you want.

Another way to view the repository is to select Visualize master's History from the Repository menu (figure 2.15). A window like that in figure 2.18 opens, showing the history of this repository. Specifically, this is the history of commits, the auditing capability that every version control system must provide.

Each line in this history view represents a commit and shows the date that I made that commit. As you click each commit, it's highlighted, and the SHA1 ID text box changes to reflect the correct ID.

2.3.2 *Using the command line to tour a Git repository*

Cloning a repository via the GUI has a concise equivalent on the command line. After you've explored this repository via the GUI, go ahead and delete this directory. Because there's no server interaction, you can do this anytime you want to start over! Once you've deleted this directory, open the command-line prompt.

To access the Git command line in Windows, you'll have to use Git BASH. For Mac and Unix/Linux users, the command line is the terminal program. Once the terminal window is open, it looks like figure 2.19.

```
Git Bash

Rick@MININT-OSPDAI2 ~
19:38 505> git --version
git version 1.8.1.msysgit.1

Rick@MININT-OSPDAI2 ~
19:39 506>
```

Figure 2.19 The Git command line

You can type `git --version` into this window, and after you press Return/Enter, it should respond with Git's version number. Please make sure you can do this. Figure 2.19 shows the output of the `git --version` command. For the Mac and Ubuntu, this output is similar.

Next, type the following:

```
git clone https://github.com/rickumali/RickUmaliVanityWebsite
```

The preceding command and the output from Git should look like the following listing.

Listing 2.5 Output of `git clone`

```
% git clone https://github.com/rickumali/RickUmaliVanityWebsite
Cloning into 'RickUmaliVanityWebsite'...
remote: Reusing existing pack: 91, done.
remote: Total 91 (delta 0), reused 0 (delta 0)
Unpacking objects: 100% (91/91), done.
```

You duplicated the steps you followed in the previous section, this time with the command line. As a result, you've made another clone of the repository you just deleted. Let's explore the command-line equivalents of our explorations with the Git GUI.

To view the files, change into the directory via the `cd` command. Once you're in the target directory, you're in what is known as the repository's *working directory*. You can get the list of files that the repository knows about by typing `git ls-files`, as shown in the following listing. (Additionally, you can list the files by using the command-line tool `ls`.)

Listing 2.6 Output of `git ls-files`

```
% git ls-files
README.txt
bio.tmpl
blog_start.tmpl
contact.tmpl
footer.tmpl
getfeed.pl
make_flickr_tmpl.sh
make_new_index.pl
make_ramblings_tmpl.sh
make_rick_index.sh
make_sports_tmpl.sh
make_tech_tmpl.sh
pictures_start.tmpl
```

Listing 2.6 is the equivalent of figure 2.16, where you viewed all the files of a repository via the Git GUI.

To see a detailed view of the README.txt file, type `git blame README.txt` (see listing 2.7). Depending on the size of your command-line window, you may have to press the spacebar to allow the output to continue scrolling. Git uses a pager to display text that's longer than the height of the screen. Press the Q key to exit this pager. (You'll learn about the pager tool in the next chapter.)

Listing 2.7 Output of `git blame`

```
22:25 514> git blame master README.txt
436cf890 (Rick Umali 2009-09-05 02:10:36 +0000  1) rickumali-index
436cf890 (Rick Umali 2009-09-05 02:10:36 +0000  2)
436cf890 (Rick Umali 2009-09-05 02:10:36 +0000  3) This is the software
436cf890 (Rick Umali 2009-09-05 02:10:36 +0000  4)
436cf890 (Rick Umali 2009-09-05 02:10:36 +0000  5) The best way to read
436cf890 (Rick Umali 2009-09-05 02:10:36 +0000  6)
436cf890 (Rick Umali 2009-09-05 02:10:36 +0000  7) min hr day month weekday
436cf890 (Rick Umali 2009-09-05 02:10:36 +0000  8) 0   10   *     *       *
436cf890 (Rick Umali 2009-09-05 02:10:36 +0000  9) 10  10   *     *       *
436cf890 (Rick Umali 2009-09-05 02:10:36 +0000 10) 20  10   *     *       *
7308dd03 (Rick Umali 2013-01-07 23:03:45 -0500 11) 25  10   *     *       *
436cf890 (Rick Umali 2009-09-05 02:10:36 +0000 12) 30  10   *     *       *
```

The output in listing 2.7 is roughly the same as figure 2.17.

To get a listing of the repository's history of commits, type `git log --oneline`. Again, depending on your terminal size, Git may display the output in the pager. You may have to press the spacebar to page through the output in the pager, and press Q to exit the pager. The following listing shows the output.

Listing 2.8 `git log --oneline` output

```
% git log --oneline
fe4e8f9 Refactored fix.
0fa9e1d Fixed processing of tech feed.
7308dd0 Updating README to reflect new tech tab.
447606a Fixing bad link in tech tab intro page.
364d2d4 Adding link to GitHub.
821d75c Adding 'tech' tab.
c4a15c5 Shortened intro text for the "Pictures" tab.
23db75c Shortened the intro text to the BLOG tab.
```

Don't worry if you're not comfortable with the command line. You can rely on this book to tell you exactly what to type. The most complicated thing you'll do outside Git is editing files, and I'll guide you in doing this properly. Be warned, though, that the command line dominates the later portions of this book.

This chapter doesn't have a lab. Ideally, you've followed all the steps in the previous section. If you haven't, and there's still time in your lunch hour, then by all means go for it. But the next chapters go through each aspect of Git's version control system much more slowly, so don't be discouraged if this section went too quickly. You'll be going through the material in more detail shortly.

2.4 Version control terminology

You'll encounter many terms when reading about version control. This section offers a brief listing of those terms and a short definition for each:

- *Branch*—Another path, or line, of development in the repository.
- *Check out*—To request a copy of a file so you can work on it; a typical feature of centralized version control systems.
- *Clone*—To make a copy of a repository that exists somewhere locally (in another directory) or remotely (on another server, or Git hosting site such as GitHub).
- *Commit*—A change that's saved to a repository, recording itself into the timeline.
- *Distributed*—A characteristic of a system such that its operations can be performed without the need of a server (as opposed to *centralized*).
- *Repository*—A storage area for files; in the context of version control, this storage area is usually a directory or folder with special operations for viewing the timeline, committing files, and branching.
- *Staging area*—A feature of Git that enables the developer to commit certain parts of files instead of the whole file.
- *Timeline*—A set of events ordered by time, from the earliest to the most recent event; also known as a *history*.
- *Version control*—The practice of keeping track of changes such that you can always go back to a known state.

3 Getting oriented with Git

The command line is where you'll interact with Git. Yes, there are GUIs for Git (for example, Git GUI), and you'll examine those, but most Git work is done by typing commands. Git was born on the command line, so it's no surprise that all of its functionality is oriented toward commands that you have to type in. You'll first learn about Git's command-line syntax, a command-line pattern that will help you understand how all Git commands are structured. You'll also orient yourself to the command line in general. If you're a command-line veteran, chances are this chapter will be straightforward!

Finally, you'll learn about the Git help system. Every command has help that you can access, and learning how to access it will enable you to become more effective with Git. Moreover, Git has longer-form documentation that is accessible from the help system. This documentation is worth reading, and this chapter shows you how to do that.

3.1 Getting set up

By now you should have already installed Git, which makes it immediately available to the command line. To open the command line on Mac or Linux, start your terminal program. In Windows, the CMD or PowerShell environments won't suffice; you'll have to start the Git BASH program, which should be a double-click away from either your desktop or your Windows Start menu, depending on how you installed it. From here on out, I'll refer to this window as the *command-line window* (or *command line* for short).

Because Git doesn't have a server, there isn't anything to start. Once you have the command line, you should be able to type `git --version`. After you press

Return/Enter (which I'm going to assume from here on out), you should see the output in figure 3.1 (though your text prompt and Git version may vary).

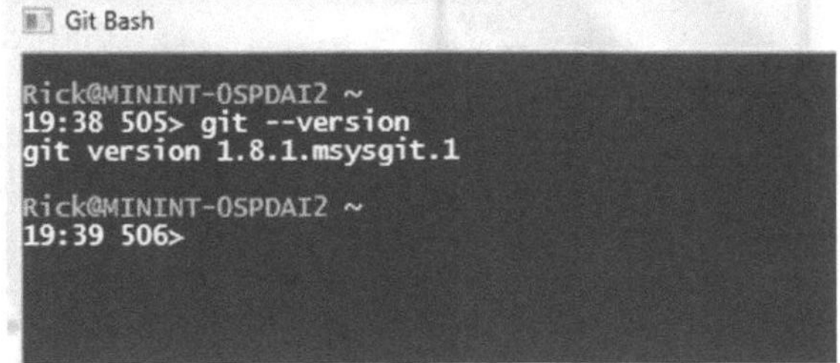

Figure 3.1 The Git command line

Git keeps track of who performs version control actions, so you must at least configure Git with your own name and email. Note that you're not connecting to a server or making a network connection, but rather configuring the Git software that's installed on your machine.

To configure Git with your name and email, type the following Git commands on the command line.

TRY IT NOW

```
% git config --global user.name "Your Name"
% git config --global user.email "Your E-mail@example.com"
```

Replace your own name and email address for the appropriate configuration settings.

To see that you've set these properly, you can type `git config user.name`, and your name will print as a response to that command. You can list all the configurations that Git knows about with `git config --list`. When I type this command, I see the output in the following listing. Notice that the `user.name` and `user.email` are shown in the list.

Listing 3.1 `git config --list`

```
% git config --list
core.symlinks=false
core.autocrlf=true
color.diff=auto
color.status=auto
color.branch=auto
color.interactive=true
pack.packsizelimit=2g
help.format=html
http.sslcainfo=/bin/curl-ca-bundle.crt
sendemail.smtpserver=/bin/msmtp.exe
diff.astextplain.textconv=astextplain
rebase.autosquash=true
user.name=Rick Umali
user.email=rickumali@gmail.com
gui.recentrepo=C:/Users/Rick/Documents/gitbook
gui.recentrepo=C:/Users/Rick/Documents/RUVW
```

TRY IT NOW Perform the following to view Git's configuration. The `git config` command can be run to show all the settings or just a specific setting.

```
% git config user.email
% git config user.name
```

3.2 Using commands

Commands are what you type into the command line. There are two kinds of commands: those supplied by Git, and common commands that perform typical operations.

3.2.1 Git command-line syntax

All Git commands follow the same convention: the word `git`, followed by an optional switch, followed by a Git command (so far, you've learned `config`), followed by optional arguments (abbreviated in figure 3.2 as args) that the command recognizes.

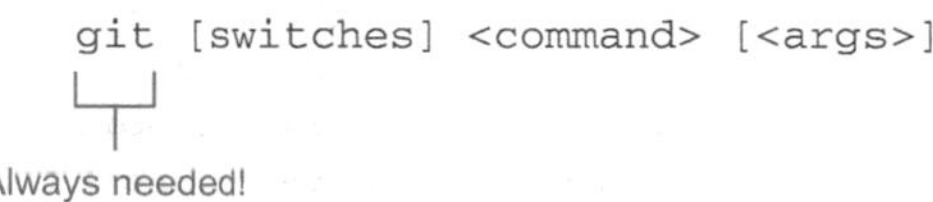

Figure 3.2 The Git command-line syntax

You've already typed a Git command, `git config`. Figure 3.3 is a diagram of a slightly more complicated version of this command, showing its structure.

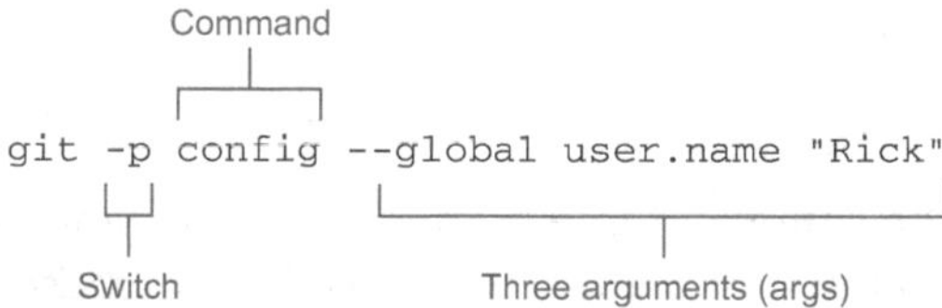

Figure 3.3 A breakdown of the Git command-line structure

The only command that's absolutely necessary is `git`. Go ahead and type `git`. You should see output that looks like listing 3.2.

In the previous section, when you typed `git --version`, the string `--version` is a switch to the `git` command itself. Notice in listing 3.2 how `--version` is but one of several optional switches you can pass into the `git` command. In figure 3.3, I used the `-p` switch, which paginates the output if needed.

(The use of one or two dashes is important. In general, switches that are a single character use one dash, and switches that are fully spelled-out words use two dashes.)

Listing 3.2 Output from typing `git`

```
usage: git [--version] [--exec-path[=<path>]] [--html-path]
[--man-path] [--info-path]
[-p|--paginate|--no-pager] [--no-replace-objects] [--bare]
[--git-dir=<path>] [--work-tree=<path>] [--namespace=<name>]
[-c name=value] [--help]
  <command> [<args>]
```

```
The most commonly used git commands are:
   add        Add file contents to the index
   bisect     Find by binary search the change that introduced a bug
   branch     List, create, or delete branches
   checkout   Checkout a branch or paths to the working tree
   clone      Clone a repository into a new directory
   commit     Record changes to the repository
   diff       Show changes between commits, commit and working tree, etc
   fetch      Download objects and refs from another repository
   grep       Print lines matching a pattern
   init       Create empty git repository or reinitialize an existing one
   log        Show commit logs
   merge      Join two or more development histories together
   mv         Move or rename a file, a directory, or a symlink
   pull       Fetch from and merge another repository or a local branch
   push       Update remote refs along with associated objects
   rebase     Forward-port local commits to the updated upstream head
   reset      Reset current HEAD to the specified state
   rm         Remove files from the working tree and from the index
   show       Show various types of objects
   status     Show the working tree status
   tag        Create, list, delete or verify a tag object signed with GPG

See 'git help <command>' for more information on a specific command.
```

The output in listing 3.2 is the same output you would get if you typed `git help`. You'll explore the Git help system in the next section, and perhaps become a little more comfortable with typing Git commands on the command line.

3.2.2 Common commands

The most important thing about the command line is knowing where you are. When you first open the command line, you're in the home directory. You might see this displayed as a tilde (~) or `$HOME` (pronounced *dollar home*). Both mean the same thing: home.

To display the name of the current directory, type `pwd`. This prints the name of the current directory. Find this directory in the graphical directory-browsing tool of your choice (Windows Explorer, Mac Finder, Ubuntu Files). I hope this is an obvious exercise. The main thing is knowing where you are.

To list the files in your current directory, type `ls`. This produces a listing of files. For now, don't worry if the output scrolls off the top of the screen, or even what its contents are.

To create an empty directory, type `mkdir my_empty_dir`. Please type the underscores. If you leave them out, you'll create three directories. This command creates a directory named my_empty_dir in the current directory. See if you can find this directory in your graphical directory-browsing tool.

To go into this newly created directory, type `cd my_empty_dir`. After that, if you type `pwd`, you'll see the name of the directory you just created. If you type `ls`, you won't see any output because there are no files in the directory. Confirm in your graphical directory-browsing tool that you don't see any files. To go back to your parent directory (the directory where you typed `mkdir`), type `cd ..`. The two dots (`..`) signify the parent directory. There's no space between these two dots. There's a space between the `cd` and the dots.

If you traverse into multiple different directories, you can go back home by typing `cd`. In the command line, it's always easy to get back home. (This book creates its repositories in the home directory for this reason. If another directory is more convenient, remember to make this substitution every time you see a reference to *home* in this book.) Git repositories are directories in your file system, which is why knowing the commands in table 3.1 is important.

Table 3.1 Common command-line commands

Command	Explanation
`pwd`	Print the current directory. (PWD is short for present, or print, working directory.)
`ls`	List files in the current directory.
`mkdir directory_name`	Make a directory with the name directory_name.
`cd directory_name`	Change the current directory to directory_name.

Because you'll be typing commands into the command line, and navigating up and down directories, it's helpful to know your way around the command line. Knowing how to list files (`ls`), print your current working directory (`pwd`), make directories (`mkdir`), and change your directory (`cd`) are the most essential commands you need to learn in any operating system.

TRY IT NOW Try the commands in table 3.1 by typing the following:

```
% pwd
% ls
% mkdir my_empty_dir
% ls
% cd my_empty_dir
% ls
% pwd
% cd ..
```

Certain users will recognize these commands as standard Unix/Linux commands, but don't worry. This is the only chapter where I specifically mention Unix/Linux. If this is your first time using these commands, congratulations! You're learning not just Git, but a little about the command line as well.

The last topic before we leave this section on the command line is how to remove files and directories. The command to do this is `rm`. As with any command that removes files, be careful! To remove a directory, use the `rm -r` command. If you've followed the TRY IT NOW steps, you're in the directory where you first typed `mkdir`. To remove my_empty_dir, type `rm -r my_empty_dir`.

TRY IT NOW Create another empty directory and then remove it. Use the following steps:

```
% mkdir another_dir
% rm -r another_dir
```

To practice removing a file, you first have to create one. Use your favorite text editor (Windows Notepad, Notepad++, Mac TextMate, Sublime Text, or Unix/Linux vi, Vim, or nano) to create a file in your home directory. The file can contain any content. Now locate this file by using a combination of the `cd`, `ls`, and `pwd` commands. Once you find the file, remove it with the `rm` command.

One thing to notice with this task is that you created a file by using a tool outside the command line. In the next chapter, you'll be creating files from the command line!

3.3 Improving command-line efficiency

Typing from the command line can get tedious. You often find yourself repeating the same commands or changing parts of the previous command that you just typed. Fortunately, the command line does have features that reduce the tedium. This section describes these essential command-line features.

The first thing to notice is that pressing the up arrow or down arrow recalls the previous or subsequent commands, respectively. The command line has a history of the commands that you've typed. You can list this history by using the `history` command.

Once you've started typing something on the command line, but before you've pressed Return/Enter, you can use the left arrow and right arrow to move around in the text. Using the Delete key (or Ctrl-D) or the Backspace key (Ctrl-H) allows you to remove characters before or after the cursor location.

Take a look at figure 3.4. The cursor is on top of the capital F. If you press Ctrl-D, the letter F disappears, and all the characters to its right move to the left. If you press Ctrl-H, the letter g to the left of the letter F disappears.

Other shortcuts on the command line include pressing Ctrl-U to erase everything to the left of the cursor, or Ctrl-K to erase everything to the right of cursor. Ctrl-A positions your cursor at the beginning of the line, and Ctrl-E positions your cursor at the end of the line.

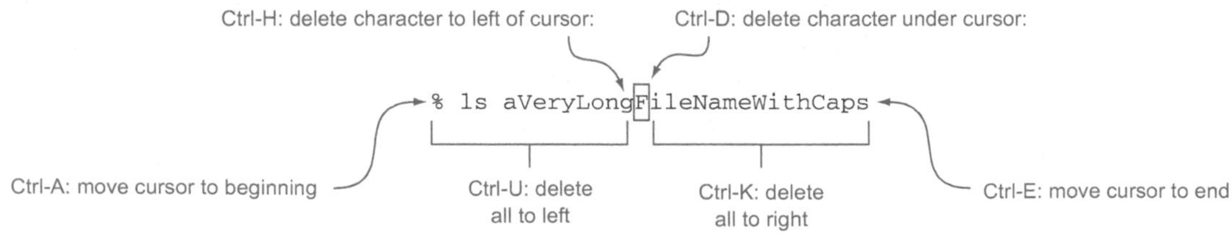

Figure 3.4 Command-line editing summarized

TRY IT NOW On the command line, press the up arrow a few times. You should see the commands you typed earlier. Then press the down arrow. The history of your session is kept in chronological order.

Now type some gibberish letters and then practice using the Delete (Ctrl-D) and Backspace (Ctrl-H) keys to remove characters. Use the arrow keys to move the cursor around on the line that you're typing. Type some more gibberish, and then test the other shortcut keys (Ctrl-U, Ctrl-K, Ctrl-A, and Ctrl-E). Again, move the cursor around on the line you're typing. More command-line techniques are available, so explore the resources in the "Further exploration" section at the end of this chapter. I also have some videos on MoreLunches.com that demonstrate some of these techniques.

Another shortcut that's incredibly helpful is tab completion. At the command line, if you partially type the name of a directory, you can press Tab, and the command line will complete the name of the directory. This completion works for filenames as well.

TRY IT NOW On the command line, create a directory with a long name:

```
% mkdir directoryWithLongName
```

Now type the following exactly:

```
% cd direc<TAB>
```

You should see the name of the directory appear. Make sure to delete this directory by using the `rm -r` command. (The `-r` switch to `rm` signifies that the argument is a directory, and you want to remove its contents recursively.) That task should be easier now that you know about directory name completion.

To see the same thing with filenames, you'll have to learn how to easily create a file with a long filename. On the command line, this is done with the `touch` command. Type the following:

```
% touch filewithsuperlongname
% ls filewith<TAB>
```

You should see the filename appear automatically. (To remove this file, use the `rm` command, without the `-r` switch.)

Typing on the command line won't be too tedious now that you can correct what you're typing and can easily complete long filenames and directories.

3.4 Using Git help

When you type `git help`, Git presents a daunting picture by suggesting that 21 commands are the most commonly used ones (see listing 3.2). Some, such as `rebase` and `reset`, have confusing descriptions. Don't worry about these commands for now! Rest assured that by the end of this book, you'll become more familiar with all the commands in this list.

The Git help system is where you should start if you're looking for answers about Git, as it has a description of every command available, along with its valid switches and arguments, guides on how to do common tasks from Git developers, and a glossary. To get a lay of the land, try this command: `git help help`. By comparing the syntax in figure 3.2, you know that the first `help` is the command, and the second `help` is the argument. You're asking Git for help about its help system.

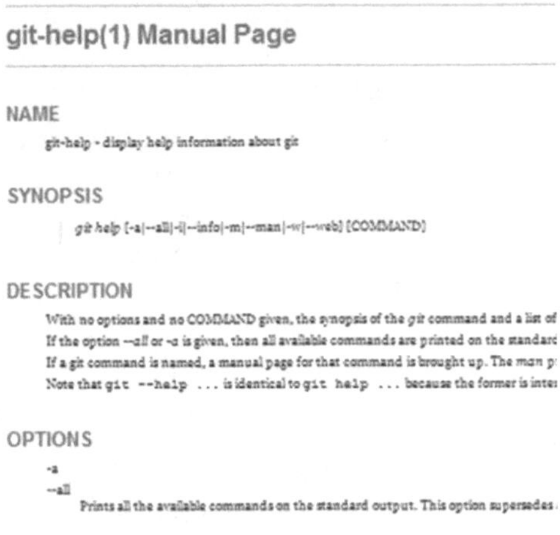

Depending on your platform, either your default browser appears, showing you a help page (figure 3.5), or your command-line window displays a help page right in the command-line window (figure 3.6).

```
git-help(1) Manual Page

NAME
    git-help - display help information about git

SYNOPSIS
    git help [-a|--all|-i|--info|-m|--man|-w|--web] [COMMAND]

DESCRIPTION
    With no options and no COMMAND given, the synopsis of the git command and a list of
    If the option --all or -a is given, then all available commands are printed on the standard
    If a git command is named, a manual page for that command is brought up. The man p
    Note that git --help ... is identical to git help ... because the former is inte

OPTIONS
    -a
    --all
        Prints all the available commands on the standard output. This option supersedes
```

Figure 3.5 A typical Git help page, displayed in a browser

In the latter case, you can page down this text by using the spacebar, and page up by using the B key. A separate tool called a *pager* enables you to move up and down a long document. You'll explore pagers in the lab.

Git's help is modeled using the classic manual page format. A brief synopsis of the command is shown, along with its valid switches and arguments. A description comes next, followed by other information, depending on the command itself.

Take some time reading the `git help help` output. Notice the `-a` switch, which enumerates all the Git commands available. When you type `git help -a`, you'll see a list of over 150 commands. It's so long that the output scrolls off the page. This puts listing 3.2 into some perspective!

Depending on your version of Git, you may have a `git help -g` switch. This switch reveals a small list of common Git guides. These guides, written by Git developers, offer a more narrative explanation of how Git works.

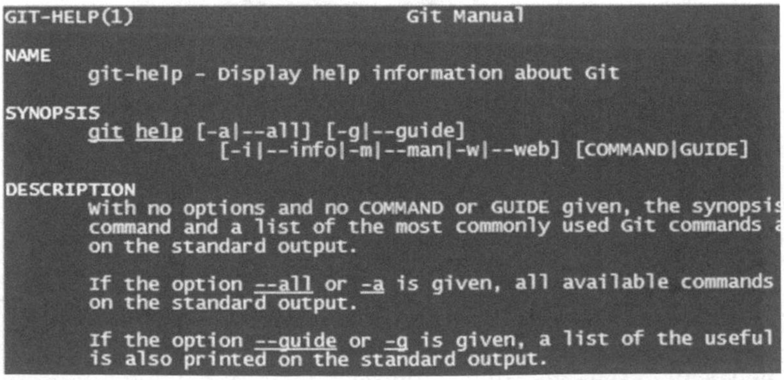

```
GIT-HELP(1)                      Git Manual

NAME
       git-help - Display help information about Git

SYNOPSIS
       git help [-a|--all] [-g|--guide]
                  [-i|--info|-m|--man|-w|--web] [COMMAND|GUIDE]

DESCRIPTION
       With no options and no COMMAND or GUIDE given, the synopsis
       command and a list of the most commonly used Git commands a
       on the standard output.

       If the option --all or -a is given, all available commands
       on the standard output.

       If the option --guide or -g is given, a list of the useful
       is also printed on the standard output.
```

Figure 3.6 A typical Git help page, displayed in the command-line window

Of special interest is the Git glossary. Type `git help glossary`. Many of the terms that you'll read in other Git help pages are described here. Keep this page in mind as you work with Git.

In section 3.1, you used `git config` to tell Git your username and email address. But notice that you used the `--global` switch. What does this switch do? Now that you know about Git's help system, you can learn about this switch by typing `git help config`.

> **TRY IT NOW** Type the help commands that you've seen so far in this section:
>
> ```
> % git help help
> % git help glossary
> % git help -a
> % git help config
> % git help -g
> ```
>
> Does your installation bring up the help in a browser or in the command-line window?

Alternatively, you can get help about any command by passing `--help`, as in `git config --help`. Who says Git isn't helpful?

It bears repeating: to get help about any Git command, type `git help command`, or `git command --help`, substituting the word *command* with the command you're interested in knowing more about. Remember to try `git help` with all the commands that you learn about in the upcoming chapters.

The documentation for Git exists on the web at many places. I often refer to www.kernel.org/pub/software/scm/git/docs/ and http://git-scm.com/documentation. The Git help manual page says that the formatted and hyperlinked help for the latest version of Git exists at http://git-htmldocs.googlecode.com/git/git.html. These are authoritative sources for Git documentation.

3.5 Controlling long output with a pager

In the previous TRY IT NOW exercise, you typed `git help -a` to see a listing of all the commands available in Git. The output of this command is so long that it scrolls past the height of the window. To see the first few lines of this output, you'd need to use the command-line window's scrollbars.

To make long output pause at each page worth of text, you can ask Git to send its output to a pager. The `-p` or `--paginate` switch to Git (see figure 3.3) tells Git to pause after it displays one full screen of text.

To display the rest of the text, press the spacebar. To go backward in the output, press B. This pager is the same mechanism used by the help text.

> **TRY IT NOW** At the command line, compare the output of these two commands:
>
> ```
> % git help -a
> % git --paginate help -a
> ```
>
> You can use the `-p` switch in place of `--paginate`.

3.6 Lab

You're now at your first lab. I hope that throughout this chapter, you've been typing the commands as I've been prompting in the text. Also, I hope you performed the TRY IT NOW exercises, especially the one at the start of section 3.2, which is a prerequisite for your future Git work. Now you're ready to follow these steps:

1. You told Git your name and email, but where does Git save that information? See if you can locate where this is.
2. Git is known as the *stupid content tracker*. What `git help` page says this?
3. What is the Git command that forward-ports local commits?
4. What does the abbreviation DAG stand for, in the context of Git?
5. Does your installation come with a Git tutorial help file?
6. The command-line commands you saw in the first TRY IT NOW are not Git commands, but rather Unix/Linux commands. Do they display anything helpful when you type them followed by the `--help` switch?
7. When Git is given the `-p` (or `--paginate`) switch, it uses the command-line tool `less`. Type `less --help` to play with this pager some more, and learn about other ways to scroll through long text (for example, instead of scrolling one page at a time, you can scroll one line at a time).

3.7 Further exploration

The command line that you've been working in is known officially as the *Unix/Linux shell*, or the *shell*. It's one part of Unix/Linux (some would argue the most visible part).

I hope working on the command line has piqued your interest in the shell. For those with an interest in learning more, seek out references for BASH, the flavor of the shell that Git is typically paired with. Because working on the command line involves a lot of typing, many of the interesting techniques used in the shell are those that reduce keystrokes. These include things like aliases, command-line history (repeating commands you've just typed), command-line editing (which we covered), and directory stacks.

It's beyond the scope of this book to discuss all of these techniques, but they're worth exploring, especially if you find yourself typing the same things over and over. If you sit with someone who's an expert at the command line, it can look like magic as commands seemingly appear out of nowhere, with just a keystroke or two. Visit these URLs to learn a bit more about these features:

- www.gnu.org/software/bash/manual/html_node/Aliases.html
- www.gnu.org/software/bash/manual/html_node/Bash-History-Facilities.html
- www.gnu.org/software/bash/manual/html_node/Command-Line-Editing.html
- www.gnu.org/software/bash/manual/html_node/The-Directory-Stack.html

The parent URL for these pages is here:

- www.gnu.org/software/bash/manual/html_node/index.html

3.8 Commands in this chapter

Starting with this chapter, the Git commands discussed in each chapter are summarized in a table, alongside a short description (see table 3.2).

Table 3.2 Commands used in this chapter

Command	Description
`git config --global user.name "Your Name"`	Add your name to the global Git configuration
`git config --global user.email "Your E-mail Address"`	Add your email address to the global Git configuration
`git config --list`	Display all the Git configurations
`git config user.name`	Display the `user.name` configuration value
`git config user.email`	Display the `user.email` configuration value
`git help help`	Ask Git for help about its help system
`git help -a`	Print all available Git commands
`git --paginate help -a`	Paginate the display of all Git commands
`git help -g`	Print all available Git guides
`git help glossary`	Display the Git glossary

4 Making and using a Git repository

You spent yesterday reviewing the command line and learning how to get help from Git. You also configured Git by telling it your name and email address. You saw in figure 1.2 of chapter 1 that Git gives each and every user the entire repository. Today you're going to create your own Git repository by using the `git init` command. You'll then add files into it by using `git add` and `git commit`. Finally, you'll learn how to get your repository's status and history with the `git status` and `git log` commands.

These are fundamental commands for working with Git! I hope you'll see that creating a repository isn't such a big deal, and that because repositories are so easy to create, it always makes sense to create one for tracking even trivial projects.

4.1 Understanding repository basics

A *repository* is a specialized storage area in which you can keep track of your work. In chapter 1, I introduced in listing 1.1 a simple example of what an ad hoc repository might look like. Let's gradually uncover this example to properly introduce the basics of a version control repository.

Let's assume you have a build process that requires you to preprocess some files. Maybe you have to rename these files or add a timestamp to them. Let's assume you're on Windows, so you'll use a Windows batch file (its suffix is BAT) to execute these steps.

You write code for this utility script, saving it to a file called filefixup.bat. The script works great, but a few days later you need to code up different preprocessing steps. The old BAT file still works, so instead of making modifications to that file, you rename it by appending the string "-01" to its name (filefixup-01.bat). This signifies version 01, and you've made a mental note that version 01 is the first version of the utility that renames and adds a timestamp to the files. This new filename also gives you the ability to keep using this old version as you work on the newer version.

You start adding the newer code to the file filefixup.bat, knowing that you can always go back to filefixup-01.bat. The file with the 01 suffix is a copy of the BAT file. The file without the 01 suffix is your working copy.

After a few hours, you reach a good stopping point with the new utility program. You realize that this would be a good point to make a new version of this program. Following your convention, you make a copy of the working file, naming it filefixup-02.bat. Once again, you make another mental note that version 02 contains your improvements.

Now your directory looks like the following listing.

Listing 4.1 An ad hoc version control system

```
C:\buildtools> dir
 Volume in drive C is GNU
 Volume Serial Number is 5101-E64D

 Directory of C:\buildtools

03/15/2014  08:22 PM    <DIR>          .
03/15/2014  08:22 PM    <DIR>          ..
03/01/2014  08:22 AM            11,843 filefixup-01.bat
03/03/2014  08:52 AM            11,943 filefixup-02.bat
03/03/2014  08:53 AM            11,943 filefixup.bat
               3 File(s)         60,066 bytes
               2 Dir(s)  467,905,187,544 bytes free
```

I hope you can see how you could get to the listing that you saw in chapter 1. You keep making changes methodically to the program, saving it to a new filename each time you get to a logical stopping point. If you're witty, you might even save a copy named filefixup-beforelunch.bat, to indicate the version that you were working on immediately before lunch. You'll throw this one away, but going to lunch is a stopping point.

Over time, you can see a makeshift history by looking at your progress via the timestamps on the version files. You might back up the entire directory to a flash drive or to a network shared drive. You might someday have to revert to an earlier version, and to do so, you copy the version that you want on top of the working file. These are the kinds of operations you could do using this ad hoc version control system.

It should come as no surprise that Git does all this, and more. Let's forget about this, or any other, ad hoc system you might have been using before, and try out Git.

4.2 Creating a new repository with git init

The command to create a new Git repository is `git init`. You can type this in any directory, and instantly a Git repository will be created in that directory. That's all there is to it.

TRY IT NOW In this section, you'll enter commands into the Git command line. Refer to section 3.1 for how to start this command line for your platform. For Windows users, you'll be typing commands into Git BASH (you can't use the DOS or PowerShell windows for these kinds of operations). For Mac and Unix/Linux users, you'll be entering commands into the standard terminal application.

These steps navigate you to the home directory, create a new directory, and then create a Git repository in that directory:

```
% cd
% mkdir buildtools
% cd buildtools
% git init
% ls
```

One thing to note is that the last command won't print anything. No files are in the repository, because you haven't added any yet!

Performing the preceding steps creates a Git repository right in the buildtools directory. There are two important things to note:

- No server was started.
- The repository is entirely local.

It's hard to overemphasize the first item: no server was started. Usually commands are available on our individual machines to let us see all the running processes. In Windows, you can visit the Task Explorer. In the Mac or on Unix/Linux, you can type `ps -e`. If you inspect these mechanisms, you won't see a Git process running. Git doesn't require a server. By not requiring a server, it's easy to decide for yourself to start version control on a particular directory because you don't have to request permission from anyone, except yourself. (And yes, you should always allow yourself to use Git!)

The second item is a corollary of the first item: the repository that you created is entirely on your machine. Note that I didn't say the repository is "running on your machine." Nothing is running. But the Git repository files have been created, and the directory buildtools has been transformed from an ordinary directory into a version control working directory.

Figure 4.1 illustrates what happens when you type `git init` after creating a working directory: you create a Git repository. The Git repository is on your machine inside the working directory, not on a server somewhere over the network! I take pains to distinguish between the working directory and the repository because they're two different things. A *working directory* is simply the place where you do your work. The *repository* is the specialized storage area in which you can save versioned files. The Git software can track files, meaning it can detect differences between files in the repository and files in the working directory.

At this point, you can start to create files in the working directory and add them to the repository.

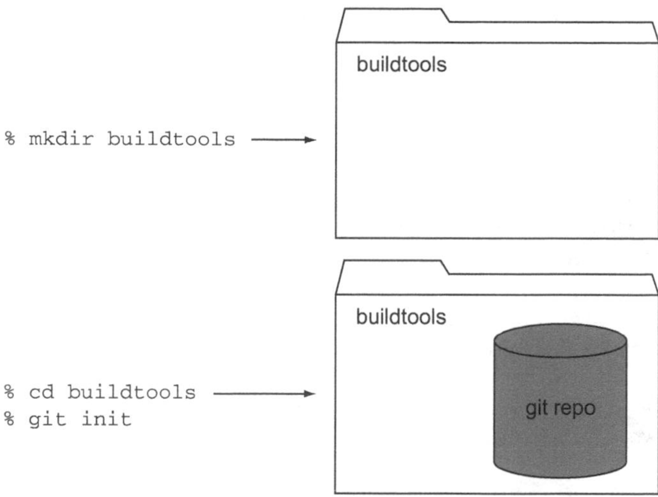

Figure 4.1
Visualizing `git init`

> **Above and Beyond**
>
> Git's repository exists in a separate hidden directory named .git. In the command line, filenames or directories that begin with a period, or dot, are considered hidden by the `ls` file- and directory-listing command. You can see the Git repository by typing `ls -a .git` in your working directory (recall that this is the directory named buildtools). You can even look around in there by typing the following:
>
> ```
> % cd .git
> % ls
> ```
>
> The nature of these files and directories is beyond the scope of this chapter. I do acknowledge that people are curious about where the repository lives, and this answers it: the repository lives inside the working directory.
>
> Be advised that you shouldn't touch any of these files! They're manipulated by Git, and modifying them manually will cause issues down the road. For the uneasily satisfied, or the eternally curious, type `git help repository-layout` to explore some more.

4.3 Tracking files with git status and git add

Your repository is empty, but it's ready for action. To see this for yourself, you'll use the `git status` command. This Git command will tell you the state of your working directory.

4.3.1 Using git status to check your repository state

The command `git status` is something you'll type often, so let's get used to it by trying it out.

TRY IT NOW These steps navigate you to the home directory and then to the newly created buildtools directory. You'll query the status of this working directory:

```
% cd
% cd buildtools
% git status
```

Notice that you can skip the first and second commands if you're already in the buildtools directory. Remember, you can find out your current directory by typing `pwd`.

The status output looks like the following listing.

Listing 4.2 The initial `git status` of an empty repository

```
# On branch master
#
# Initial commit
#
nothing to commit (create/copy files and use "git add" to track)
```

The status message enumerates information such as the branch name and the current commit identifier (this is the initial commit); these can be ignored for now. The last line is the one to focus on. It announces that there's nothing to commit. It then says how to start tracking files: use `git add` to track. Let's create a file that Git can track. You'll stick with the command line and create a file by using the `echo` command.

TRY IT NOW These steps navigate you to the home directory and then to the newly created buildtools directory. You'll create a file by using the `echo` command:

```
% cd
% cd buildtools
% git status
% echo -n contents
% echo -n contents > filefixup.bat
% git status
```

The first `echo` command takes the string argument `contents` and prints it onscreen. You'll notice that it doesn't include the newline, so the prompt prints immediately after the word `contents`. The second `echo` command takes the string argument and prints it to a file that we call `filefixup.bat`.

Both times you use the `-n` switch to suppress the newline. This avoids an end-of-line warning message that you'll examine more thoroughly in the lab at the end of the chapter. (If you do see "warning: LF will be replaced by CRLF in new_file. The file will have its original line endings in your working directory," check that you added the `-n` switch to `echo`. You can ignore this warning.)

Your `git status` output looks something like the following listing.

Listing 4.3 `git status` before adding a file

```
# On branch master
#
# Initial commit
#
# Untracked files:
#   (use "git add <file>..." to include in what will be committed)
#
#       filefixup.bat
nothing added to commit but untracked files present (use "git add" to track)
```

As with the `git status` output in listing 4.2, you can ignore the branch and commit identifier. But unlike listing 4.2, Git has now detected a new file in the working directory, and points out that it's untracked. In the last line, it offers its suggestion of what to do next: use `git add` to track.

4.3.2 Using git add to add a file to your repository

Whenever you want to introduce a new file to a Git repository, you must use `git add` on that file first. Git can only keep track of files that it has been told about. Let's run `git add` in the next exercise.

> **TRY IT NOW** This exercise, and all future ones in this chapter, assume you're already in the buildtools directory (the working directory). Refer to the earlier TRY IT NOW exercises to see how to navigate to this directory via the command line. You might also refer to the previous chapter on how to use the Tab key to autocomplete long filenames and directories.
>
> ```
> % git add filefixup.bat
> % git status
> ```

The output of `git status` tells you the same boilerplate information (your branch and your commit number), as you can see in the following listing. But it now tells you that there's a new file, and that it can be committed.

Listing 4.4 `git status` after adding a file

```
# On branch master
#
# Initial commit
#
# Changes to be committed:
#   (use "git rm --cached <file>..." to unstage)
#
#       new file:   filefix.bat
#
```

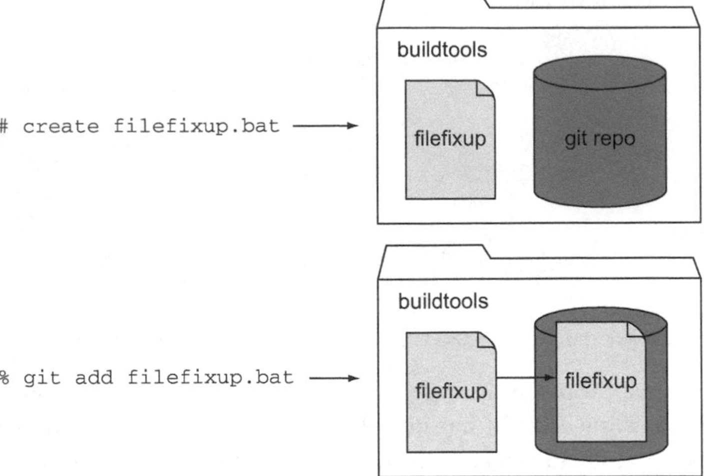

Figure 4.2
Visualizing `git add`

As you can see in figure 4.2, typing `git add` adds the file to the repository. You'll see in the next section that the command adds the file to a special part of the repository (and you'll learn more about this area in chapters 6 and 7).

4.4 Committing files with git commit

Git now knows about your file (filefixup.bat) and tracks changes to it. But you haven't yet committed this file to the repository. Committing files to the repository creates the timeline discussed back in section 2.1. If you look at figure 4.2 in the previous section, the file has been copied into the repository, but no history has been recorded.

To create a timeline event, you have to commit the file to the repository, using `git commit`. The commit step is what stores the file and records a timeline entry in the repository. Remember, this whole time you've been working on the file in the working directory, and the only thing Git has been doing is tracking it. Let's run `git commit` in the following TRY IT NOW exercise.

> **TRY IT NOW** This exercise assumes you're already in the buildtools directory. See the previous TRY IT NOW sections on how to access this directory. The next command commits your change to the repository, along with a message, which is in quotes after the `-m` switch (`-m` stands for *message*). Now type the following:
>
> ```
> % git commit -m "This is the first commit message"
> ```
>
> You should see the message in listing 4.5.

Listing 4.5 A successful Git commit message

```
[master (root-commit) 5308add] This is the first commit
 1 file changed, 1 insertion(+)
 create mode 100644 filefixup.bat
```

The string `[master (root-commit) 5308add]` is significant in that it's announcing that this is the root commit, which means this is the first time you've run `git commit` in this repository. The word `master` indicates the branch. Every repository opens a default branch, so don't worry about it for now. The `5308add` is a SHA1 ID, and it's the unique identifying number for this commit. Every commit has a unique SHA1 ID.

Skipping the commit message for a moment, the last two lines of the commit output report the number of files that have changed (in this case, one file) and the nature of those changes (in this case, one insertion). It also reports the mode for the newly created file in your repository (`100644`). The *mode* is a number representing a file's permission. For the purposes of this book, don't worry about a file's permissions, but do note that Git tracks this.

Once again, Git gives you a lot of information to look at, but the most important item here is the commit message: `This is the first commit.`

In figure 4.3, I've split the repository into two sections. When you use `git add`, the file is stored in a kind of waiting room, the staging area (the lower half of the repository in the figure). It's only when you commit the file with `git commit` that Git starts tracking the file's history (the upper half of the repository figure). You'll learn more about the staging area in chapter 7.

Recall your ad hoc version control system back in section 4.1? Every time you saved the spreadsheet with a special suffix, you had to make a mental note of what that suffix indicated. I don't know about you, but my mental note capacity is limited! In that example, the 01 suffix represented the spreadsheet with just the data, the 02 suffix

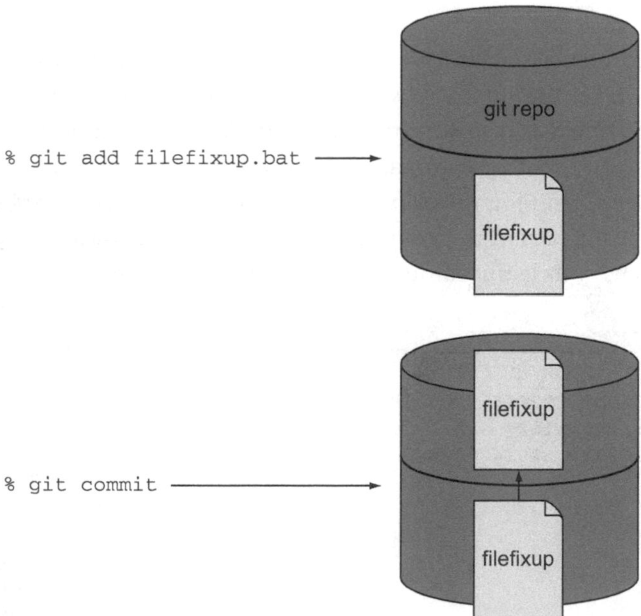

Figure 4.3
`git add` **and then** `git commit`

represented the spreadsheet with formulas down the first column, and so on. With Git's commit messages, you can write these mental notes directly into the repository!

In the next chapter, you'll learn how to make commits by using Git GUI. For now, bask in the fact that you've made your first commit into a Git repository!

4.5 Viewing the repository with git log and ls-files

The repository now has your file, and your working directory is in a clean state. To confirm this, you'll use the `git status`, `git log`, and `git ls-files` commands.

> **TRY IT NOW** This exercise assumes you're already in the buildtools directory:
>
> ```
> % git status
> % git log
> ```
>
> The first command should announce that the working directory is clean, and that there's nothing to commit. The second command displays the log.
>
> The output of `git log` appears in a pager. You learned how pagers work in the previous chapter. To page through the output, press the spacebar. To exit the pager and return to the command-line prompt, press Q.

If all goes well, you'll see output that looks like the next listing.

Listing 4.6 `git log` output

```
commit 5308adddb9a1526dbf12928f51f7c5328730d38b
Author: Rick Umali <rickumali@gmail.com>
Date:   Wed Apr 16 22:12:07 2014 -0400

    This is the first commit message
```

The `git log` command displays all the commits that have been made to the repository. It turns out that this is the typical way most people look at the repository: as a series of commits. In listing 4.6, you'll notice that the Author field is the user and email specified by our `git config` commands from chapter 3. The date should be your local date, and the commit log message is the message that you entered earlier.

To see the files that are a part of the commit, pass in the `--stat` switch to `git log`. When you use this switch, your listing should look like the following.

Listing 4.7 `git log --stat` output

```
commit 5308adddb9a1526dbf12928f51f7c5328730d38b
Author: Rick Umali <rickumali@gmail.com>
Date:   Wed Apr 16 22:12:07 2014 -0400

    This is the first commit message!
 filefixup.bat | 1 +
 1 file changed, 1 insertion(+)
```

To list the files in the repository, use `git ls-files`. Because you've committed the working directory contents to the repository, `git ls-files` will display the same output as the command `ls` (to list files in the current directory).

4.6 Lab

Creating a Git repository is a lightweight operation. Adding a file to the repository requires three steps: creating the file, telling Git to track the file (via `git add`), and then committing the file to the repo (via `git commit`). This lab reinforces these steps.

If you haven't yet performed the TRY IT NOW sections on your own, please do so now. If you've tried them, now is the perfect time to do them again, perhaps with a different directory name or a different filename.

You can encounter a lot of strange errors with Git. But if you're careful and deliberate, you can usually spot where things went wrong. In this lab, you're going to deliberately perform Git operations in the wrong order so you can see the strange error messages. Think about why you get the error message and then see if you can continue.

1. Create a new directory. Run `git init` and then run `git log`.
   What is the error that you receive? Why might you get this error?
2. Follow these step carefully:

   ```
   % mkdir twoatonce
   % cd twoatonce
   % git init
   % echo -n contents > file.txt
   % git add file.txt
   % echo -n newcontents > file.txt
   % git status
   ```

   What is the output of the `status` command? Do you see file.txt twice in the status message? Once to be committed and once to be added?
3. Create another file for a current Git repository, this time using `echo contents > new_file.txt`.
   Note that this time you didn't use the `-n` switch in the `echo` command. Now try `git add` on this file. If you're on Windows, you should see the warning message "warning: LF will be replaced by CRLF in new_file.txt. The file will have its original line endings in your working directory."

   Read the help for `git config` (type `git config --help`), and look at the `core.safecrlf` and `core.autocrlf` settings. This warning is Git saying that it will try to be careful with your text file's line endings.

   The point of this lab is to observe that Git is sensitive to end-of-line issues, the kind that has bedeviled text-file interoperability between Unix/Linux and non-Unix/Linux machines.

4.7 Commands in this chapter

Table 4.1 Commands used in this chapter

Command	Description
`git init`	Initialize a Git repository in the current directory
`git status`	Display status of current directory, as it relates to Git
`git add FILE`	Start tracking FILE in Git; adds FILE to the staging area
`git commit -m MSG`	Commit changes to the Git repository, with a message (in quotes)
`git log`	Display the log (history) of the Git repository
`git log --stat`	Display the log with the files that were modified
`git ls-files`	List the files in the repository

5 Using Git with a GUI

In Git, you'll spend the majority of your time on the command line. As I said in the previous lunch, Git was born and bred on the command line. But now is a good time to explore how to interact with Git via its GUI. That's right. Git ships with a GUI!

Command-line aficionados often look down their noses at the GUI (the somewhat derogatory abbreviation for *graphical user interface*, pronounced *gooey*). They take the stance that any tool without a command-line interface is a tool not worth using. But it's worth exploring because the GUI does allow for a much richer user experience with Git by offering a better visualization of what's happening in the repository. The `git gui` command makes it easy to visualize the state of the working directory. Making commits and visualizing the status is a one-window operation with this tool, and for some people, that's a more comfortable way of working.

This chapter focuses on interacting with a few Git GUI features, such as creating a repository, adding and committing files, and viewing your repository's history. You'll learn the other features of Git GUI later, as you learn more about Git itself.

Finally, the GUI that you'll be using is the one that ships with Git itself. Plenty of other GUIs are available for Git, and you'll explore two of these in chapter 19, but Git GUI is the official GUI for Git. It's designed to be portable, which means it can be run from any of the three major platforms (Windows, Unix/Linux, or Mac) and it generally runs the same way. When you install Git, you get Git GUI for free.

5.1 Starting Git GUI

By design, to start Git GUI, you should first be on the command line. Follow the TRY IT NOW to see Git GUI on your screen.

TRY IT NOW These steps navigate you to the home directory and then start Git GUI:

```
% cd
% git gui
```

At this point, you should see a window on your screen, as shown in figure 5.1. Behold: Git GUI!

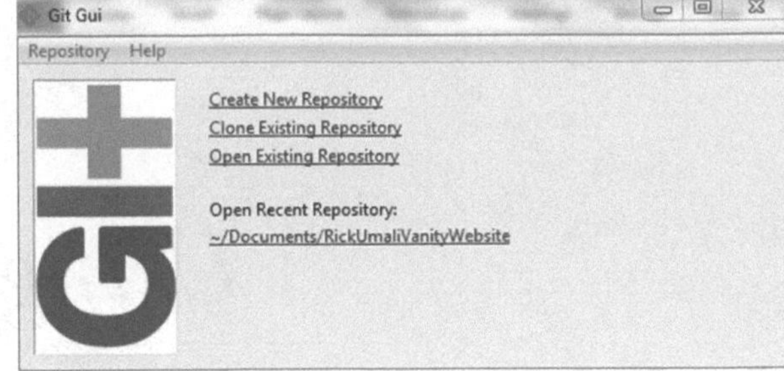

Figure 5.1 The initial screen for Git GUI

In the preceding exercise, I purposely had you run Git GUI in your home directory (remember, typing `cd` by itself switches you to the home directory). The small window that you see is the default screen offering you some options. You won't do anything with this window, but for curiosity's sake, go ahead and look at the Repository and Help menus at the top of Git GUI.

Depending on how you installed Git, and your system (Windows, Unix/Linux, or Mac), you may have a menu item that will start Git GUI. Look for this option, but I've found this reliably only on Windows machines.

5.1.1 Starting Git GUI in Windows

Windows users have an additional way to start Git GUI. When Windows users install Git, the default installation creates three context menus for Git, which users can access by right-clicking any folder in Windows Explorer.

These three context menus are Git Init Here, Git GUI, and Git BASH, as shown in figure 5.2. They do just what you might expect. If you click a directory that isn't a Git

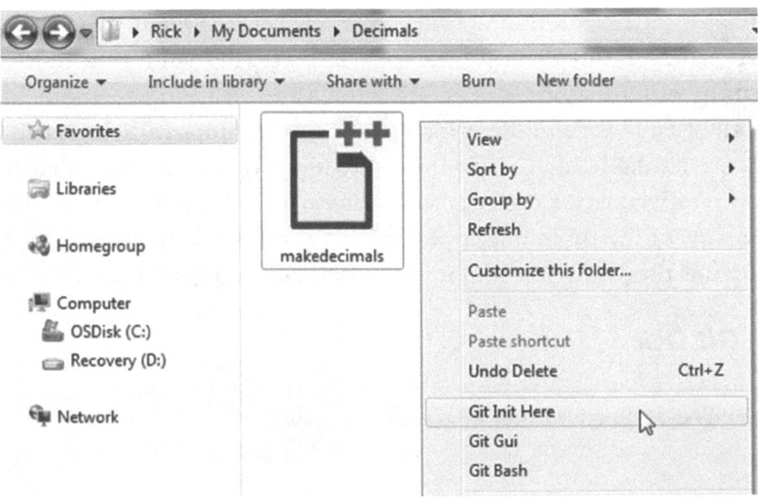

Figure 5.2 Git's context menu (in Windows)

working directory already, Git Init Here will create a repository in that directory. If you select Git GUI or Git BASH, you'll open either Git GUI or the Git command line with the selected directory as the working directory.

5.2 Creating a repository with Git GUI

Creating a repository by using Git GUI can be done from the initial Git GUI screen. Let's create a new repository in the next TRY IT NOW section. You'll use this repository in the rest of the chapter, so please follow along.

TRY IT NOW These steps position you in your home directory and start Git GUI. Remember, you may have a menu item to start Git GUI, and you can use that instead.

```
% cd
% git gui
```

Git GUI appears. Click the Create New Repository link. This prompts you for a directory name. Click the Browse button, and you'll see a familiar file browser window.

Create a new directory called `newrepo` in your home directory.

WINDOWS

If you're on Windows, you should be able to create a directory in this browser, as shown in figure 5.3.

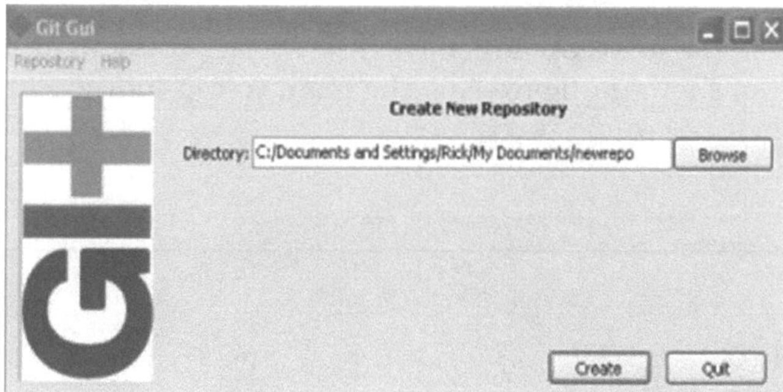

Figure 5.3 Creating newrepo in Windows

MAC

On the Mac, use Finder to create the path shown in figure 5.4. (You'll use your own home directory.)

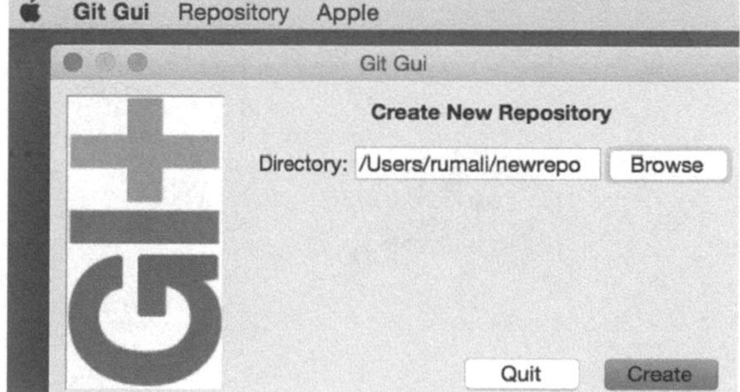

Figure 5.4 Creating newrepo on the Mac

UNIX/LINUX

On Unix/Linux, create the newrepo directory by typing `newrepo` in the dialog box that appears when you click Browse (see figure 5.5).

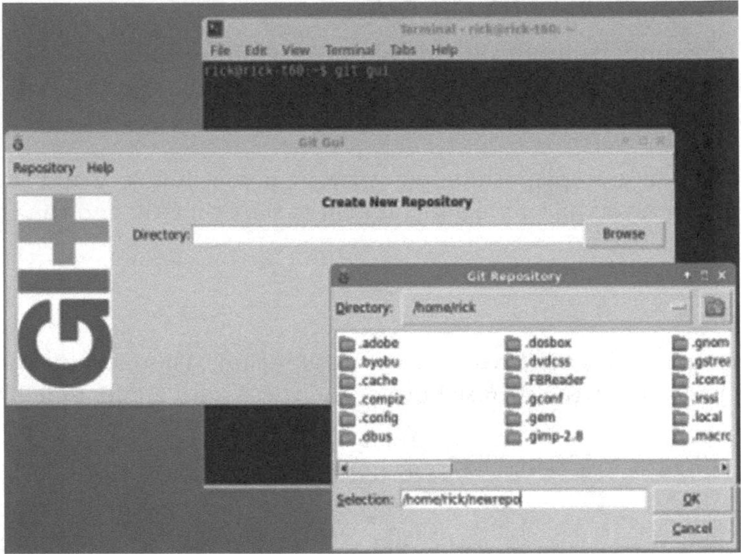

Figure 5.5 Creating newrepo on Linux

After you've entered the newrepo directory, click Create. You'll then see the main Git GUI screen shown in figure 5.6. The main screen indicates, via the title bar, that it's in the newrepo repository.

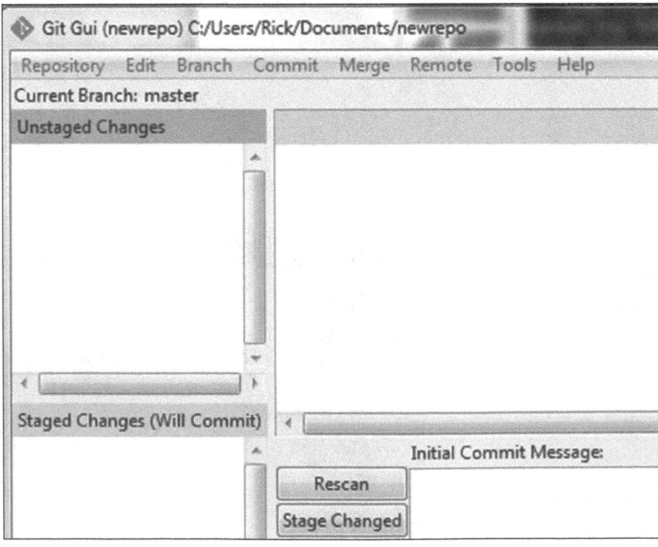

Figure 5.6 The newrepo repository open in Git GUI

Git GUI is designed to have the same look and feel across all three platforms. From here on out, I'll mostly show the Windows platform, unless important differences apply to a specific platform.

At this point, the repository is created. Access the Repository menu (see figure 5.7), and then click Explore Working Copy to see the empty working directory in your system's file browser.

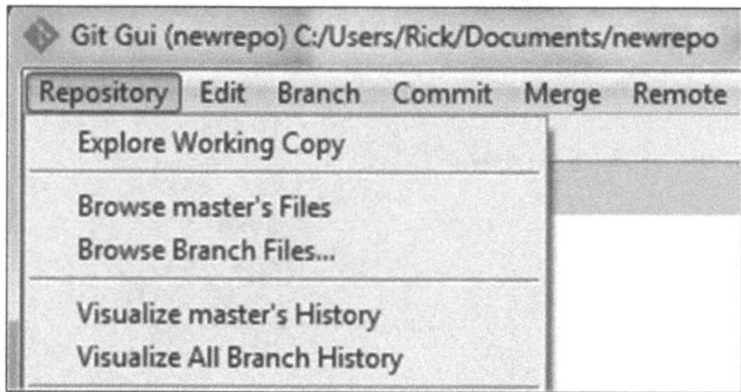

Figure 5.7 Explore Working Copy

As you learned in yesterday's lunch, Git doesn't need to communicate to a server during the repo initialization process. The repository is created entirely on your local machine and doesn't need a network connection to complete.

Git GUI's advantage is the ability to browse the file system by using your system's file-browsing tool. If you're not used to working in the command line and keeping track of where you are (via the `pwd` command), creating a repository should be a piece of cake in the GUI.

The main Git GUI screen reminds you of the working directory via the title bar. Also, using the menu item Explore Working Copy (under the Repository menu), you can open your system's browsing tool to see the directory and its contents.

5.3 Adding a file into the repository via Git GUI

The Git repository has been created in your newrepo working folder. Let's add a file to the repository, as you did yesterday.

TRY IT NOW The first part of this TRY IT NOW opens the newrepo directory with `git gui`. If Git GUI is still up and running from the last TRY IT NOW, you're already in the right spot. You can proceed to the text after figure 5.9.

You can use `git gui` to open the newrepo directory in two ways.

First, you can open the Git command-line window and then type in a few commands:

```
% cd
% cd newrepo
% git gui
```

Note that you can replace the first two commands with `cd $HOME/newrepo`. After you type `git gui`, it will appear as in figure 5.9.

The second, and possibly easier, way is to open Git GUI as in section 5.2. You open the Git command-line window and then immediately type the following:

```
% git gui
```

Git GUI's first window appears. Select the newrepo folder via the file system browser (see figure 5.8). Git GUI will open a working directory only if it's a Git repository.

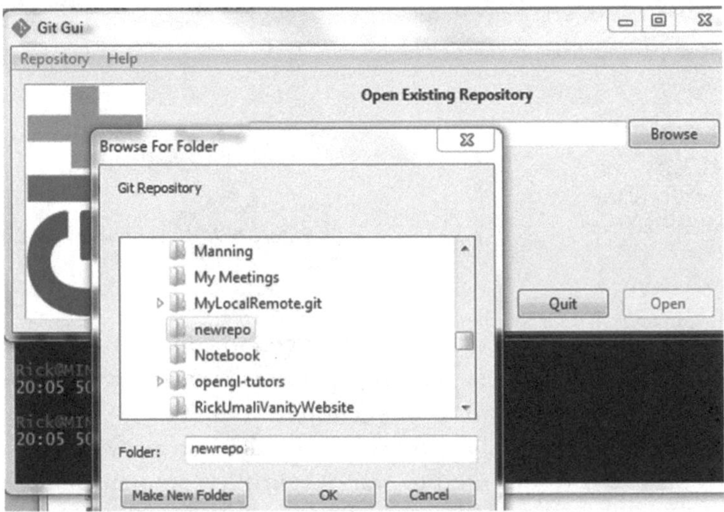

Figure 5.8 Opening an existing Git repository via Git GUI

Your Git GUI displays the screen in figure 5.9.

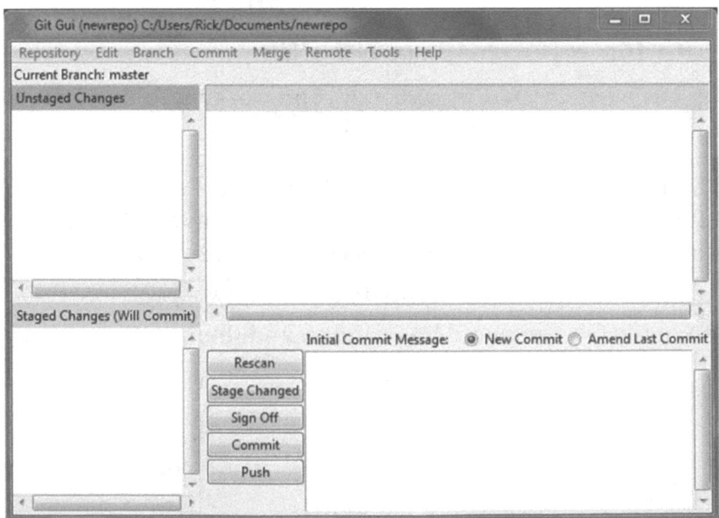

Figure 5.9 Git GUI open in the newrepo working directory

The multiple text windows capture the status of the working directory. Recall from yesterday that you learned the `git status` command. This main window shows the same information as `git status`. Let's now add a file to the repository.

TRY IT NOW Open the Repository menu and select the Explore Working Copy menu. This brings up your system's file browser. It should show an empty directory. Use your file browser to add a new text file.

WINDOWS

Figure 5.10 shows how to do this in Windows.

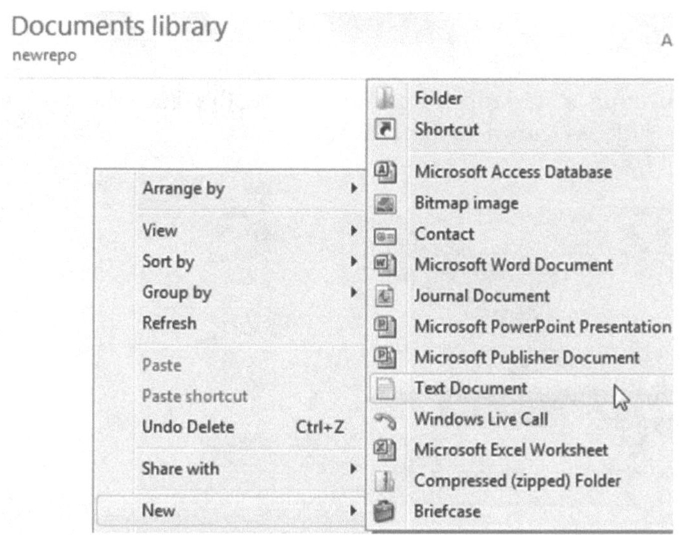

Figure 5.10 Create a new text file here.

You should name this file `sample.txt` to follow along in this section. What you've done is similar to what you did on the command line with the `echo` command. You could open the text file now by using your favorite editor and type in some text.

MAC

For the Mac, you can use TextEdit to create a new file in the newrepo directory. Open the File menu of TextEdit, as shown in figure 5.11, and click Save to save the file in the newrepo directory.

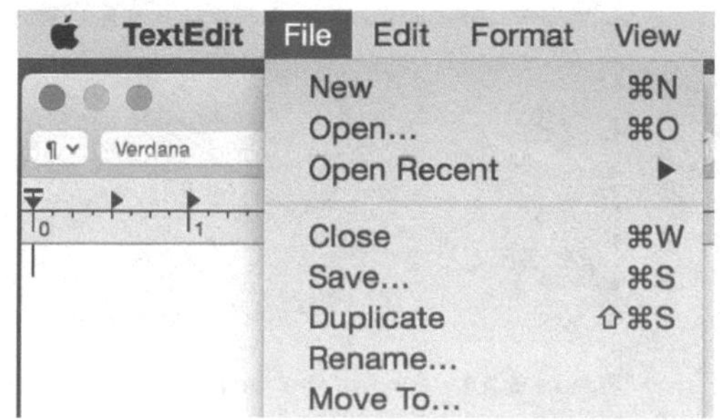

Figure 5.11 On the Mac, save your text file to the newrepo directory.

The Save dialog box in TextEdit prompts you for the location, as you can see in figure 5.12.

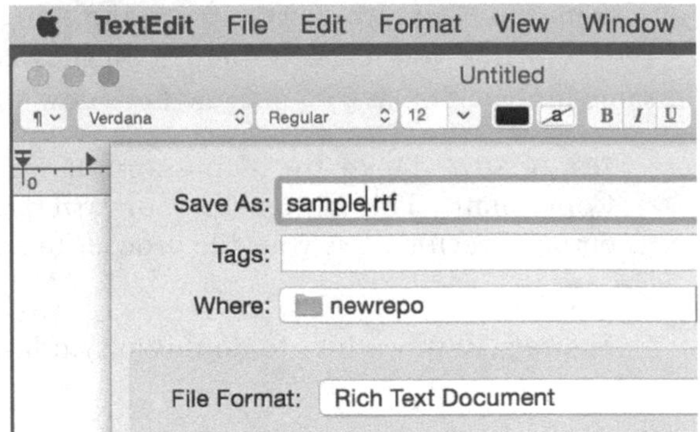

Figure 5.12 Mac's TextEdit saves files only with the RTF extension. This is OK. Just make sure the file is in the newrepo directory.

UNIX/LINUX

In a system running a desktop Unix/Linux, use the File Manager tool to introduce a new file, as shown in figure 5.13.

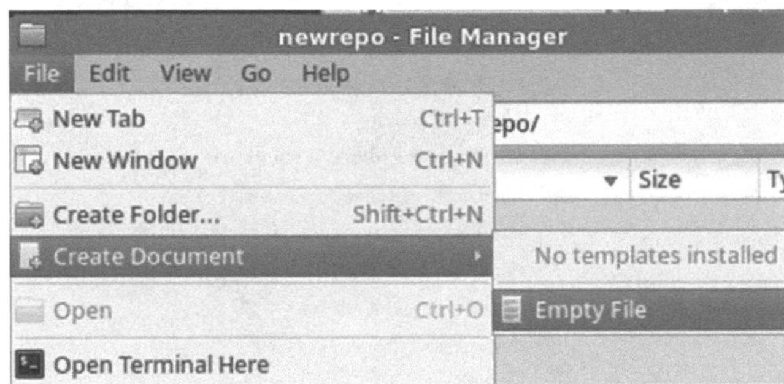

Figure 5.13 Create a new empty file by using the Linux File Manager.

Make sure to save the file in the newrepo directory, as shown in figure 5.14.

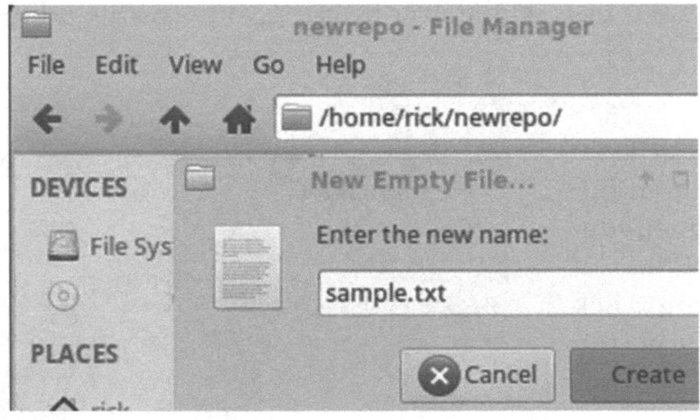

Figure 5.14 Saving the new file in the newrepo directory using Unix/Linux

Once you can see your empty file in the newrepo working directory, go back to Git GUI and click Rescan. This button appears highlighted in figure 5.15. This is also a menu item under the Commit menu.

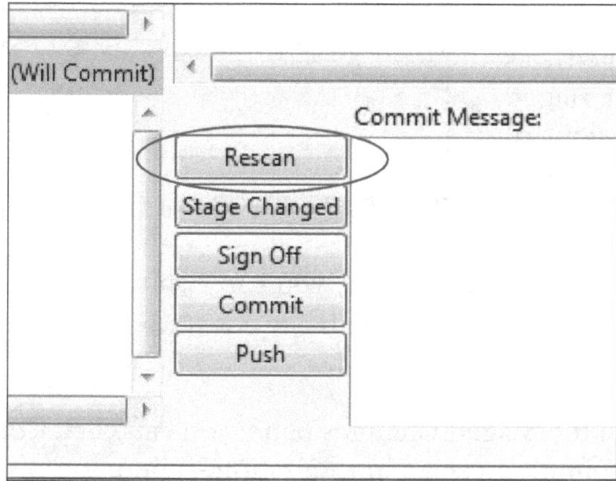

Figure 5.15 The Rescan button

Clicking Rescan makes Git GUI recognize this new file, and it now appears in the upper-left pane of Git GUI. Clicking the file in this pane makes its status appear in the upper-right pane. Figure 5.16 shows what Git GUI looks like.

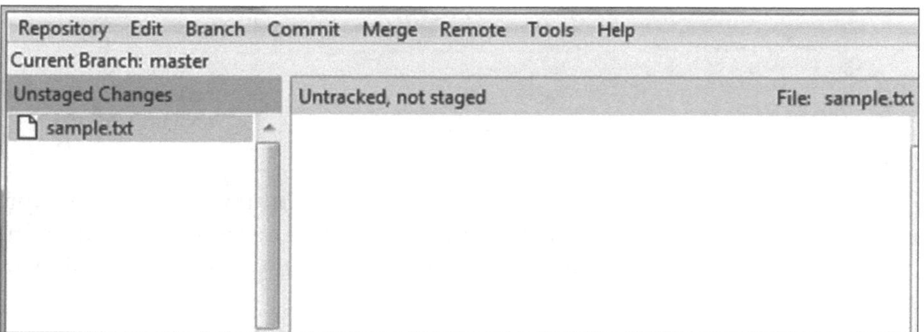

Figure 5.16 Your file in Git GUI

This file is untracked. Git recognizes that there's a new file, but announces that it's "untracked" and "not staged." To commit this file into the repository, you must first introduce it to the repository. You'll now use the equivalent of the `git add` command in Git GUI.

TRY IT NOW To begin tracking the file, first select the file by clicking it. Then open the Commit menu and select the Stage to Commit menu item. The file moves from the top-left pane (Unstaged) to the bottom-left pane (Staged), as in figure 5.17. You've just staged this file! This is the equivalent of typing `git add sample.txt`, which you did in the preceding chapter.

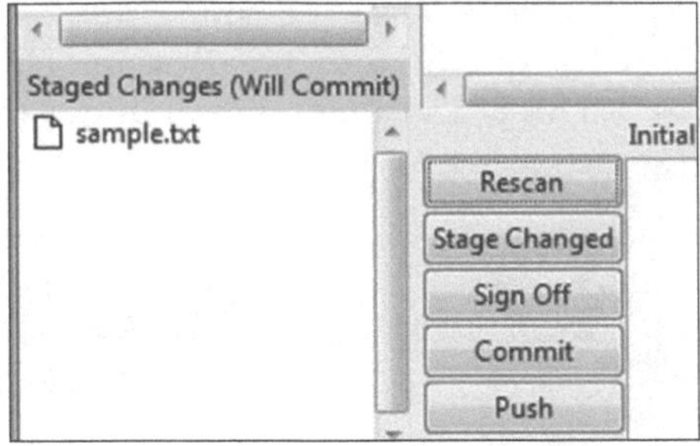

Figure 5.17 Staging your file via Git GUI

Now that the Git repository knows about this file, you can commit it. The file is now "staged" (in the lower-left pane), and you can probably guess what button you'll press next!

TRY IT NOW With the file in the Staged Changes pane, you can click Commit. But first you must enter a commit message in the bottom-right pane.

Just as you did in the preceding chapter, let's enter one line in the text pane, as shown in figure 5.18: `This is the first commit.`

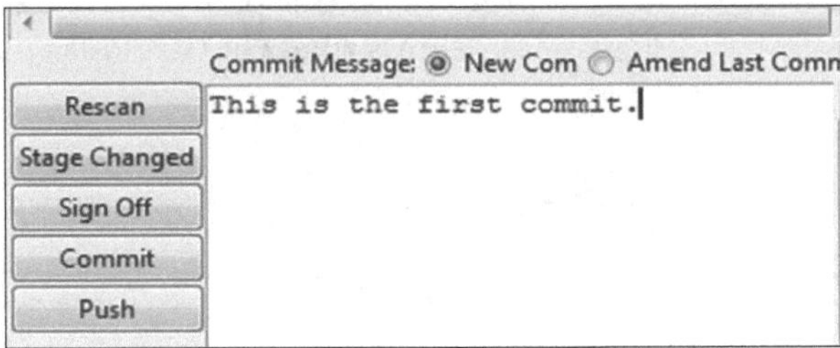

Figure 5.18 The first commit message

After entering this message, click the Commit button. You'll notice that your file disappears from the lower-left text pane, and your commit message appears in the status bar at the bottom of the Git GUI window, as shown in figure 5.19.

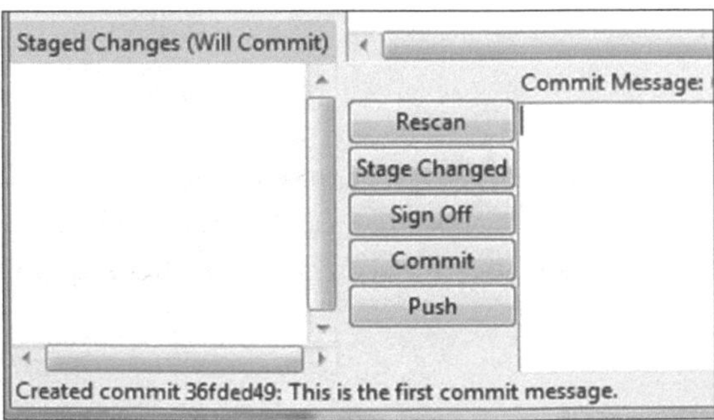

Figure 5.19 How Git GUI looks after the Commit button is clicked

Your file is now in the repository after this commit. If you're comfortable with command-line editors, bringing up a UI to enter a commit log message may be overkill!

5.4 Looking at your history

At the end of the preceding chapter, you were able to examine the contents of your repository by running the `git log` command. Let's check out the equivalent in Git GUI.

TRY IT NOW From the Repository menu, select Visualize Master's History, as in figure 5.20. Recall from yesterday that every repository opens a default branch called *master*.

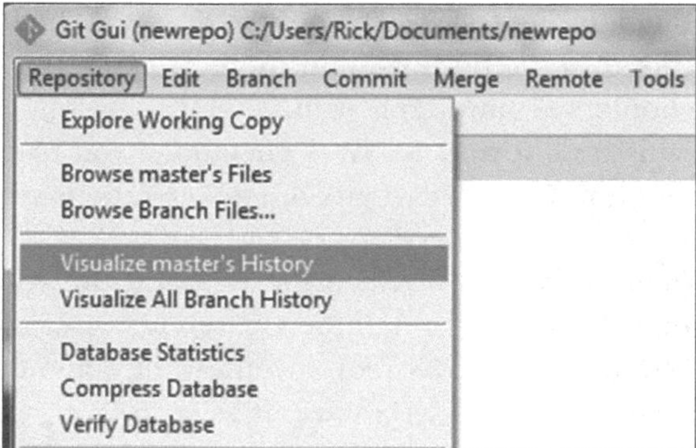

Figure 5.20 Visualize the master's history.

Clicking this item brings up another window that should look like figure 5.21.

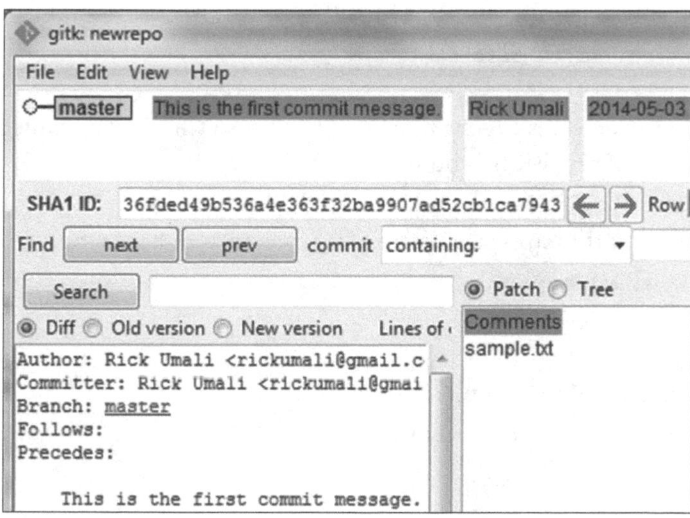

Figure 5.21 `gitk`, the equivalent of `git log`

Observe that the window that appears has the word *gitk* in the title, next to the repository name (newrepo). This window is a separate program from the Git GUI window you've been using so far in this chapter.

The gitk window contains a lot of text panes, but the key information is the ribbon of text in the upper-left pane. This is your log message. The text panes to the right show the committer's ID and date of this commit. The SHA1 ID is displayed underneath this commit information.

Because your repository contains only one commit, most of the other buttons and menu items aren't of any interest at the moment. Let's try a quick lab!

5.5 Lab

Let's exercise Git GUI a bit, but first make sure you've completed the TRY IT NOW sections if you didn't attempt them during your reading of the chapter. The flow of creating a repository, adding a file to the repository, and then checking the repository's contents is something you should be comfortable with.

If you've already done this during your lunch, I encourage you to perform the steps once again. Think about how Git GUI and gitk mimic operations you saw in the previous chapter.

In this lab, you'll intermix operations between the GUI and the command line (from yesterday). This section demonstrates that the Git repository can be accessed from either the GUI or the command line. Moreover, you can do some operations in the GUI, and some from the command line, as you see fit.

Follow the steps carefully:

1. From the command line, type the following:

   ```
   % cd
   % mkdir labrepo
   % cd labrepo
   % git init
   % echo -n contents > file.txt
   % git citool
   ```

   In what text pane does file.txt appear? (For those who have problems bringing up the `git citool`, visit this book's forum for help.)
2. What is `git citool`? (Hint: read its help documentation.)
3. Exit the Git GUI that appeared after you typed `git citool`; then from the command line, type this:

   ```
   % git add file.txt
   % git citool
   ```

   In what text pane does file.txt appear?
4. Exit the Git GUI that appeared after you typed `git citool`. Now type this:

   ```
   % echo -n more > file.txt
   % git citool
   ```

   Do you see file.txt in both the upper-left and the lower-left text panes?

5 How do you think you commit both of these changes?
6 Type `gitk` at the command line. Are you surprised that this ends up being a separate command? Does `gitk` have a `--help` switch?

5.6 Further exploration

You spent a lot of time with Git GUI in this chapter. Graphical user interfaces are a great jumping-off point for two areas of further exploration.

5.6.1 Other GUIs for Git

At least a dozen GUI clients are available for Git, across all the major platforms. Search around and try them out! Attempt the operations that you did in this chapter (create a repository and then add a file to it), to see how they approach things.

As you continue to learn more about Git via the command line, you'll also touch on how to complete those same operations with Git GUI and gitk, so you haven't seen the last of these tools.

5.6.2 Tcl, Tk, and Wish

A tantalizing area of exploration is the underlying technology behind Git GUI and gitk. Both tools are implemented in the computer language Tcl, which stands for *Tool Command Language*. Tcl is a dynamic, interpreted language invented in 1988 by John Ousterhout. Tk, a toolkit of GUI controls (a.k.a. widgets), was added to the language not long after. To learn more about the Tcl/Tk, visit www.tcl.tk.

Both Git GUI and gitk are written in Tcl/Tk. To get a taste for how to interact with the windowing shell and create a GUI entirely on the fly, open the command line and type `wish`. A window should appear on your screen, as in figure 5.22.

In Windows, two windows appear, one titled Console, and the other titled Wish. For Mac and Unix/Linux, only the Wish window appears. The console is on the command line, indicated by the % prompt.

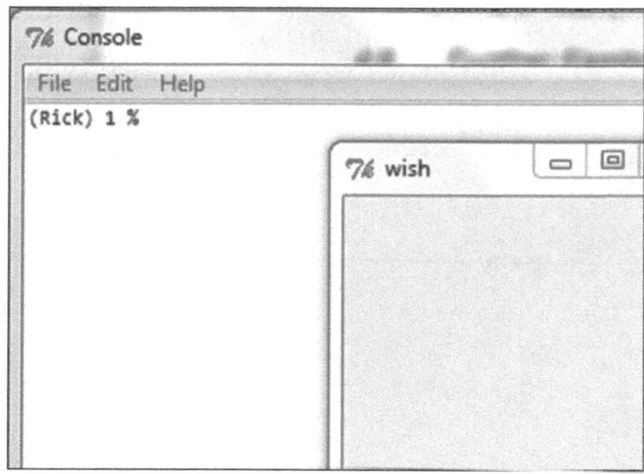

Figure 5.22 The Wish GUI window

In the console, type the lines in the following listing.

Listing 5.1 A simple wish GUI program

```
button .submit -text "git" -command { catch { exec git --version }
➥ results ; puts $results}
pack .submit
```

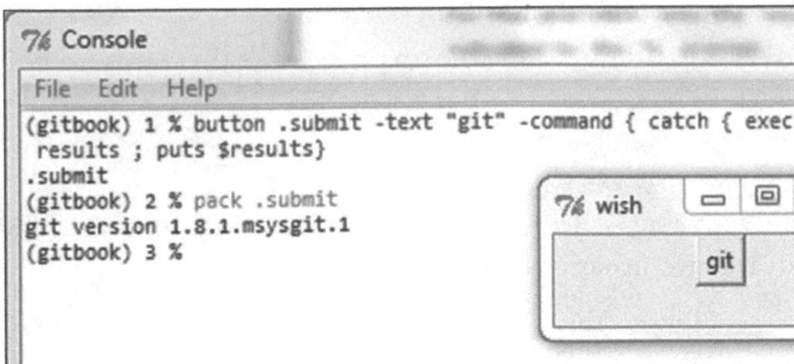

Figure 5.23 A small GUI implemented in Wish

After you type the `pack .submit` line, you'll see the Wish window change. It becomes a single button with Git as its label. Pressing this button displays the output of `git --version` in the Console window, as in figure 5.23.

As you can see, you've made a window with a button that prints out the Git version when you click it. How easy was that? As I said, this is an interesting area to explore, and when you install Git, you have the capabilities right on your machine.

5.7 Commands in this chapter

Table 5.1 Commands used in this chapter

Command	Description
`git gui`	Start Git GUI
`git citool`	Start Git GUI to commit changes
`gitk`	Start gitk (git log viewer)

6 Tracking and updating files in Git

You're now the proud owner of a Git repository. You've successfully added a single file into it. You can view the contents of your repository. It's now time to dive a little deeper into how to update and keep track of changes that you make to files in your repository. This is clearly one of the most important features of any version control system!

In this chapter, you'll create a new repository and then add files by using `git add` and `git commit`. You'll learn about the `git diff` command, which will help you keep track of what you've changed in your repository as you're working. You'll then dive into the staging area, one of the special features of Git that enables you to commit parts of your changes into the repository. This is a use case that comes up often, and Git's support for this is worth learning. Finally, you'll see how Git GUI provides these same operations.

6.1 Making simple changes

In this chapter, you'll be working in a new repository in which you'll build a simple program that adds two numbers.

6.1.1 Creating a new repository

Let's make an entirely new repository and put one file in it.

> **TRY IT NOW** You'll create a working directory called math, and then create a Git repository in that directory. You'll next create a small program called math.sh that initially contains one comment. This will be added into the repository.
>
> Given what you've learned over the preceding two chapters, you already know how to do this via Git GUI. This section describes the steps using the

command line. To that end, the first step is to open the command line. In Windows, this is Git BASH, but in Mac or Unix/Linux, this is the terminal.

Next, you can execute all the commands shown here:

```
% cd
% mkdir math
% cd math
% git init
% echo "# Comment " > math.sh
% git add math.sh
% git commit -m "This is the first commit."
% git log
```

These commands give you a new repository with a new file in it. Remember that the first `cd` command puts you in your home directory. Notice that you use the `git commit -m` switch, which you first saw in chapter 4. The result of using these commands is shown in figure 6.1.

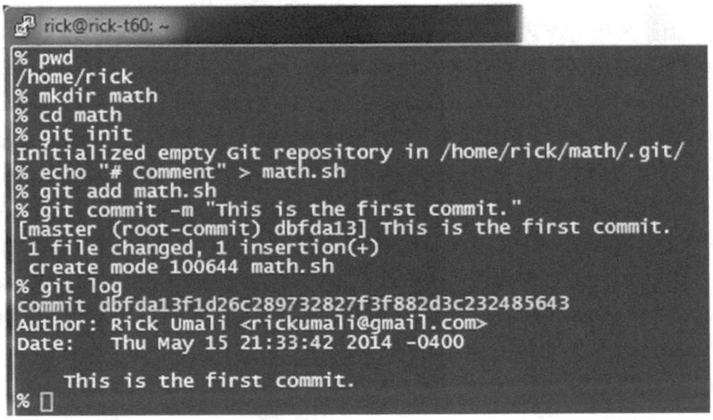

```
rick@rick-t60: ~
% pwd
/home/rick
% mkdir math
% cd math
% git init
Initialized empty Git repository in /home/rick/math/.git/
% echo "# Comment" > math.sh
% git add math.sh
% git commit -m "This is the first commit."
[master (root-commit) dbfda13] This is the first commit.
 1 file changed, 1 insertion(+)
 create mode 100644 math.sh
% git log
commit dbfda13f1d26c289732827f3f882d3c232485643
Author: Rick Umali <rickumali@gmail.com>
Date:   Thu May 15 21:33:42 2014 -0400

    This is the first commit.
% 
```

Figure 6.1 Our entire TRY IT NOW exercise

For Windows users, because you're not using the `-n` switch to `echo`, you may see the warning message "warning: LF will be replaced by CRLF in math.sh. The file will have its original line endings in your working directory." For now, this can be ignored. (See the lab in chapter 4 for more details about this warning.)

6.1.2 Telling Git about changes

Now let's make a change to this math.sh file and see how Git tracks changes.

TRY IT NOW Add one line to your math.sh program:

```
a=1
```

You can use your favorite text editor, but let me introduce a command-line technique to append a line to a file. Type this:

```
% echo "a=1" >> math.sh
```

The >> symbols say to add the text in quotes to the end of the file. After typing this, math.sh will have one more line in it. To see the entire file on the command line, type this command:

```
% cat math.sh
```

This command shows you the two lines of your file. Because you added this file to Git (via `git add`), you can ask Git if the file has changed by typing this command:

```
% git status
```

You should then see the following output.

Listing 6.1 `git status`

```
On branch master
Changes not staged for commit:
  (use "git add <file>..." to update what will be committed)
  (use "git checkout -- <file>..." to discard changes in working directory)

        modified:   math.sh

no changes added to commit (use "git add" and/or "git commit -a")
```

Notice how `git status` reports that there are `Changes not staged for commit`. It instructs us to use `"git add <file>" to update what will be committed`. It also tells you how to discard the changes: use `"git checkout -- <file>..."` to discard changes in the working directory.

The important thing here is that in order to commit this change, you have to add it to Git (via `git add`). You'll explore this in the next section, but first, let's take a look at what's different.

6.1.3 Seeing what's different

The `git status` command tells you that a file has been changed, but can you find out exactly what that change is? Yes.

TRY IT NOW In the math directory, type the following:

```
% git diff
```

You should see the following code listing.

Listing 6.2 `git diff` output

```
diff --git a/math.sh b/math.sh
index 8ae40f7..a8ed9ca 100644
--- a/math.sh
+++ b/math.sh
@@ -1 +1,2 @@
 # Comment
+a=1
```

The output of the `git diff` command demonstrates that Git knows exactly how the file has changed. Let's concentrate on these five lines:

```
--- a/math.sh
+++ b/math.sh
@@ -1 +1,2 @@
 # Comment
+a=1
```

The output begins with a two-line header that shows which two files are being compared. The first line (`a/math.sh`) is the original file. Remember, the Git repository has the original file! The second line (`b/math.sh`) is the new file in your working directory. You can make changes only to files in the working directory.

The string `@@ -1 +1,2 @@` says how to interpret the next two lines. These next two lines are affectionately known as a *hunk*, and a hunk shows one area of the two files that is different. Our change adds one line, so the line starting with `@@` is roughly saying take the original file (`a/math.sh`) at the first line (`-1`) and then apply the contents of the new file (`b/math.sh`) starting at line 1 and going for two lines (`+1,2`).

The real meat of the diff is the hunk:

```
 # Comment
+a=1
```

You'll recognize the first line (`# Comment`) from when you created the file. The second line is what you added to the file in the working directory. It has a `+` prepended to it, to indicate that it's an addition. More complicated diff outputs can include multiple hunks, and they each are prefaced by `@@` lines.

Git can use this difference output to transform a file from one version to another. (You can learn more about this format by searching for *unified format* or *unidiff* on the web.) For our purposes, it's a helpful record of what has changed. As you progress with Git, you'll learn to read this difference output, but as you can see from this section, there's a certain intuition about it.

6.1.4 Adding and committing changes to the repo

Let's add this file to the repository by following the instructions from `git status`.

> **TRY IT NOW** To commit this to the repository, you have to add this change via `git add` and then type `git commit`.
>
> The `git status` command offers an alternative way to commit this change to the repository: use the `git commit` command with the `-m` switch and the `-a` switch.
>
> Type the following:
>
> ```
> % git commit -a -m "This is the second commit."
> ```
>
> By using the `-a` and `-m` switches together, you've avoided the `git add` step *and* entered a message at the same time.

The Git command from the previous TRY IT NOW should give an output that looks like the following listing.

Listing 6.3 `git commit -a -m` output

```
[master e9e6c01] This is the second commit
 1 file changed, 1 insertion(+)
```

Performing the `git add` at the same time as `git commit` is a common shortcut. You do have to add the file first (with an initial `git add`) before this shortcut can work. The `-a` switch to `git commit` says to automatically stage (run `git add`) any files that Git knows about.

6.2 Thinking about git add

One thing that might begin to bother you about Git is that you have to use `git add` at least once for every file that you want to commit into the repository. Then, after every change you make to that file, you *still* have to use `git add` on the changed file before you can commit the file to the repository.

What's up with that? After the first `git add`, doing it again for every change seems redundant. The last section ended with a helpful shortcut (`git commit -a`), which saves you from having to type `git add` before typing `git commit`, but why is this step necessary to begin with?

6.2.1 An analogy to introduce the staging area

One analogy is to pretend that your code is an actor in a theater production. The dressing room is the Git working directory. The actor gets a costume and makeup all prepared for the upcoming scene. The stage manager calls the actor and says that the scene is just about to start. The actor (the code) doesn't jump in front of the audience right then. Instead, it waits in an area behind the curtain. This is our Git staging area. Here, the actor might have one last look at the costume or makeup. If the actor (the code) looks good, the stage manager opens the curtains, and the code commits itself to its performance. You might visualize this as in figure 6.2.

It must be said that most actors will spend only the briefest of moments behind the curtain before appearing in front of the audience. But what if you had a last-minute change in costume? Naturally, you leave the staging area, perhaps even go back to the

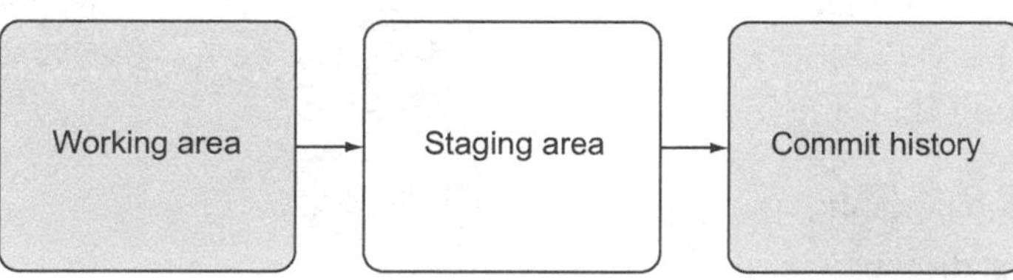

Figure 6.2 Three stages in the commit

dressing room, and make more changes. Then you rush back to the staging area before you commit.

The only thing that gets committed must pass through the staging area first!

6.2.2 *Adding changes to the staging area*

The Git staging area contains the version of the Git working directory that you want to commit. Git acknowledges with the shortcut `git commit -a` that most of us will typically put the changes in our working directory into the repository. But Git does allow you to make changes to what is staged (to go back to the dressing room, if you will).

> **TRY IT NOW** Let's add some more code to the math.sh program. Again, using your favorite editor, add these two lines:
>
> ```
> echo $a
> b=2
> ```
>
> I recommend that you use your favorite editor to add these two lines. On the command line, you could echo these two lines as you did earlier:
>
> ```
> echo "echo \$a" >> math.sh
> echo "b=2" >> math.sh
> ```
>
> To see this code running, type the following:
>
> ```
> % bash math.sh
> ```
>
> You should see the number 1 after you press Return/Enter. You're printing the value of the variable `a`.
>
> Now type `git status` and `git diff`. You should see that Git recognizes a change has been made, and in order to put these changes into the staging area (stage for commit), you should type `git add`. Let's do this now:
>
> ```
> git add math.sh
> ```

Your command-line session should look roughly like figure 6.3.

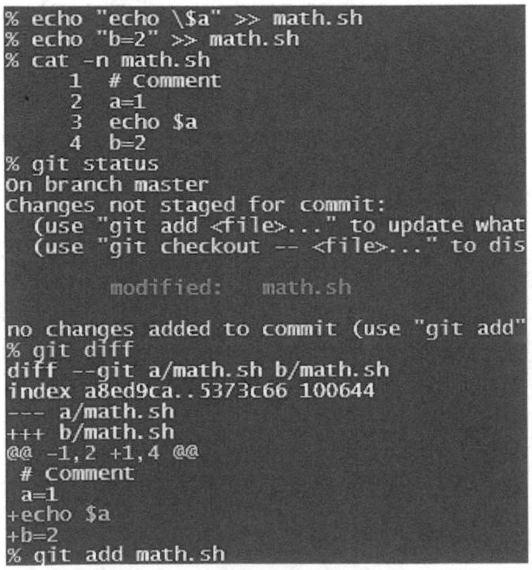

```
% echo "echo \$a" >> math.sh
% echo "b=2" >> math.sh
% cat -n math.sh
     1  # Comment
     2  a=1
     3  echo $a
     4  b=2
% git status
On branch master
Changes not staged for commit:
  (use "git add <file>..." to update what
  (use "git checkout -- <file>..." to dis

        modified:   math.sh

no changes added to commit (use "git add"
% git diff
diff --git a/math.sh b/math.sh
index a8ed9ca..5373c66 100644
--- a/math.sh
+++ b/math.sh
@@ -1,2 +1,4 @@
 # Comment
 a=1
+echo $a
+b=2
% git add math.sh
```

Figure 6.3 The working session to add your latest change

6.2.3 *Updating the staging area*

In the previous section, you added some code into the staging area. You can now commit this change. But wait! Suddenly you realize that you no longer want that `echo $a` line! You originally added it to confirm the contents of the variable `$a`, but you don't want to commit that line into the Git repository.

To change that, let's edit the math.sh file in the working directory and add this corrected version to the staging area.

TRY IT NOW Your file has four lines. You've already added this code to the staging area via `git add`. Can you update it one more time before doing a commit? Yes!

First, use your favorite editor to remove the line `echo $a`. (Any editor will do, provided you can open the file and delete lines with it.) After you've saved the file, type this:

```
git status
```

You'll have the puzzling output that follows.

Listing 6.4 Slightly confusing output

```
On branch master
Changes to be committed:
  (use "git reset HEAD <file>..." to unstage)

        modified:   math.sh

Changes not staged for commit:
  (use "git add <file>..." to update what will be committed)
  (use "git checkout -- <file>..." to discard changes in working directory)

        modified:   math.sh
```

Git has listed math.sh twice, showing that it's been modified twice. Let's see why.

The first section of the status output says that math.sh is "to be committed." This means it's already staged. The second section of the status output says that the math.sh isn't yet staged. How can there be two copies of math.sh? Simple: one is in the staging area, and one is in the working directory!

6.2.4 *Understanding the staging area*

In the previous section, the file math.sh has gone through the changes in figure 6.4.

Figure 6.4 shows your working area (the directory in which you're making edits), a staging area (where files are stored when you use the `git add` command), and finally the commit history itself: the final permanent record of your file.

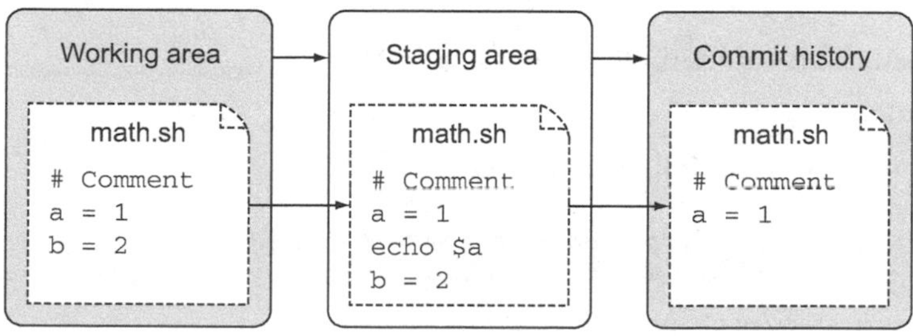

Figure 6.4 `git add`, **followed by** `git commit`

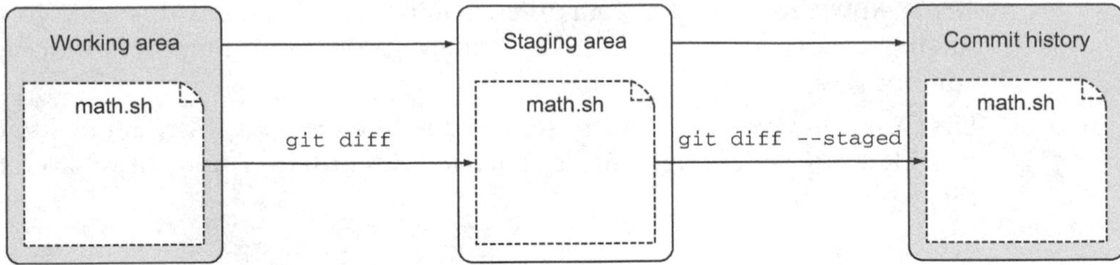

Figure 6.5 Understanding the confusing `git status` output

When you perform the `git status` from the TRY IT NOW in section 6.2.3, Git simultaneously checks the math.sh file in the staging area with the math.sh that is already committed, and the math.sh file in the working directory. These two checks are shown in figure 6.5. Remember, the version in the staging area is what will be committed when you run `git commit`.

> **TRY IT NOW** Let's look at two commands that demonstrate the three different math.sh files. At the command line, type the following:
>
> ```
> % git diff
> ```

When you type `git diff`, you should see the following output.

Listing 6.5 `git diff` (between working area and staging area)

```
diff --git a/math.sh b/math.sh
index 964c002..5bb7f63 100644
--- a/math.sh
+++ b/math.sh
@@ -1,4 +1,3 @@
 # Comment
 a=1
-echo $a
 b=2
```

By looking at figure 6.5, you should see how this `diff` output is formed: when you run `git diff`, you're comparing the copy of the file in the working directory with the copy of the file in the staging area. Let's account for the differences.

> **TRY IT NOW** At the command line, type the following:
>
> ```
> git diff --staged
> ```

This command produces the following output.

Listing 6.6 `git diff --staged` (between the staging area and the repository)

```
diff --git a/math.sh b/math.sh
index a8ed9ca..964c002 100644
--- a/math.sh
```

```
+++ b/math.sh
@@ -1,2 +1,4 @@
 # Comment
 a=1
+echo $a
+b=1
```

Again, look at figure 6.5 to convince yourself that `git diff --staged` is comparing the copy of math.sh that you committed earlier with the version of the math.sh that you staged (via `git add`).

This situation may be easier to see in the Git GUI tool, in figure 6.6.

> **TRY IT NOW** Start Git GUI in the math directory that you've been working in. The simplest way to do this on the command line is via the following commands:
>
> ```
> % cd
> % cd math
> % git gui
> ```
>
> Remember, the first `cd` command changes you to the home directory. Refer to the previous chapter for other ways to invoke this. Then after Git GUI appears, select math.sh in the Unstaged Changes pane (the upper-left pane). Observe the difference pane (the larger, upper-right pane).
>
> Next select math.sh in the Staged Changes pane (at the lower left). Observe the output in the difference pane.

In figure 6.6, you have the equivalent of the `git diff` output. Thanks to Git GUI's colors, you can see that you've just removed the `echo` statement between the `a=1` and `b=1` lines.

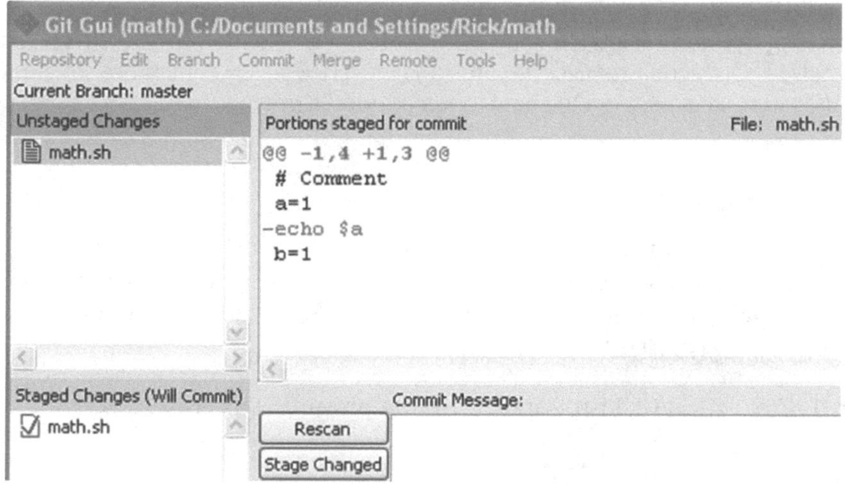

Figure 6.6 `git diff`

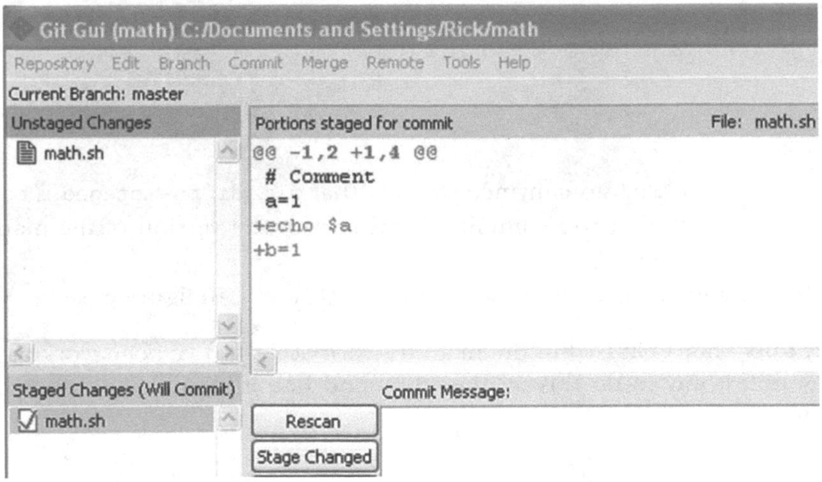

Figure 6.7 `git diff --staged`

In figure 6.7, you have the equivalent of the `git diff --staged` output. Again, thanks to Git GUI's colors, you can see what you're doing to the file that you've committed: you're adding two lines.

> **Above and Beyond**
>
> One question that might have crossed your mind at this stage is how Git might work with binary files, such as spreadsheet files or image files. The short answer is that Git works pretty well with these files, except for the case of taking a difference between one version of a binary file and another. A longer and more proper treatment is unfortunately beyond the scope of this book.
>
> Searching the web will lead to some standard answers for how Git handles binary files (namely, a combination of Git attributes and the `git config` command). To start learning these techniques, read the help for Git attributes by typing `git help attributes`.

6.2.5 *Committing changes*

To commit the most recent change (the file in your working directory), you should type `git add` at the command line. Refresh your memory by typing `git status` at the command line (you should see listing 6.4). In the next TRY IT NOW section, you'll do the add by using Git GUI.

TRY IT NOW If Git GUI isn't open from the previous TRY IT NOW section, type the following:

```
% cd
% cd math
% git gui
```

This opens Git GUI. You'll see the two math.sh files, one in the staged pane and one in the unstaged pane. Select math.sh in the Unstaged Changes pane (upper left). Then, instead of accessing the Commit menu's Stage to Commit item, right-click the hunk itself. In the resulting context menu, choose Stage Hunk For Commit, shown in figure 6.8.

This GUI step is roughly equivalent to typing `git add`.

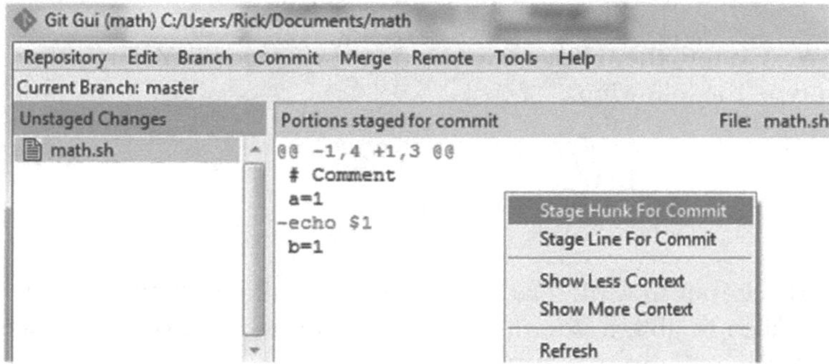

Figure 6.8 Staging a hunk

After staging this hunk, you can commit the file. This time, enter this message in the Commit Message pane (lower right): `Adding b variable`. Then click the Commit button, as shown in figure 6.9. Your Git's status bar will read, "Created commit SHA1: Adding b variable." These steps are the equivalent of typing `git commit -m "Adding b variable."`

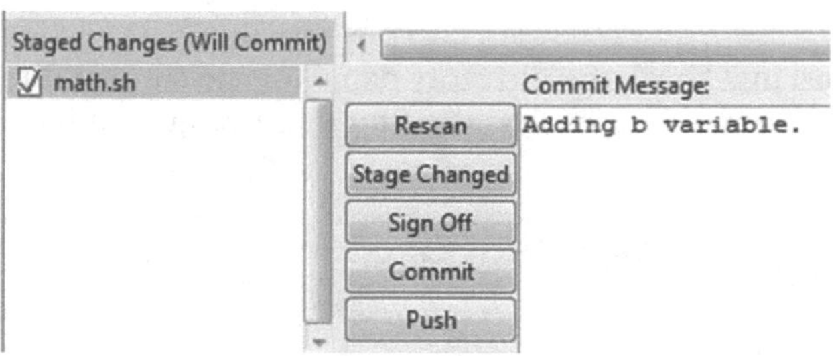

Figure 6.9 Committing the change

To check that everything is correct, exit Git GUI and then type `git log --oneline`. You should see the following output.

Listing 6.7 `git log --oneline` output

```
d4cf31c Adding b variable.
e9e6c01 This is the second commit
dbfda13 This is the first commit.
```

This listing shows that three commits have taken place. The `--oneline` switch is a more compact listing of the Git log output.

6.3 Adding multiple files

Throughout this chapter, you've been operating on one file. But software typically consists of multiple files. How might you add multiple files into a Git repository? To exercise this functionality, let's first create lots of multiple files. In the command line, `touch` is a command that can create multiple empty files.

TRY IT NOW Go into the math directory (see the earlier TRY IT NOW sections) and then type the following:

```
% touch a b c d
```

Now type this:

```
% ls
```

You should see four new files: a, b, c, d. They are all empty files (0-byte files). You should also be able to see this in your file-browsing tool. To see what Git thinks, type the following:

```
% git status
```

This indicates that these four files are untracked.

You could add these new files one at a time with four individual `git add` commands. But `git add` does take a directory name that will add all the files in that directory (including the untracked ones).

TRY IT NOW Before you run `git add` and pass in a directory name, let's find out what `git add` would do. In the math directory, type the following:

```
% git add --dry-run .
```

That period is important. That's the directory name for the current directory. When you run this command, it displays the files that it would add, as in the following listing.

Listing 6.8 `git add --dry-run .` output

```
add 'a'
add 'b'
add 'c'
add 'd'
```

The `dry-run` switch does what you'd expect from that descriptive name: it does a dry run, showing you what it would have done. Using the period as an argument to `git add` causes Git to add all the files in the current directory that it doesn't yet know about (in addition to any files that have changed).

To add these four empty files at once, type the following:

```
% git add .
% git status
```

The `git add` command won't produce any output, but `git status` shows that you now have four new files to commit. Let's commit these files. You have at least two ways to do this:

- `git commit -m`
- `git gui`

The first option allows you to pass in your commit message as an argument to the `-m` switch. The last option opens up Git GUI, where you can confirm the four staged files (in the lower-left pane). Enter a message in the Commit Message pane and then click the Commit button, as shown in figure 6.10.

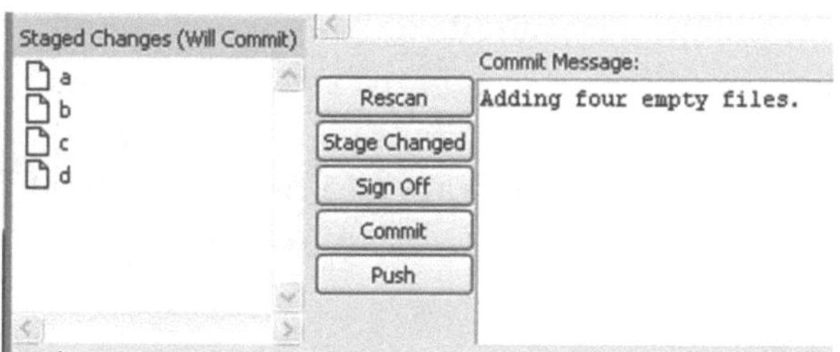

Figure 6.10 Committing four files

TRY IT NOW Use one of the preceding methods to commit these four files. When you're finished, type `git log` or look at gitk. Your commit history should look like the following listing.

Listing 6.9 Current state of repository

```
commit 6f51fb1d4528f11e3c9936ec68e6fa03a1f236a0
Author: Rick Umali <rickumali@gmail.com>
Date:   Wed May 21 21:06:13 2014 -0400

    Adding four empty files.

commit d4cf31c0506d5207f8c6ef410c6506e820fe87b5
Author: Rick Umali <rickumali@gmail.com>
Date:   Tue May 20 21:45:01 2014 -0400

    Adding b variable.

commit e9e6c019ca153eb12da3a5e878f0dff30b2d2b44
Author: Rick Umali <rickumali@gmail.com>
Date:   Thu May 15 22:51:19 2014 -0400

    This is the second commit

commit dbfda13f1d26c289732827f3f882d3c232485643
Author: Rick Umali <rickumali@gmail.com>
Date:   Thu May 15 21:33:42 2014 -0400

    This is the first commit.
```

6.4 Lab

This chapter was about adding and tracking files, and learning that there's a staging area that all files must pass through before they're committed. Also, along the way, I've shown you nuances and alternative forms of various Git commands, as well as some new command-line commands.

6.4.1 Understanding command-line nuances

To further explore the command line, answer the following questions:

1. What is another way to call `git diff --staged`?
2. What is the short form of `git add --dry-run`?
3. How do you display line numbers to your file via the `cat` command?
4. The `--oneline` switch that you passed to `git log` is shorthand for a longer `git log` command. What is it?
5. The `-a` switch to `git commit` (to automatically pass files to `git add`) has a longer alternative switch that is surprisingly not `--add`. What is it?

6.4.2 Getting out of trouble

When you realized you had done a `git add` for a change that you didn't want to add, you backed out of it by plowing ahead: you added your fix on top of the staging area! You could have backed out another way. How? Try it now!

6.4.3 Adding your own file

Add a new file to this math directory. Call this file `readme.txt`. You can add anything to this file (or even leave it empty). Now add this to the repo. The output of `git log --shortstat --oneline` should match the following listing.

Listing 6.10 Adding one more file

```
0c3df39 Adding readme.txt
 1 file changed, 0 insertions(+), 0 deletions(-)
6f51fb1 Adding four empty files.
 4 files changed, 0 insertions(+), 0 deletions(-)
d4cf31c Adding b variable.
 1 file changed, 1 insertion(+)
e9e6c01 This is the second commit
 1 file changed, 1 insertion(+)
dbfda13 This is the first commit.
 1 file changed, 1 insertion(+)
```

6.5 Further exploration

The simple program that you've been building in this chapter is in BASH. BASH stands for Bourne-Again Shell. You'll be building on this little program in the coming chapters, so it might be worthwhile to explore it further on your own. The official starting point for BASH is www.gnu.org/software/bash/manual/.

6.6 Commands in this chapter

Table 6.1 Commands used in this chapter

Command	Description
`git commit -m "Message"`	Commit changes with the log message entered on command line via the `-m` switch
`git diff`	Show any changes between the tracked files in the current directory and the repository
`git commit -a -m "Message"`	Perform a `git add`, and then a `git commit` with the supplied log message
`git diff --staged`	Show any changes between the staging area and the repository
`git add --dry-run .`	Show what `git add` would do
`git add .`	Add all new files in the current directory (use `git status` afterward to see what was added)
`git log --shortstat --oneline`	Show history using one line per commit, and listing each file changed per commit

7 Committing parts of changes

The Git staging area, introduced in the previous chapter, is a powerful but, at the same time, confusing feature of Git. This chapter builds on what you've learned about the staging area, helping you to understand it better.

The staging area is used to add, delete, and rename files before they are committed into the repository. Any change to the repository goes through the staging area, which means you must come to grips with it. The commands that manipulate the staging area include `git add`, `git rm`, `git mv`, and `git reset`.

The staging area also provides the ability to commit parts of files to the repository. If you add debugging code or print statements, and don't want to commit these into the repository, you can leave these out of the staging area without having to delete them from the file. This technique enables you to make more-refined commits and take control of your repository.

7.1 Deleting files from Git

In Git, adding files to the repository is achieved via the `git add` and `git commit` commands. But what if you want to delete files from the repository? It turns out you follow a similar two-step pattern from the preceding chapter: `git rm` and `git commit`.

In the previous chapter, after you completed the lab, your working directory contained six files: a, b, c, d, math.sh, and readme.txt. The a, b, c, and d files were empty files that you created with the `touch` command.

To delete the first file, a, you can use your operating system's file browser to manually delete the file. On the Git command line, you would use `rm a`. In both cases, because you're deleting a file that you've previously committed in the Git repository, you can always recover this file. You'll see this in the next chapter, but for now, let's try deleting files and see what happens.

> **TRY IT NOW** On the command line, let's change to the math repository that you created in the preceding chapter. Then let's delete the first file. Use these commands:
>
> ```
> cd
> cd math
> rm a
> ```
>
> Now let's use the helpful `git status` command to see what Git considers to be the status. Type the following:
>
> ```
> git status
> ```

You should see output like the following listing.

Listing 7.1 Output from `git status`

```
On branch master
Changes not staged for commit:
  (use "git add/rm <file>..." to update what will be committed)
  (use "git checkout -- <file>..." to discard changes in working directory)

        deleted:    a

no changes added to commit (use "git add" and/or "git commit -a")
```

As you saw in the preceding section, you've updated your working directory, but Git needs to have this update added to the staging area. To make a delete to the staging area, you must use `git rm`.

> **TRY IT NOW** In the math directory, delete the file from the staging area by typing this:
>
> ```
> git rm a
> ```
>
> This reports back a line that reads `rm 'a'`. Now type this:
>
> ```
> git status
> git gui
> ```
>
> Observe that the deleted file is already staged in the lower-left pane (you should see something like figure 7.1). Select the Repository menu and click Quit to exit Git GUI. Now type the following:
>
> ```
> git status
> ```

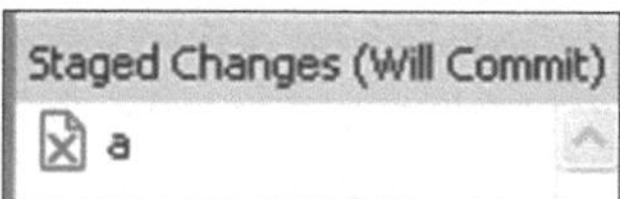

Figure 7.1 `git rm` **in Git GUI**

You should see the following output from the `git status` command.

Listing 7.2 Output after deleting a file and running `git status`

```
On branch master
Changes to be committed:
  (use "git reset HEAD <file>..." to unstage)

        deleted:    a
```

It may seem counterintuitive to tell the staging area that a file has been deleted after you delete it from the working directory. But remember that the staging area represents the content that you want to commit into the repository! If you want to delete it from the repository, you have to delete it from the staging area first, and then commit what is in the staging area.

You can delete files from the working directory and the staging area at the same time by using the `git rm` command. This is a handy shortcut to remember. Let's use this to delete the b file.

> **TRY IT NOW** In the math directory, delete the b file by typing the following:
>
> ```
> git rm b
> ```

Notice that this is the same thing you would have had to type if you did a regular deletion of the file! You now have two ways to do the same thing, but this method is faster (one step versus two).

To finally delete these two files (a and b), use `git commit`, as you've done before.

> **TRY IT NOW** In the math directory, to commit the two deletions that you performed in the previous TRY IT NOW sections, you can type this:
>
> ```
> git commit -m "Removed a and b"
> ```

This should give you the following output.

Listing 7.3 Output from `git commit` after deleting the two files

```
[master 38ac358] Removed a and b
 2 files changed, 0 insertions(+), 0 deletions(-)
 delete mode 100644 a
 delete mode 100644 b
```

7.2 Renaming files in Git

To rename a file on the command line, you must use the `mv` command. `mv` stands for *move*, as in "move c to the new file named renamed_file." To Git, a file rename consists of two steps: copying the original file to a new file and then deleting the original file. Let's explore this.

> **TRY IT NOW** On the command line, in your directory, type this command:
>
> ```
> ls
> ```
>
> You should see at least four files (if you've been following along): c, d, math.sh, and readme.txt.

Now type this:

```
mv c renamed_file
```

Confirm by typing `ls` that you no longer have a file c, but you have a file renamed_file. Now type the following:

```
git status
```

You should see output like the next listing.

Listing 7.4 `git status` after renaming a file

```
On branch master
Changes not staged for commit:
  (use "git add/rm <file>..." to update what will be committed)
  (use "git checkout -- <file>..." to discard changes in working directory)

        deleted:    c

Untracked files:
  (use "git add <file>..." to include in what will be committed)

        renamed_file

no changes added to commit (use "git add" and/or "git commit -a")
```

Look carefully at this status. Git considers the rename as two distinct actions:

1. Deleting a file (the file c)
2. Adding a file (the untracked file renamed_file)

To properly record this action into the staging area, use the steps in the next TRY IT NOW.

TRY IT NOW In the math directory, first remove the original c file by typing this command:

```
git rm c
```

Add the untracked file by typing this:

```
git add renamed_file
```

Now type the following:

```
git status
```

From `git status`, you should see the same output as the following listing.

Listing 7.5 `git status` after renaming a file (and after `git rm` and `git add`)

```
# On branch master
# Changes to be committed:
#   (use "git reset HEAD <file>..." to unstage)
#
#       renamed:    c -> renamed_file
#
```

At this point, Git realizes that you were renaming c to renamed_file. Is there a better way to announce to Git that you're renaming a file? Yes: `git mv`. If you're renaming a file, this command saves you the additional steps of `git rm` and `git add`.

TRY IT NOW The last file that you have is d. Let's rename it by typing this:

```
git mv d another_rename
```

Now check the status:

```
git status
```

Your `git status` output should look like this listing.

Listing 7.6 Output of `git status` after using `git mv`

```
# On branch master
# Changes to be committed:
#   (use "git reset HEAD <file>..." to unstage)
#
#       renamed:    c -> renamed_file
#       renamed:    d -> another_rename
#
```

You didn't need to use `git rm` or `git add`! If you were to examine this working directory in Git GUI, you'd see the two steps exposed as from the earlier `git status` (see figure 7.2).

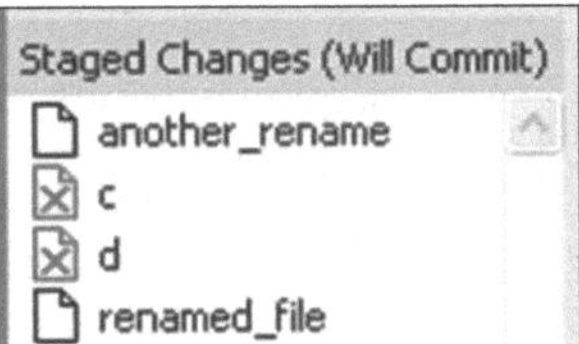

Figure 7.2 `git mv` in Git GUI

TRY IT NOW In the same directory, type the following:

```
git gui
```

From Git GUI, let's perform the commit. Type the commit message `Renaming c and d.` in the lower-right Commit Message pane, as shown in figure 7.3.

Click the Commit button. The status bar reports the commit information.

Exit Git GUI by selecting Quit from the Repository menu.

From the command line, type `git log --oneline`.

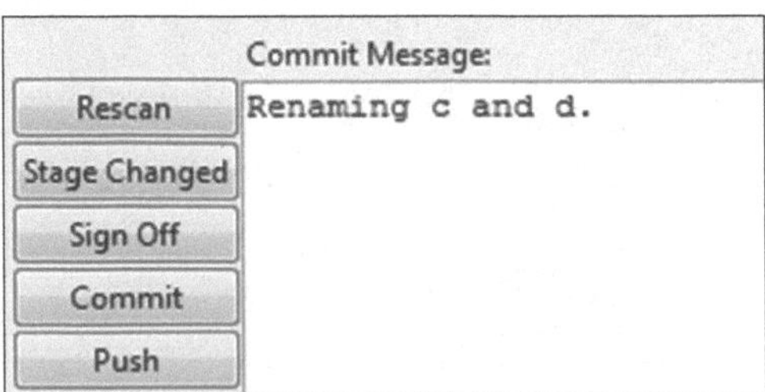

Figure 7.3 Committing the renames of c and d

The git log output should look roughly like the following listing.

Listing 7.7 git log --oneline output

```
3e39fec Renaming c and d.
4d2d662 Removed a and b
0c3df39 Adding readme.txt
6f51fb1 Adding four empty files.
d4cf31c Adding b variable.
e9e6c01 This is the second commit
dbfda13 This is the first commit.
```

7.3 Adding directories into your repository

A lot has been happening in your little repository. You've been adding files, deleting files, and renaming files. You saw that you can use the standard command-line programs (rm to delete, mv to rename) or even your OS file browser, provided that you inform Git by using git add, git rm, and git mv. Even better, you can use the git rm and git mv commands directly to save a step. Whenever you use these commands, you affect the staging area, and it's this staging area that you commit into the repository.

You can see this more clearly by using Git GUI.

TRY IT NOW In the math directory, start Git GUI by typing the following:

```
git gui
```

In Git GUI, click the Repository menu and select Explore Working Copy. This brings up your computer's file browser, which should look roughly like figure 7.4.

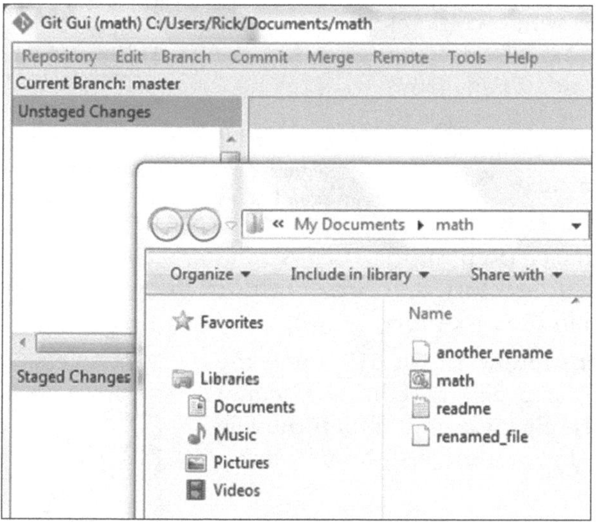

Figure 7.4 The working directory of your repository

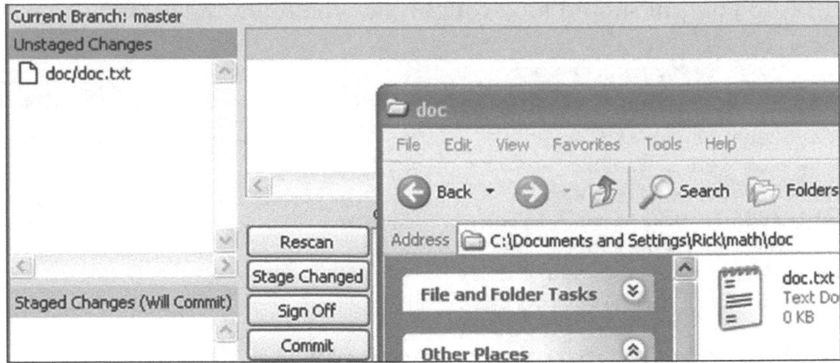

Figure 7.5 Adding a new directory to the repository

In this file browser, create a directory named `doc`. In that new directory, create an empty file called `doc.txt`. (Git will add only directories that already have files in them.)

In Git GUI, click the Rescan button. You should see the new file and directory in the Unstaged Changes pane (upper left), as in figure 7.5.

Before you go through the stage and commit steps, let's confirm that Git doesn't know about this new directory. In Git GUI, click the Repository menu and click Browse Master's Files.

You should see a small window that shows all the files that the Git repository knows about. Notice that the newly created directory and file aren't present yet (see figure 7.6).

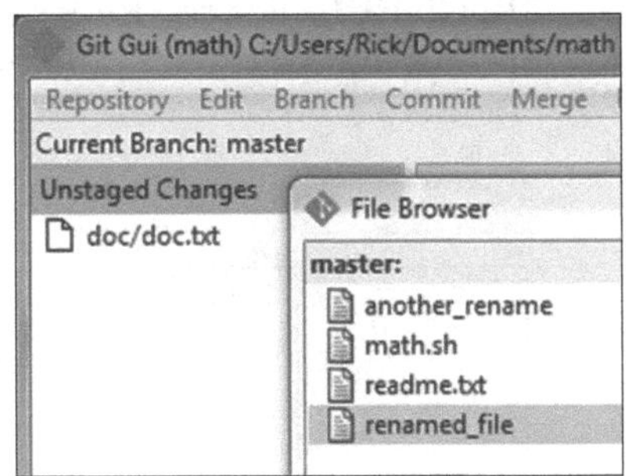

Figure 7.6 The new directory and file aren't present in the repository yet.

Close this File Browser window. In Git GUI, click doc/doc.txt from the Unstaged Changes pane. Then, from the Commit menu, select Stage to Commit. You should be prompted by a dialog box with the text "Stage 1 untracked files?" Click Yes.

Now open the File Browser (Repository > Browse Master's Files) and confirm once again that the new directory isn't present. It's in the staging area only, but not yet committed in the repository.

Close this File Browser window. In Git GUI, enter the string `Adding new doc dir and file` in the Commit Message pane and then click Commit. Open the File Browser and confirm that you can see the new directory, as in figure 7.7.

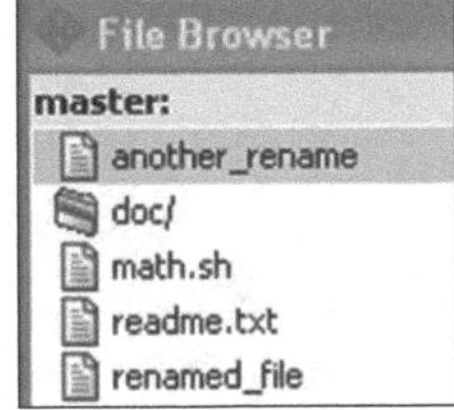

Figure 7.7 The new doc directory

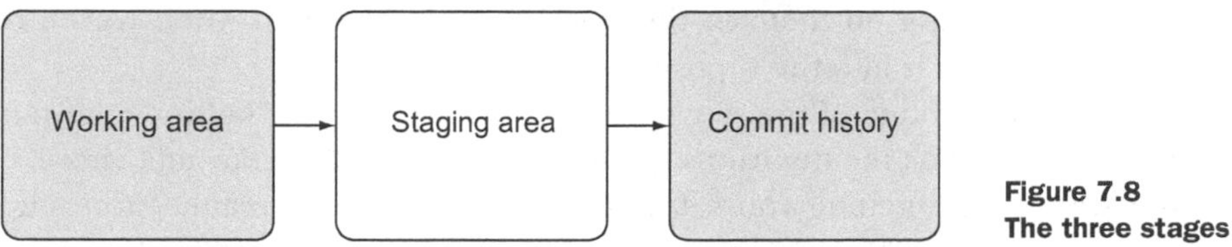

Figure 7.8 The three stages

The previous chapter introduced a diagram (repeated here in figure 7.8) showing the three stages that a file must complete before being committed into the repository.

The next figure shows the three stages using the windows that you've just explored in this section. To see files in the working area, you use your computer's file browser. To see files in the staging area, you can use the lower-left pane of Git GUI. And to see files that were committed into the repository, you can look at Git GUI's File Browser (figure 7.9).

7.4 Adding parts of changes

You've been adding, deleting, and renaming directories and files in your repository, and each operation has you going through the staging area in a fairly straightforward, though seemingly superfluous, manner. In this section, you'll see a powerful technique that the staging area offers.

7.4.1 Reconsidering the stage analogy

If you remember the analogy from the previous chapter, you might begin to think of the staging area as something entirely unnecessary. After all, you only pause to take a breath in that small area behind the curtain. This is also the case with your files. You

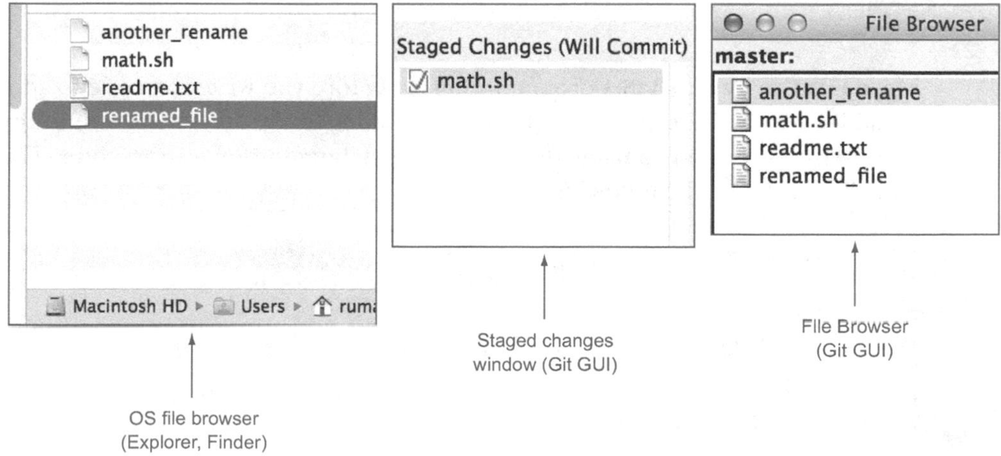

Figure 7.9 The three stages using the GUI

spend time working on them in the working area, but then when you're ready, you commit the file as is into the repository.

Instead of thinking of the staging area as a narrow space behind a curtain, what if you thought of it as the luxurious green room of a modern-day talk show? Here, you would spend time getting yourself ready before the performance. You might review notes, watch some TV, or have a snack. Same thing with your files! After you've finished your work in the working directory, maybe you want to polish up the file before you commit it to history.

This turns out to be a pretty common need. As you do development, you might place things in your file that you don't want to commit into the project's history, such as comments that say TODO, debugging code meant for your development environment, and more. The Git staging area lets you check in just the parts of the file that you want. With the staging area, you can be precise about what you commit.

7.4.2 Considering when to commit

You've been slowly developing our little example math program. Right now, its contents are modest. In listing 7.8, you can see that you have only three lines.

Listing 7.8 Your initial file

```
# Comment
a=1
b=1
```

Let's pretend you've been asked to add the values contained in these two variables. You make a change to the comment in line 1. Your file will look like the following listing.

Listing 7.9 Changing the first comment

```
# Add a and b.
a=1
b=1
```

Maybe you're new to BASH programming, so before you write the code to add the two variables, you write code to print out their values to the screen. You dutifully write a comment to remind yourself that this is just debugging or testing code. This makes your file look like the following listing.

Listing 7.10 Adding some debugging lines

```
# Add a and b.
a=1
b=1
#
# These are for testing
#
echo $a
echo $b
```

If you were to run this code, you'd see the values of those two variables, as shown in figure 7.10.

The new code from listing 7.10 is purely for testing.

```
% pwd
/home/rick/math
% cat -n math.sh
     1  # Add and b
     2  a=1
     3  b=1
     4  #
     5  # These are for testing
     6  #
     7  echo $a
     8  echo $b
%
% bash math.sh
1
1
% 
```

Figure 7.10 Running your program so far

Now you're on a roll. You change your program one last time, making it look like the next listing.

Listing 7.11 Adding code to sum up two variables

```
# Add a and b
a=1
b=1
#
# These are for testing
#
echo $a
echo $b
let c=$a+$b
echo $c
```

If you run this code, you'll see not only the value of the two variables, but the sum of those two variables, as shown in figure 7.11.

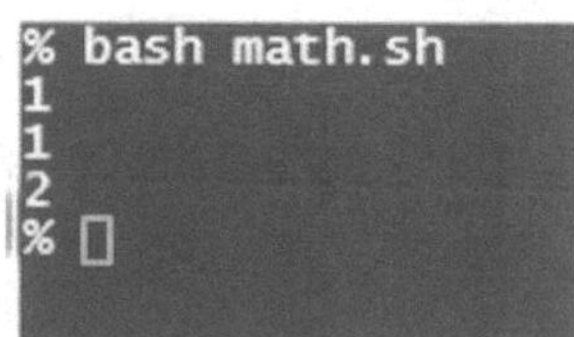

Figure 7.11 Your program is finally doing some math.

Your program grew incrementally. In figure 7.12, take a look at the growth of the program. It should be easier to pick out which lines you added. The leftmost listing is the last commit you made to your local repository.

```
# Comment
a=1
b=1
```

→

```
# Add a and b
a=1
b=1
```

→

```
# Add a and b
a=1
b=1
#
# These are for testing
#
echo $a
echo $b
```

→

```
# Add a and b
a=1
b=1
#
# These are for testing
#
echo $a
echo $b
let c=$a+$b
echo $c
```

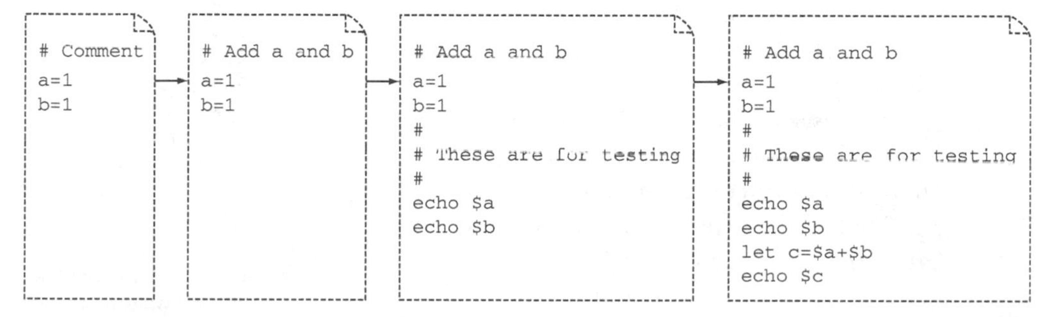

Figure 7.12 Side-by-side listing of your program

At this point, your program is done. Now is a good time to commit it into the repository. You've run `git commit` many times already, but let's consider the act of committing more closely.

When you commit code into the repository, you're leaving a record. You're preserving an artifact that you can return to. When should you commit? It makes sense to commit to the repository under any of these conditions:

- Adding or deleting a file
- Renaming a file
- Updating a file to a known good working state
- When you anticipate being away from the work
- When you introduce some questionable code

One of the overriding refrains you'll hear about Git is that it enables you to commit often. The list I enumerated should cover practically every hour of your working day. Committing something into the repository is a local act, because Git doesn't require a server to run.

Committing should be treated like punctuation. Add it at the end of every sentence, and between every break. Committing should reflect the trail of thought you followed to get to the present code.

I've encountered organizations that have guidelines and rules about how and when to commit to the repository. With Git, you'll still have guidelines and rules, and we'll get to that, but it's important to keep in mind that Git is distributed. How your commits appear in your local repository has no bearing on how you ultimately share your code with others. (You'll learn how to share your code in chapter 13.) Commit as often as you'd like!

7.4.3 Committing parts of a file by using Git GUI

Your code is ready to be committed, but let's pretend that you want to commit the code that adds the numbers. You might do this because you don't want to commit your debugging code. There might be rules or guidelines about committing tests in your organization. Regardless of your reason, what you want to do is commit only the highlighted changes in figure 7.13, and keep the rest in your working directory.

In this section, you'll handpick lines that you want to commit into the repository, something that Git lets you do because of its staging area.

You can edit your file to delete the comment and the lines that echo variables `a` and `b`. That might suffice for this example, but consider your own work. Haven't you written code that was

```
# Add a and b
a=1
b=1
#
# These are for testing
#
echo $a
echo $b
let c=$a+$b
echo $c
```

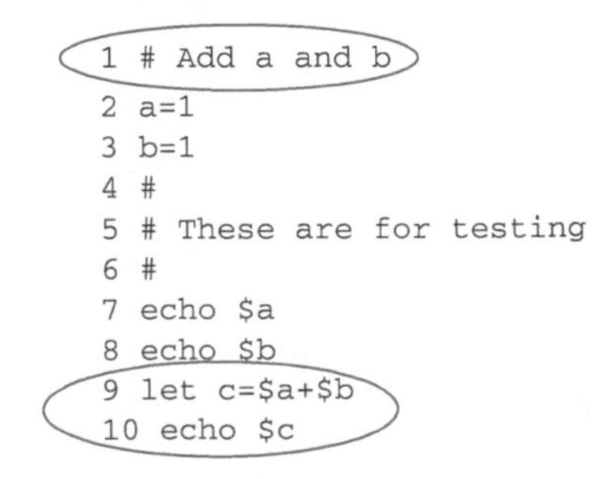

Figure 7.13 The code that is circled is what you want to commit. You'll handpick these lines.

meant to help you? Code that you probably wouldn't want to commit, but nonetheless was helpful for your own day-to-day development work? Git supports partial file commits for precisely this use case!

TRY IT NOW First, update the math.sh file to look like the following listing. Use your favorite editor for this operation. You're going make the edits as shown in figure 7.12.

Listing 7.12 math.sh contents

```
# Add a and b
a=1
b=1
#
# These are for testing
#
echo $a
echo $b
let c=$a+$b
echo $c
```

After you've edited your file, you'll look at it with Git GUI, because it's easier to visualize:

```
cd
cd math
git gui
```

In Git GUI, you should see the diff pane (at the upper right) displaying the output in figure 7.14.

Stop and consider why the picture looks the way that it does. Observe that the original lines (`a=1` and `b=1`) don't have a `+` or `-`, because the file has always contained these lines.

The diff pane shows that you edited the first line and then added all the lines after `b=1`. Figure 7.15 is what you saw in the earlier side-by-side diagram

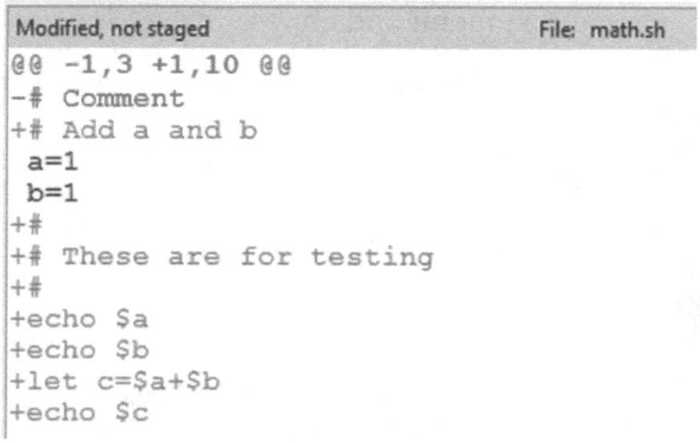

```
Modified, not staged                          File: math.sh
@@ -1,3 +1,10 @@
-# Comment
+# Add a and b
 a=1
 b=1
+#
+# These are for testing
+#
+echo $a
+echo $b
+let c=$a+$b
+echo $c
```

Figure 7.14 Your diff pane

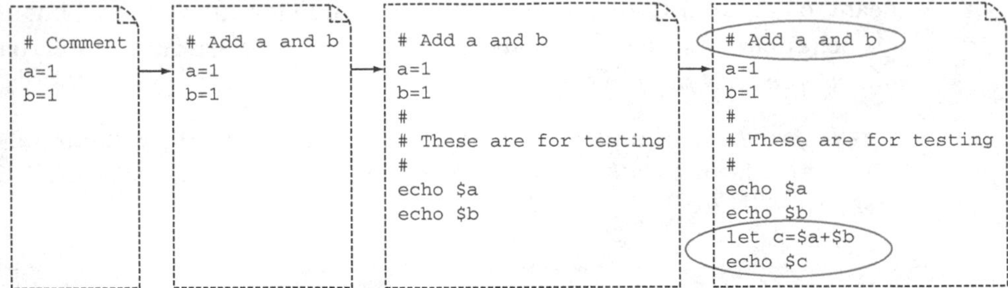

Figure 7.15 Your editing session, with the lines you're going to handpick circled in the far-right listing

(figure 7.12), reproduced here, with the lines that you're going to handpick for your commit highlighted.

If your file looks like figure 7.16, you'll need to modify how your editor treats the end of the line. Each line has to be on its own separate line. Resave the file by using a different end-of-line format, and then reopen in Git GUI.

```
Modified, not staged
@@ -1,2 +1 @@
-a=1
-b=1
+a=1b=1# # These are for testing#echo $aecho $blet c=$a+$becho $c
\ No newline at end of file
```

Figure 7.16 If your Git GUI display looks like this, examine the end-of-line configuration in your editor.

To save the lines that are specified in figure 7.15, position your cursor over the first diff line (-# Comment), as in figure 7.17.

```
Modified, not staged
@@ -1,3 +1,10 @@
-# Comment
+# Add a and b
 a=1
 b=1
```

Figure 7.17 Position the cursor over the first diff line.

Next, right-click to bring up a context menu and choose Stage Line For Commit, as shown in figure 7.18.

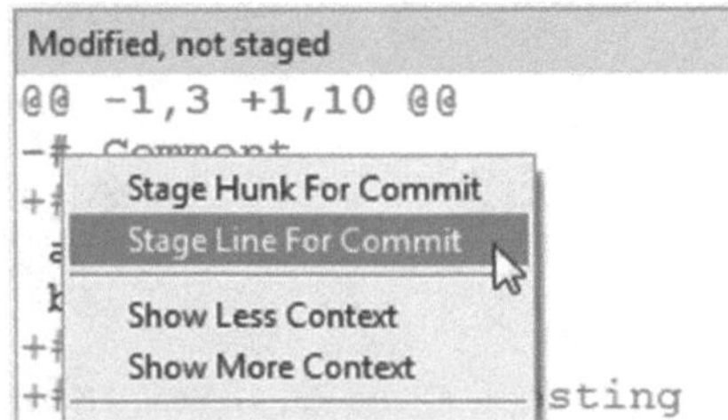

Figure 7.18 Choose the Stage Line For Commit option.

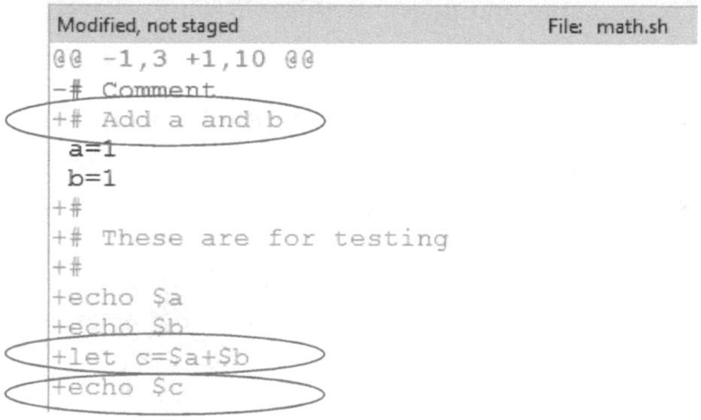

```
Modified, not staged                              File: math.sh
@@ -1,3 +1,10 @@
-# Comment
+# Add a and b
 a=1
 b=1
+#
+# These are for testing
+#
+echo $a
+echo $b
+let c=$a+$b
+echo $c
```

Figure 7.19 Stage the three circled lines.

To complete this exercise, do the same for the highlighted three lines in figure 7.19.

As you perform these steps, you'll notice that the diff pane changes as more lines are added to the staging area. When you're finished, the status line at the top of the diff pane will read, "Portions staged for commit." To see those, click math.sh in the Staged Changes pane (at the lower left). You'll see your diff pane look like figure 7.20.

```
Portions staged for commit
@@ -1,3 +1,5 @@
-# Comment
+# Add a and b
 a=1
 b=1
+let c=$a+$b
+echo $c
```

Figure 7.20 Your final diff pane

At this point, quit Git GUI. Then in the command line, type the following:

```
git diff --staged
```

The output should look like the following listing. Remember from section 6.2.4 that `git diff --staged` compares what is in the staging area with the version that was last committed.

Listing 7.13 `git diff --staged` output

```
diff --git a/math.sh b/math.sh
index 5bb7f63..dab42fb 100644
--- a/math.sh
+++ b/math.sh
@@ -1,3 +1,5 @@
-# Comment
+# Add a and b
 a=1
 b=1
+let c=$a+$b
+echo $c
```

Take stock of what you just did: you handpicked individual lines to commit into the repository. You've left out your personal debugging lines and are staging only the good stuff. The only thing left is to commit your code. You know two ways to do it: `git commit` and `git citool` (which is the same thing as `git gui`).

TRY IT NOW Before you commit the changes, type the following:

```
git status
```

What do you think you'll see? (To refresh your memory, look at listing 6.4 from chapter 6.)

Now commit the changes by typing the following:

```
git citool
```

This brings up the Git GUI. Enter this commit message: `Adding two numbers`. Then click the Commit button.

7.4.4 Committing parts of a file using git add -p

The TRY IT NOW steps that you performed in the previous section have their corresponding functionality on the command line. Unlike the point-and-click functionality you saw with Git GUI, selecting which lines to include does require some command-line editor know-how.

In this section, you'll handpick lines to commit, this time without using a GUI tool. For some, this may be a preferred method of interacting with the staging area! You're going to add the contents of the circled lines of the working directory into the staging area by using `git add -p`, as shown in figure 7.21.

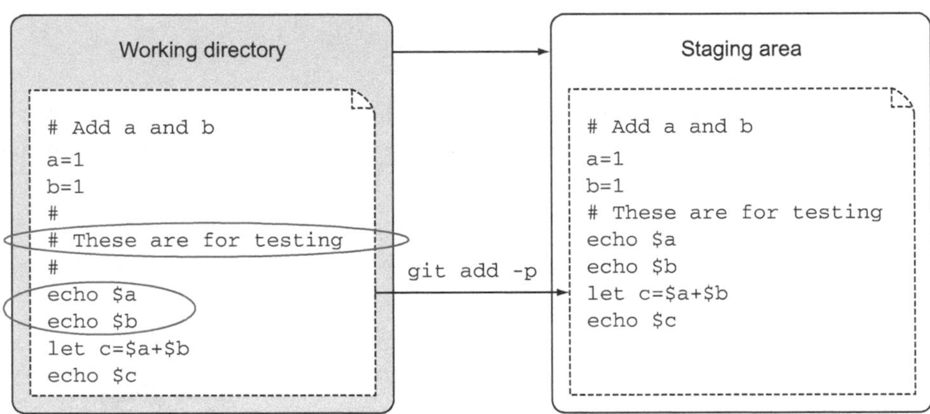

Figure 7.21 Changing the staging area by using `git add -p`

TRY IT NOW First, type the following:

```
git status
```

This should report that you have changes that have yet to be staged. How can you figure out what these changes are? That's right—type the following:

```
git diff
```

This compares the working directory with the staging area. (Remember from the previous section you didn't commit every change from the working directory. You handpicked what lines you wanted to commit.) This shows you output like the following listing.

Listing 7.14 `git diff` output

```
diff --git a/math.sh b/math.sh
index 135274b..d187591 100644
--- a/math.sh
+++ b/math.sh
@@ -1,5 +1,10 @@
 # Add a and b
 a=1
 b=1
+#
+# These are for testing
+#
+echo $a
+echo $b
 let c=$a+$b
 echo $c
```

The lines that are prefixed with the + symbol are new to the file. To stage this, you would type `git add`. But you now know you can pick exactly which lines to stage. Let's do that using the command line to delete just the # characters before and after the line `These are for testing`.

To initiate the same staging area editing functionality that you had with Git GUI, type this:

```
git add -p
```

You'll be presented with a diff, plus a prompt, as in the following listing.

Listing 7.15 `git add -p` output and prompt

```
diff --git a/math.sh b/math.sh
index 135274b..d187591 100644
--- a/math.sh
+++ b/math.sh
@@ -1,5 +1,10 @@
 # Add a and b
 a=1
 b=1
+#
+# These are for testing
+#
+echo $a
+echo $b
 let c=$a+$b
 echo $c
Stage this hunk [y,n,q,a,d,/,e,?]?
```

The listing repeats the output from the git diff command, but the line that starts with Stage this hunk? is a prompt asking you what to do. From listing 7.15 (and from your screen—you're trying this now, aren't you?), the question "Stage this hunk?" has eight possible responses: y, n, q, a, d, /, e, ?. To find out these what these responses mean, you can enter ?.

Type it now:

```
?
```

You should see a brief sentence for each of these responses, including a few responses that weren't given to you, as shown in the following listing.

Listing 7.16 git add -p prompt help

```
y - stage this hunk
n - do not stage this hunk
q - quit; do not stage this hunk nor any of the remaining ones
a - stage this hunk and all later hunks in the file
d - do not stage this hunk nor any of the later hunks in the file
g - select a hunk to go to
/ - search for a hunk matching the given regex
j - leave this hunk undecided, see next undecided hunk
J - leave this hunk undecided, see next hunk
k - leave this hunk undecided, see previous undecided hunk
K - leave this hunk undecided, see previous hunk
s - split the current hunk into smaller hunks
e - manually edit the current hunk
? - print help
```

Type this:

```
e
```

This brings up the vi editor, which is the default editor that Git is configured to use. Don't be afraid! This section walks you through how to use vi to choose the right lines from this hunk. (See chapter 20 for how to use git config to change the default editor from vi to something you're more familiar with.)

The text that you're editing is the diff hunk from listing 7.15. In comments (lines starting with #), the file explains how to edit this hunk in a line-by-line fashion. You should see instructions that look like the following listing.

Listing 7.17 Instructions for editing the diff file

```
# ---
# To remove '-' lines, make them ' ' lines (context).
# To remove '+' lines, delete them.
# Lines starting with # will be removed.
#
# If the patch applies cleanly, the edited hunk will immediately be
```

```
# marked for staging. If it does not apply cleanly, you will be given
# an opportunity to edit again. If all lines of the hunk are removed,
# then the edit is aborted and the hunk is left unchanged.
```

If you're familiar with the vi editor, delete the two highlighted lines marked in figure 7.22 (the ones with the empty comments). In the file that you're editing, the highlighted lines are lines 6 and 8.

```
# Add a and b
a=1
b=1
#
# These are for testing
#
echo $a
echo $b
let c=$a+$b
echo $c
```

Figure 7.22 Remove the two lines containing the single # marks.

Don't worry about getting this next sequence exactly right. In the next section, you end up discarding these changes, so if you're not familiar with the vi editor and want to jump ahead, then type `:wq` to exit the editor.

If you're not familiar with vi and want to delete the two lines, carefully type in all of these letters in the exact sequence you see here:

```
1
G
jjjjj
dd
j
dd
:wq
```

This sequence will be fully demonstrated as a screencast, which you can find on the book's website. The end result of this editing will be the deletion of these two empty comment lines.

After typing `:wq`, the program returns to the command-line prompt.

The preceding steps make a change to the staging area. After you type `:wq`, the edited hunk will be staged, just as if you had typed `git add`. In the next section, you'll learn how to reset the staging area, which removes the changes that you've staged.

7.4.5 Removing changes from the staging area

The change you made in the last section was artificial. It was meant to introduce how to use the command-line environment to make a detailed change to the staging area. To undo a staging area change, you have to use `git reset`, which is the opposite of `git add`.

TRY IT NOW Take stock of the situation with the staging area by typing the following three commands:

```
git status
git diff --staged
git diff
```

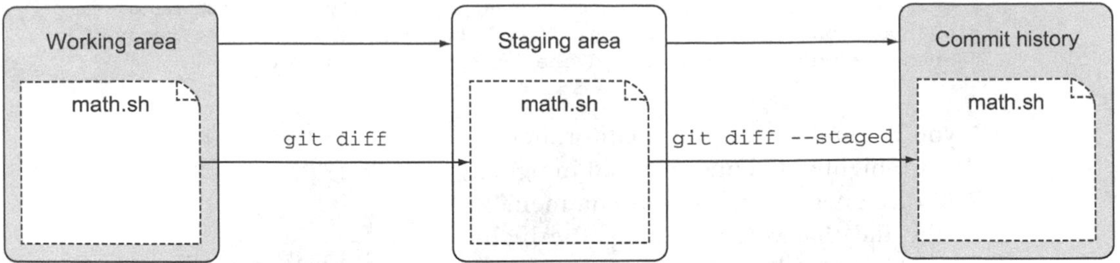

Figure 7.23 How `git diff` does its comparisons

If you followed the vi steps from the previous section, you should see two changes with these two uses of `git diff`. Figure 7.23 shows the two files that `git diff` uses for its comparison.

To undo the staging area change that you made, type the following:

```
git reset math.sh
```

The edits that you added to the staging area are removed. More important, the working directory is untouched: your debugging code is still in the math.sh file.

To visualize what just happened, look at figure 7.24.

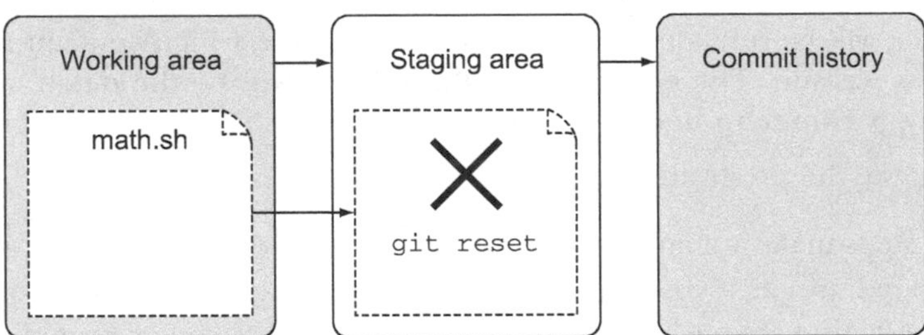

Figure 7.24 Resetting the staging area

In figure 7.24, anything in the staging area is removed from the staging area. Keep in mind that the working area always has the staging area changes. You'll learn more about this point in section 7.4.7.

7.4.6 Resetting a file to the last committed version

At this point, you may decide to remove this debugging code after all. As before, you could use your editor to remove the lines, but because the current code in the repository doesn't have this debugging code, you can more simply check out the latest version of math.sh (see figure 7.25).

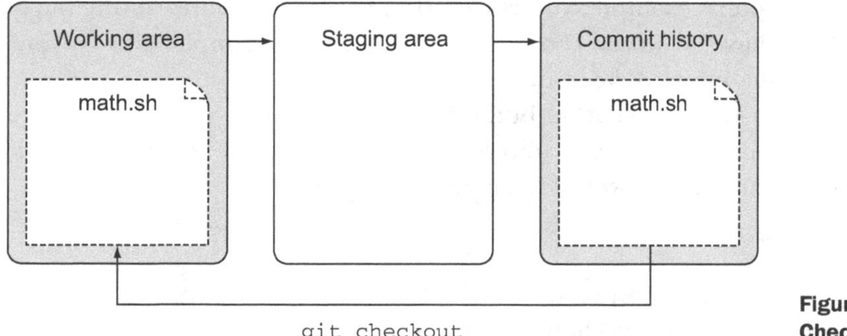

Figure 7.25 Checking out a file

When you check out a file in this manner, you overwrite the file in the working directory.

To check out a file, follow this last TRY IT NOW.

> **TRY IT NOW** In the repository you've been working in, examine the current state by typing the following:
>
> ```
> git status
> git diff --staged
> git diff
> ```
>
> Now type this:
>
> ```
> git checkout -- math.sh
> bash math.sh
> ```

At the end of this TRY IT NOW, math.sh will no longer have any debugging code. You have successfully checked out the latest committed version of math.sh. Its contents are shown in the following listing.

Listing 7.18 Contents of math.sh

```
# Add a and b
a=1
b=1
let c=$a+$b
echo $c
```

7.4.7 Understanding consequences of partial commits

When you handpick individual lines to commit, as you've done in sections 7.4.3 and 7.4.4, you should be careful that the rest of the code still works even without the lines you've omitted. It's easy to think that your partial commit will produce working code because your working directory still has all the code (including the lines you've chosen to omit). But other people who collaborate with the code won't have the code in your working directory.

In this chapter's example, you removed only comment lines and debugging code, so you don't have to worry about this issue for this example. But as you begin to embrace the staging area, keep this point in mind.

One way to run an accurate build/test cycle is to use `git stash` to completely remove the omitted code. This allows you to test the code as others will see it. (The `git stash` command is covered in chapter 9.)

7.5 Lab

In this chapter, you've done a lot of detailed work with the staging area. In the exercises that follow, you'll take slight variations of these steps.

7.5.1 Working with multiple hunks

To set up the repository for this lab, follow these steps. You should be able to do all of these without specific instructions, but if you need help, please visit the book's website.

1. Create a new directory named bigger_file, and make a Git repository.
2. Download the zip file named LearnGit_SourceCode.zip from the book's website. Unzip this file, and then copy the lorem-ipsum.txt file from the ZIP This Git repository.
3. Commit this file into the repository.
4. Copy lorem-ipsum-change.txt from the zip file of the previous step, and add it to the directory. (You'll need this again, so perhaps save this file to a good location.)
5. Rename (or copy) this new file to lorem-ipsum.txt.

By following the preceding steps, you've created a change in the lorem-ipsum.txt file. Git can detect these changes. Now, some questions and tasks:

1. How many hunks does this change have? (Hint: Each hunk is delimited by a line that starts with `@@`. You can see these lines with `git diff`.)
2. Using `git add -p`, stage and commit the change to just the second hunk.
3. Now check out the latest code (removing the changes in the working directory). (Hint: use `git status` to see the syntax for how to do this.)
4. Once again, copy the lorem-ipsum-change.txt file into the directory, and then rename (or copy) this new file on top of the lorem-ipsum.txt file.
5. Now how many hunks does this change have?
6. Commit the entire file.

To exercise this, delete this repository, and perform the exercise again, but this time use Git GUI for step 2. (You may notice that `git gui` produces fewer hunks than `git diff`.)

7.5.2 *Changing your mind with a delete*

In the previous lab, you had a file named lorem-ipsum.txt. Delete this file from the working directory by using git rm lorem-ipsum.txt. For this lab, bring this file back, following the instructions from the `git status` command.

7.5.3 *Reading assignments*

Read the Git help pages for `git checkout` and `git reset`. Notice that the commands have different forms. In this chapter, you used these commands:

- `git checkout -- math.sh`
- `git reset math.sh`

For each of these commands, what form are you using?

7.6 *Commands in this chapter*

Table 7.1 Commands used in this chapter

Command	Description
`git rm file`	Remove *file* from staging area
`git mv file1 file2`	Rename *file1* to *file2* in the staging area
`git add -p`	Pick parts of your changes to add to the staging area
`git reset file`	Reset your staging area, removing any changes you've added with `git add`
`git checkout file`	Check out the latest committed version of the file into your working directory

8 The time machine that is Git

When you make a commit into the repository, as you've been doing all along, you're making a mark on a timeline. Each commit says that on this date, at this time, I made this content change. Each commit is a version of your entire project. In this chapter, you'll learn that Git lets you visit any version of your project via the `git checkout` command. You'll also learn how to bookmark past versions of your project by using `git tag`. Learning this gives you the capability to go back in time with your code, which is important for making bug fixes to released software. Ultimately, Git is a time machine for your code!

8.1 Working with git log

In the repository that you've been working on so far, you've made eight commits. You can examine the timeline of these commits by typing `git log`. This produces a listing that looks roughly like the following.

Listing 8.1 `git log` listing for your math repository

```
commit 934e62e6a56843e4c6a859cb3e85e7901b007c2b
Author: Rick Umali <rickumali@gmail.com>
Date:   Fri Jun 13 21:15:58 2014 -0500

    Adding two numbers.

commit 595b6786212c9b329bb09fef81ff50ccc1208caf
Author: Rick Umali <rickumali@gmail.com>
Date:   Thu Jun 12 20:15:58 2014 -0500

    Renaming c and d.
```

```
commit 9289ea1d30a7fc9a2799edd9c5cb2a9f457a6814
Author: Rick Umali <rickumali@gmail.com>
Date:   Thu Jun 12 19:15:58 2014 -0500

    Removed a and b.

commit 5ecc3d2efebdd8763d6948e3bd712aa947da0198
Author: Rick Umali <rickumali@gmail.com>
Date:   Wed Jun 11 21:15:58 2014 -0500

    Adding readme.txt

commit 8a9a8bd631d7a2eacc0afe8490b91d9f86d3d31d
Author: Rick Umali <rickumali@gmail.com>
Date:   Tue Jun 10 21:15:58 2014 -0500

    Adding four empty files.

commit 874a7942a1ab43ee6d6b01a6b12f312ee2ee3b63
Author: Rick Umali <rickumali@gmail.com>
Date:   Mon Jun 9 21:15:58 2014 -0500

    Adding b variable.

commit 90d1dda323e79ad70c669f27a8083d2d236428de
Author: Rick Umali <rickumali@gmail.com>
Date:   Mon Jun 9 19:15:58 2014 -0500

    This is the second commit.

commit 96bfa4e220dcf74313e6ecf7cc8b41a11bd17198
Author: Rick Umali <rickumali@gmail.com>
Date:   Mon Jun 9 17:15:58 2014 -0500

    This is the first commit.
```

The listing is in reverse chronological order: the most recent commit is shown first, and then the second most recent, and so forth, all the way back to the first commit. As you step backward, all the way to the beginning, the commit log message should help you recall the various changes that were made at each commit.

8.1.1 Working with the SHA1 ID

The first thing to point out about the `git log` output is that each commit has an ID. This ID is unique to this commit even if you share your repository with a different server. In the `git log` output in listing 8.1, the ID is displayed next to the word `commit`. This ID is the SHA1 ID.

You first see the SHA1 ID when you use `git commit`. Every time you run the `git commit` command, it displays a status message and a new SHA1 ID. The `git commit` output in the following listing shows an abbreviated SHA1 ID: `96bfa4e`.

Listing 8.2 Sample `git commit`

```
[master (root-commit) 96bfa4e] This is the first commit.
 1 file changed, 1 insertion(+)
 create mode 100644 math.sh
```

Notice that the SHA1 ID is abbreviated in listing 8.2, but the default `git log` output (shown in listing 8.1) shows the full SHA1 ID string (which is 40 characters). In Listing 8.2, the abbreviated SHA1 ID is `96bfa4e`, but the full SHA1 ID is `96bfa4e220dcf74313e6ecf7cc8b41a11bd17198`. Either form serves as the commit's name, and is the equivalent of a version number that you might see in another version control system.

The SHA1 IDs are usually unique even if you were to compare only their first six or seven characters. Therefore, many Git commands display or accept a SHA1 ID that is shorter than the 40 characters.

The SHA1 IDs are cryptographically unique; they're guaranteed not to repeat between any file or any server. This is a powerful property that enables Git's ability to be distributed. Because no other file or server could ever generate this ID, you're free to share this commit with anyone without fear that this ID will be repeated by another commit.

8.1.2 Exploring meta information

Each commit contains, at a minimum, the committer's name and the date and time at which the user made the commit. These come from the `git config user.email` and `git config user.name` settings that you created previously, in section 3.1.1.

Each commit also contains the commit log message. This is the message that you entered with the `-m` switch of `git commit` or `git gui`. (You may also be aware that you can enter a Git commit log message using Git's default editor, though you've avoided doing this.)

One thing that's not so obvious in the `git log` output is that every commit has a parent, except for the first commit. The parent commit can be revealed by using the `git log --parents` switch.

TRY IT NOW Let's look at the parents of your commits. Go to the command line and type the following:

```
cd
cd math
git log --parents
```

The first two `cd` commands change your directory, first to the home directory and then to the math directory. The second command outputs your commit history. Git sends this output to the pager, which you learned about in detail in section 3.4. Remember that to page through the output, you press the spacebar, and to quit the pager (and return to the prompt), you press the Q key. The output may be a bit unwieldy because of the display of the full SHA1 IDs. Now type the following:

```
git log --parents --abbrev-commit
```

You should see a listing that looks like the following.

Listing 8.3 Partial output from `git log --parents --abbrev-commit`

```
commit cef45ff 29c7e58
Author: Rick Umali <rickumali@gmail.com>
Date:   Sat Jun 14 18:34:58 2014 -0500

    Adding two numbers.

commit 29c7e58 e5f8486
Author: Rick Umali <rickumali@gmail.com>
Date:   Fri Jun 13 17:34:58 2014 -0500

    Renaming c and d.

commit e5f8486 50534f8
Author: Rick Umali <rickumali@gmail.com>
Date:   Fri Jun 13 16:34:58 2014 -0500

    Removed a and b.

...
```

In the excerpted listing 8.3, for the first entry, you see the commit's abbreviated SHA1 ID, `cef45ff`, and its parent's SHA1 ID, `29c7e58`. The `abbrev-commit` switch shows only the first few characters of the SHA1 ID. Notice that the second entry in listing 8.3 is the commit with the ID `29c7e58`. It too has a parent, which is immediately before that one, and so on.

Figure 8.1 illustrates the proper way to think about commits. They're a list of objects, each commit object pointing to its parent. Each commit contains the full complete working directory at the time of the commit. When you do a `git add` and then a `git commit`, you're saving that working directory into this timeline. In figure 8.1, the first commit (`96bfa4e`) doesn't have a parent, because it's the root commit.

Commits are almost always shown in reverse chronological order because as a developer, you'll typically review your most recent work first. The questions "What did I do just now?" or "... an hour ago?" or "... yesterday?" are more common than "How did I start this repository?" This is especially the case in repositories with long histories.

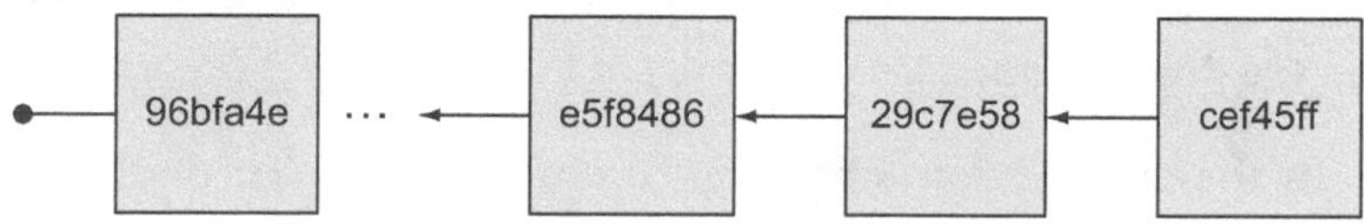

Figure 8.1 Each commit always points back to its parent.

8.1.3 Using gitk to view the commit history

The gitk GUI offers a great alternative to viewing the commit history, especially for repositories with lots of history. You explored gitk in chapter 5.

TRY IT NOW In the command window, assuming you're in the math directory, type this:

```
gitk
```

You should see a window like figure 8.2.

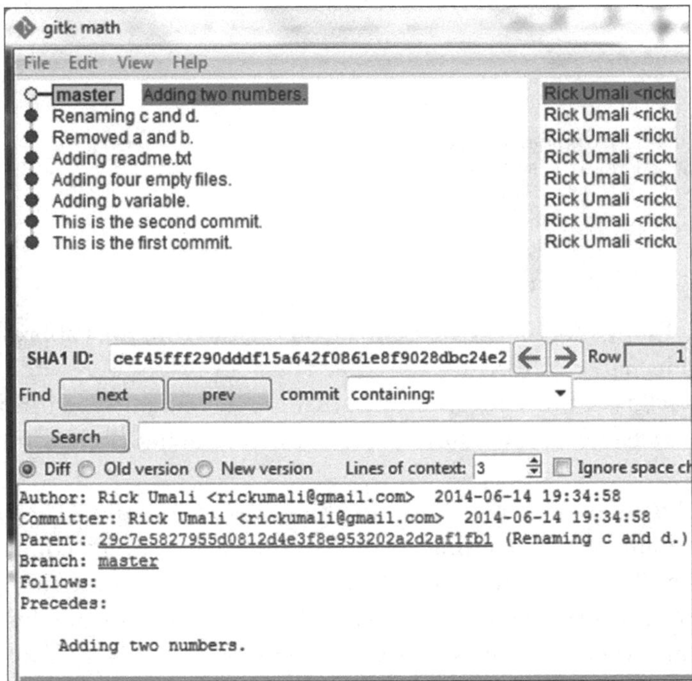

Figure 8.2 The gitk window

You should observe that the upper-left pane shows the commit log, in the same reverse chronological listing as the `git log` command-line output. The selected commit log (`Adding two numbers.`) is used to populate the rest of the panes. Your SHA1 IDs will be different, but in our example, the last commit's abbreviated SHA1 ID was `cef45ff`.

The gitk window lets you view any commit by clicking any commit line in the upper-left pane. Try this for yourself.

TRY IT NOW In the gitk window, click any commit line. Notice that the two bottom panes show more detailed information. Right above the pane on the lower-right, you'll see a toggle for either Patch or Tree, as shown in figure 8.3.

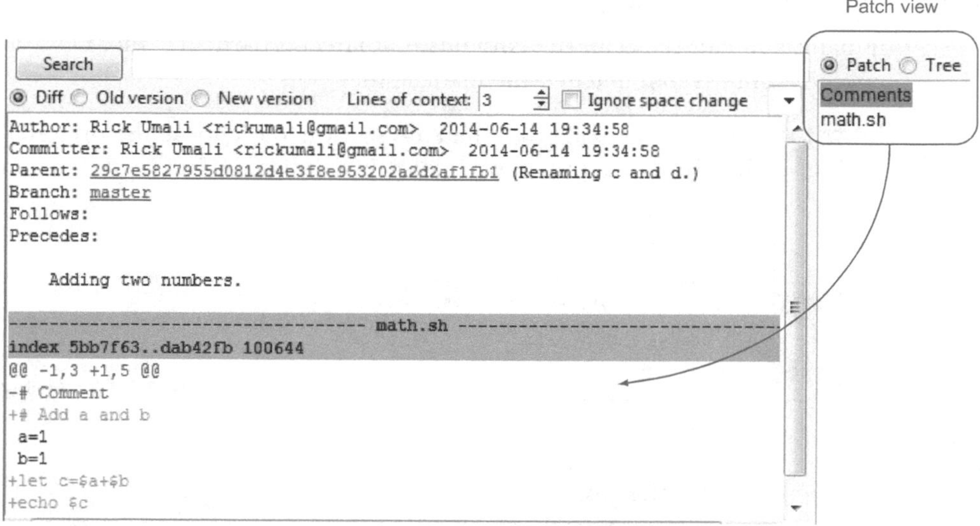

Figure 8.3 The patch view

In Patch mode, this window lists the files that were patched by the selected commit. The lower-left pane is a patch viewer of sorts. Clicking any file (here, you have only one, math.sh) will show how thc filc was changed (patched) to get to this commit.

Now select Tree mode, as shown in figure 8.4. The lower-right pane lists all the files that existed in the staging area for that commit. This directory listing, or tree of files, is associated with every commit. Clicking any file displays that file in the lower-left pane, which is now a file viewer.

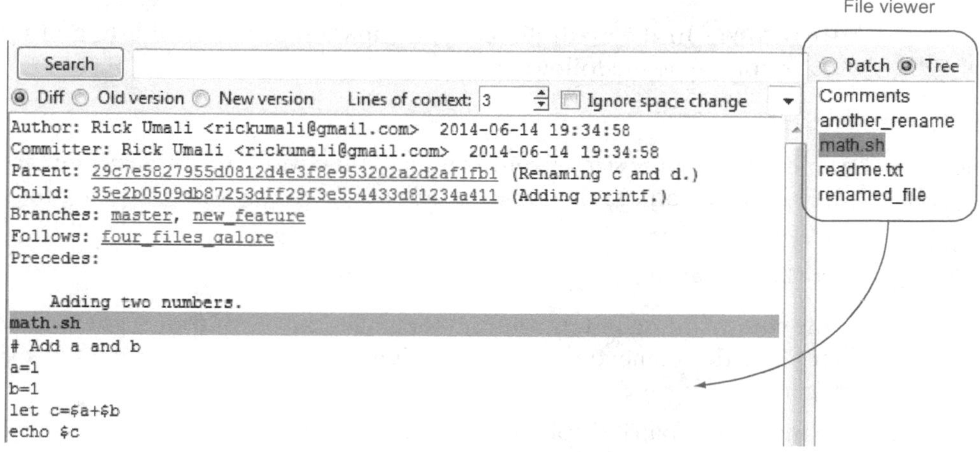

Figure 8.4 The tree view

8.1.4 Finding all commits by file

Finally, you can select all the commits that affect a particular file. Let's do this to see all the commits that touched the math.sh file.

TRY IT NOW Using gitk, go to the Find section and choose the option Touching Paths from the pull-down menu, as shown in figure 8.5. Then type `math.sh`.

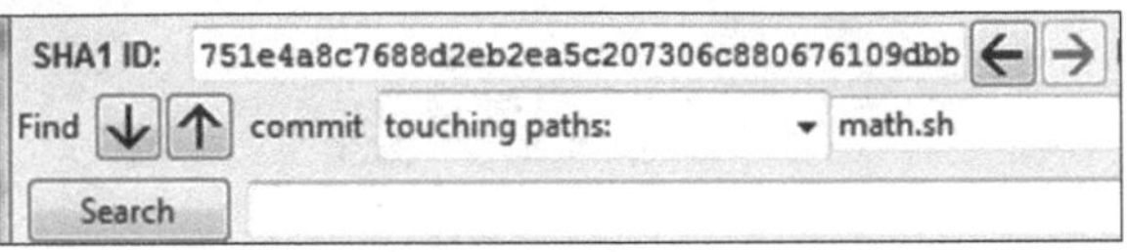

Figure 8.5 The Find section

The pane in the upper-left corner should change, boldfacing each commit log that touches math.sh, as shown in figure 8.6.

Use the up and down arrow buttons (shown previously in figure 8.5) to visit the three commits, and observe the patch pane. You should be able to see the various stages of your math.sh file!

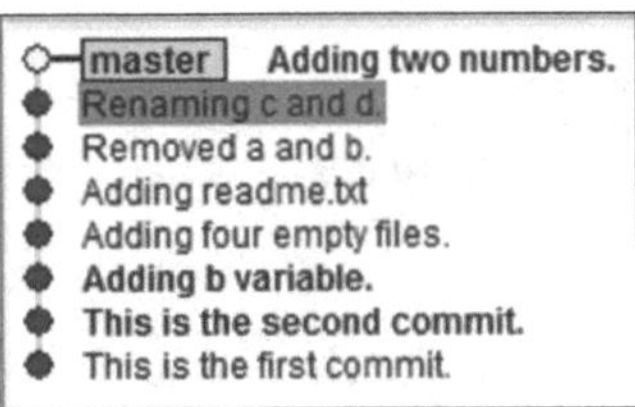

Figure 8.6 Finding all the math.sh commits

8.1.5 Using variations of git log

The `git log` command-line tool is capable of doing all the operations that you saw in the previous sections with gitk. Unlike gitk, the `git log` tool's output is limited to the size of your command window, but don't let that deter you from trying out these commands.

TRY IT NOW In the math directory, to view a more concise listing of the `git log` output, type the following:

```
git log --oneline
```

Depending on the size of your command-line window, you may need to press the spacebar to page through the output. To view a listing showing the patch information, type the following:

```
git log --patch
```

This command definitely requires the pager, as the output is going to be longer than the command window's height. Remember to press Q to exit the pager.

To match the patch display from gitk, type the following:

```
git log --stat
```

With this command, `git log` will output what has changed from the point of view of the files in each commit. Both of these commands can be combined by typing the following:

```
git log --patch-with-stat
```

Finally, to view the commits that pertain to just math.sh, pass the filename as a command-line argument. Type this:

```
git log --oneline math.sh
```

This lists all the commits that touch math.sh. Again, remember to press Q to exit the pager that `git log` uses.

The `git log` command is a powerful tool that makes quickly looking at commit history easy. The gitk GUI does offer multiple views of your repository's state, but if you know what you're looking for, `git log` can usually give you those same views concisely.

8.2 Making proper commit log messages

When you run `git log --oneline`, you'll see output that looks like the following listing as well as the upper-left pane shown previously in figure 8.2.

Listing 8.4 Output from `git log --oneline`

```
cef45ff Adding two numbers.
29c7e58 Renaming c and d.
e5f8486 Removed a and b.
50534f8 Adding readme.txt
bcaa6e2 Adding four empty files.
80f0ccc Adding b variable.
ea91623 This is the second commit.
8c31e35 This is the first commit.
```

This terse listing is easy to understand right now, because you've been looking at these changes for the past few days. But what happens a week from now? Or a month from now? Or even a year from now? Will these short sentences help you remember why you performed these changes?

One of the best ways to improve the utility of `git log` is to create proper commit messages. The `git commit` help page includes a DISCUSSION section that's worth reading. It states, "Though not required, it's a good idea to begin the commit message with a single short (less than 50-character) line summarizing the change, followed by a blank line and then a more thorough description."

Using this format will make a huge difference in your history messages.

TRY IT NOW The DISCUSSION section referenced in the preceding text can be seen by typing the following:

```
git help commit
```

Scroll or page down to see the DISCUSSION. Bookmark this to read later.

Next, let's make a change to math.sh and make a commit that follows the convention for a proper commit message. Using your favorite editor, open the math.sh file. Replace the last line (`echo $c`) with this line:

```
printf "This is the answer: %d\n" $c
```

Exit your editor and then run this new version of the math.sh program:

```
bash math.sh
```

You should see that the output is different. Let's commit this change. Remember that you have to add this change to the staging area:

```
git add math.sh
git citool
```

I covered `git citool` in chapter 5. If you have problems bringing up this window, visit the book's forum (www.manning.com/umali) for suggestions. In the window that pops up, enter the following text exactly:

```
Adding printf.

This is to make the output a little more human readable.

printf is part of BASH, and it works just like C's printf()
function.
```

Figure 8.7 shows what your commit message should look like. Now click the Commit button.

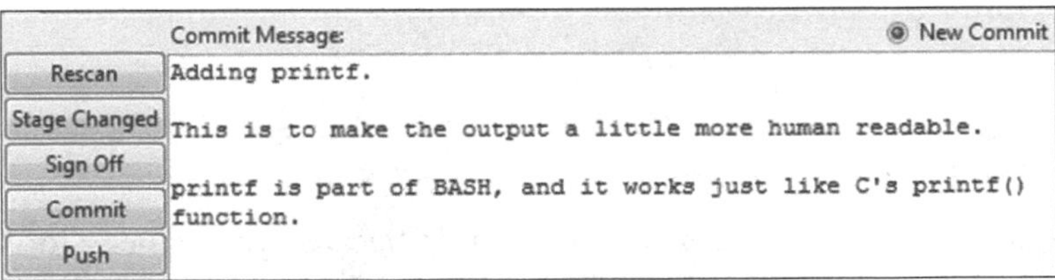

Figure 8.7 A proper commit message

You've added one more commit to the end of your work, but this commit message is different from the previous messages. Instead of a one-line message, it contains multiple lines.

The `git log --oneline` output looks only at that first line of your commit, as in the following listing.

Listing 8.5 Output from `git log --oneline`

```
cbcd3e3 Adding printf.
cef45ff Adding two numbers.
29c7e58 Renaming c and d.
e5f8486 Removed a and b.
...
```

To see how this full commit looks, use `git log`. You'll see output that looks like the next listing.

Listing 8.6 Excerpted output from `git log`

```
commit cbcd3e3d61aed114c695b5c308fa0ca4e869bf5c
Author: Rick Umali <rickumali@gmail.com>
Date:   Mon Jun 16 22:26:12 2014 -0400

    Adding printf.

    This is to make the output a little more human readable.

    printf is part of BASH, and it works just like C's printf()
    function.

commit cef45fff290dddf15a642f0861e8f9028dbc24e2
Author: Rick Umali <rickumali@gmail.com>
Date:   Fri Jun 13 21:15:58 2014 -0500

    Adding two numbers.

commit 29c7e5827955d0812d4e3f8e953202a2d2af1fb1
Author: Rick Umali <rickumali@gmail.com>
Date:   Thu Jun 12 20:15:58 2014 -0500

    Renaming c and d.
...
```

What you've done with this latest commit is add a little bit of a story to it. The core change is that you've replaced the `echo` statement with the `printf` statement. This is the bare-bones summary of the change. But the text that you've added elaborates on the change.

When you're writing a commit log message, anything that isn't in code or a code comment is a candidate for documentation in the commit message. The moment when you're committing code is the best time to write the reasons that you're making the change. At code commit time, you're the best and most experienced person (and your memory is freshest) at that moment. Take advantage of the moment by writing a commit message that will help you or a fellow developer in the future.

If you've finished making a commit but need to give it more text, you can amend your commit via the `git commit --amend` command. This helpful command lets you edit your most recent commit. You'll try this out in an exercise!

8.3 Checking out a specific version

When you were exploring gitk in section 8.1.3, you observed that the tree view shows a directory listing for every commit. Each commit contains the entire staging area (or working directory) for that commit. In figure 8.8, Git has clearly kept track of what files go with what commit.

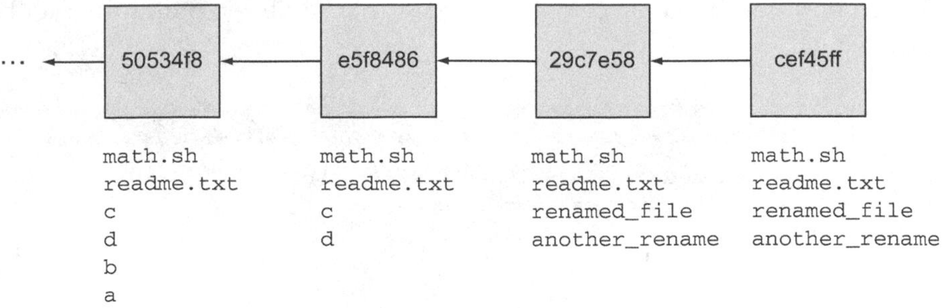

Figure 8.8 Git commits reference an entire tree of files.

Even better, Git can quickly and efficiently regenerate any tree of files just by asking for the commit by its SHA1 ID. This is the Git time machine! What distinguishes Git from other version-control systems is that you can go backward in time without any help from a server. Your repository has everything it needs for every commit in the `git log`.

8.3.1 *Understanding HEAD, master, and other names*

Before you start using your time machine, it makes sense to know how to get back to the present. What good is a time machine if you can take only a one-way trip?

Figure 8.9 is a diagram of your commit history. It's a line of commits, each commit pointing to its parent, as depicted by the arrows.

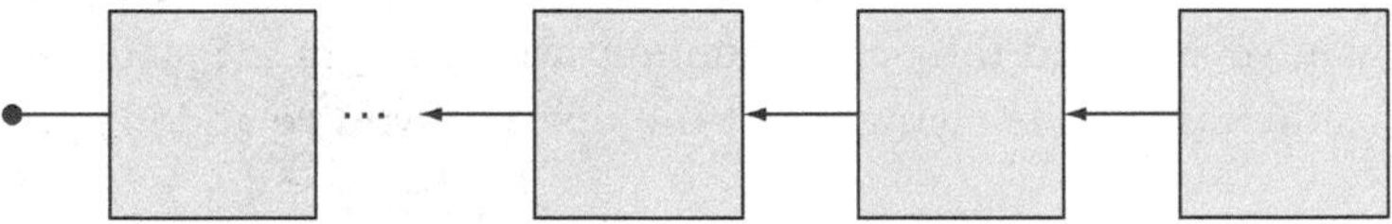

Figure 8.9 Your line of commits, each pointing back to its parent

Your line of commits is also known as a *branch*. When you ran `git init`, Git made a default branch for you to put your commits into. You'll spend tomorrow's lunch on branches, but in order to get back to the present, you should know that this default branch has a name, and that name is *master* (see figure 8.10).

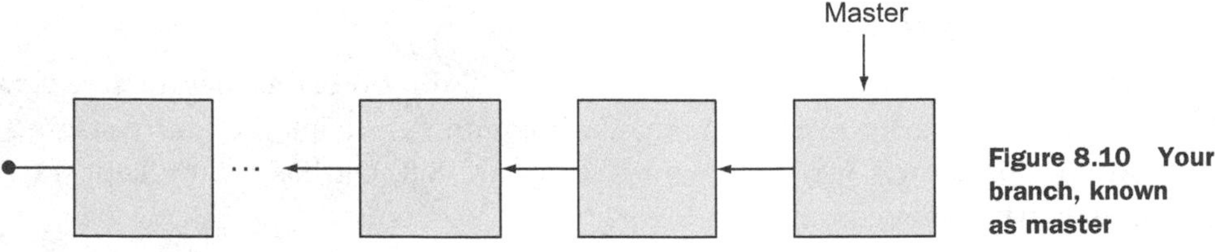

Figure 8.10 Your branch, known as master

The master branch represents the entire line of development, but in figure 8.10, master is also a reference, or a pointer, to the last commit made on this branch. This is why I drew master with an arrow pointing to the last commit. Git always keeps master pointed at this last commit, and this is how you can safely travel back in time: you know that master points to the last commit (which is your present)!

The last piece that you need for time travel is the device that does the travel. It turns out that Git supplies this device by default: it's your HEAD (see figure 8.11).

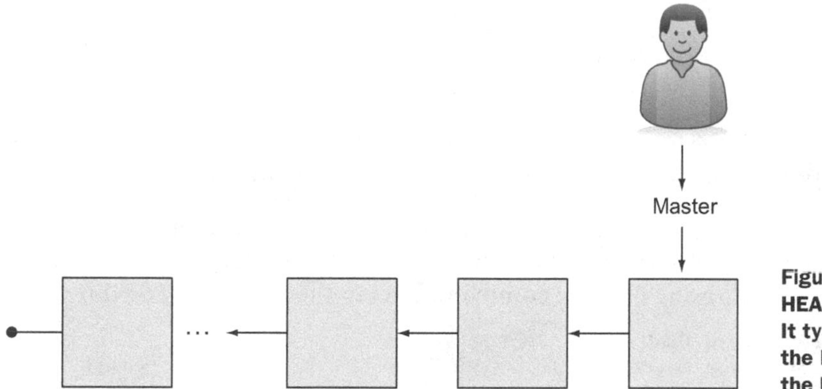

Figure 8.11 The HEAD of your branch. It typically points to the last commit of the branch.

At its simplest, the HEAD is the current branch. This HEAD is analogous to the laser in a CD/DVD player, the needle of a record player, the tape player head in a cassette player, or that marker on a progress bar that you can move with your finger to any part of a song on your music-playing device. HEAD is also your actual head. What is your head (you) looking at right now? At the moment, you're looking at the present, but in just a few paragraphs, you'll move your head to another point in the timeline.

When you read Git documentation, HEAD, our time-travel device, is always capitalized. The final official picture looks like figure 8.12, but if you find it easier to think of a smiling face, please do so!

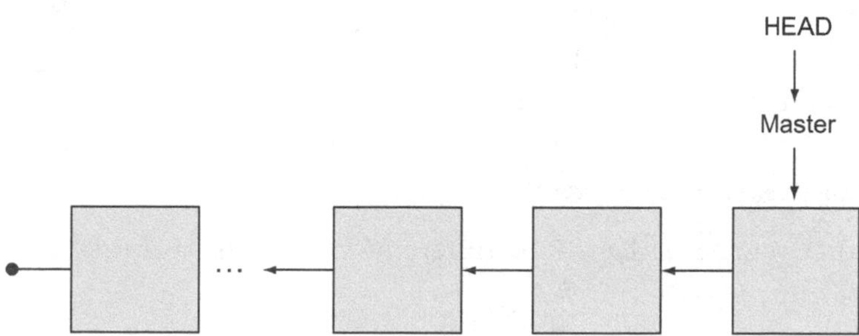

Figure 8.12 The HEAD and master

Let's see how these pictures work in your repository. The following listing shows your commits via `git log --oneline`.

Listing 8.7 Your commits on the master branch

```
cbcd3e3 Adding printf.
cef45ff Adding two numbers.
29c7e58 Renaming c and d.
e5f8486 Removed a and b.
50534f8 Adding readme.txt
bcaa6e2 Adding four empty files.
80f0ccc Adding b variable.
ea91623 This is the second commit.
8c31e35 This is the first commit.
```

You know that this list of commits is a branch called master. You can refer to the entire master branch by its latest commit (`cbcd3e3`). Let's confirm this.

> **TRY IT NOW** In this exercise, you'll use a command called `git rev-parse`. This command translates branch names to their corresponding SHA1 IDs. Type the following into the command line, in the math working directory:
>
> ```
> git rev-parse HEAD
> git rev-parse master
> ```

Both commands produce the same output: the SHA1 ID of the latest commit.

8.3.2 Going back in time with git checkout

If you wanted to go back to the version of your repository when you added the four empty files, you can go back in time to that specific version by typing `git checkout bcaa6e2`. Let's carefully try this.

> **TRY IT NOW** In the math directory, you must first obtain the SHA1 ID of the commit to which you added the four empty files. Type the following:
>
> ```
> ls
> git log --oneline
> ```
>
> The first command lists the files in the directory. This should return this list: `another_rename`, `math.sh`, `readme.txt`, `renamed_file`, and `doc`.
>
> The second command returns the list of commits. Look for the SHA1 ID for the commit that reads `Adding four empty files.` From listing 8.7, this ID is `bcaa6e2`:
>
> ```
> bcaa6e2 Adding four empty files.
> ```
>
> Remember that your SHA1 ID will be different! Once you have this ID, type the following:
>
> ```
> git checkout YOUR_SHA1ID
> ```

where `YOUR_SHA1ID` is the ID of the commit in which you added the four empty files. You can specify the entire 40-character SHA1 ID, or you can use the first four characters. (Git will complain if it can't find a match with a shortened SHA1 ID.)

Next you'll see a big warning. Ignore this for now, and type this:

```
ls
```

You should see this list: `a`, `b`, `c`, `d`, `math.sh`.

The following listing shows the warning that you see from Git after performing the checkout.

Listing 8.8 The warning about the detached HEAD state

```
Note: checking out 'bcaa6e'.

You are in 'detached HEAD' state. You can look around, make experimental
changes and commit them, and you can discard any commits you make in this
state without impacting any branches by performing another checkout.

If you want to create a new branch to retain commits you create, you may
do so (now or later) by using -b with the checkout command again. Example:

  git checkout -b new_branch_name

HEAD is now at bcaa6e2... Adding four empty files.
```

There's a lot going on in this note from Git! There's a mention of branches, a HEAD that is detached, and another switch to `git checkout` (the `-b` switch). No wonder some people claim Git is hard to learn!

Figure 8.13 illustrates the detached HEAD state.

Normally, your HEAD is associated with a branch. When you move your HEAD around by using `git checkout`, you disassociate your HEAD from your current branch, which Git considers detached.

A more detailed diagram of your repository and HEAD is shown in figure 8.14. Compare this figure with figure 8.8. Whenever you move the HEAD (what you're

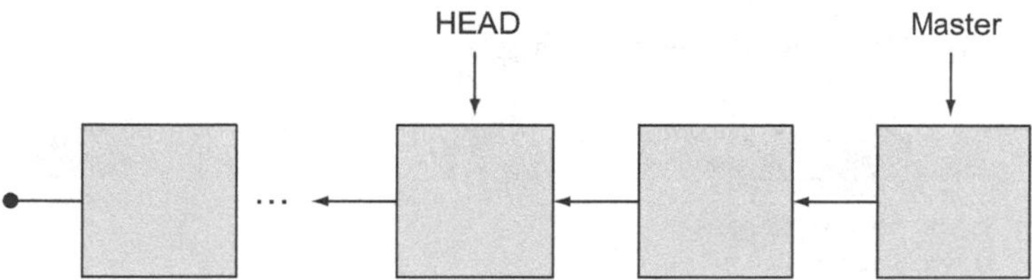

Figure 8.13 The detached HEAD. Compare this to figure 8.12, when HEAD pointed to the same place as master.

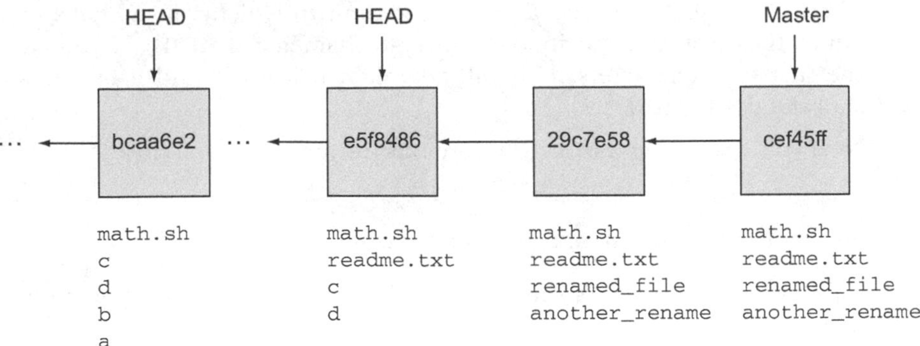

Figure 8.14 The detached HEAD state in Git

looking at) to a new commit, your working directory changes to match that commit's contents. The master pointer (reference) doesn't move. It always points to the last commit of the branch.

The warning from Git does say you can look around, so let's do that next.

TRY IT NOW Read the DETACHED HEAD section of the `git checkout` documentation page. Type this:

```
git help checkout
```

In the math directory, in your detached state, confirm that the math.sh file has indeed reverted to an earlier version:

```
cat math.sh
```

You should see that the file contains only your first comment and the two variables. Run this program by typing the following:

```
bash math.sh.
```

Notice how it doesn't return anything. This is because the program has reverted to its older functionality, back when you did this commit. Now type this:

```
git log --oneline
```

You should see output like the following listing.

Listing 8.9 `git log` output after you went back in time

```
bcaa6e2 Adding four empty files.
80f0ccc Adding b variable.
ea91623 This is the second commit.
8c31e35 This is the first commit.
```

You've gone back in time, and one consequence is that `git log` in your repository now stops at this version. It's as if the future (your present) didn't even happen. Stop and think about this: you're in an earlier state! But it's only temporary.

To get back to your present state, you use `git checkout` again, except this time you'll check out master.

TRY IT NOW In the math directory, type the following:

```
git checkout master
```

You should confirm that math.sh contains your current settings, and that `git log` returns the entire history of your work.

Think back to all those time-travel movies, where the hero is in the past. The hero can remember the future but can affect only the present. In Git, when you travel around your timeline, you're also limited to seeing and affecting wherever your HEAD is.

Figure 8.15 shows a simplified timeline. Each commit is a box labeled with a letter (instead of a SHA1 ID). Your HEAD is pointing at master, which is pointing at the last commit. This is your present. If you were to run `git log`, you'd see all the commits: A, B, C, D, E, F, and G.

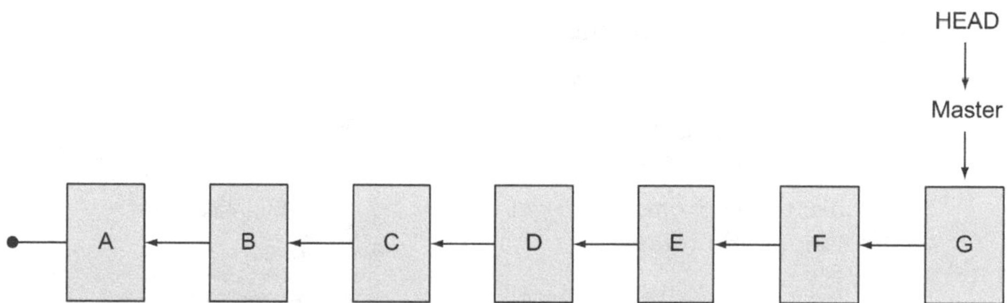

Figure 8.15 HEAD pointing to master: this is your present.

If you ran `git checkout D` (to move HEAD to the commit labeled D), `git log` would show you only commits A, B, C, and D. Your HEAD would be detached as well, as shown in figure 8.16.

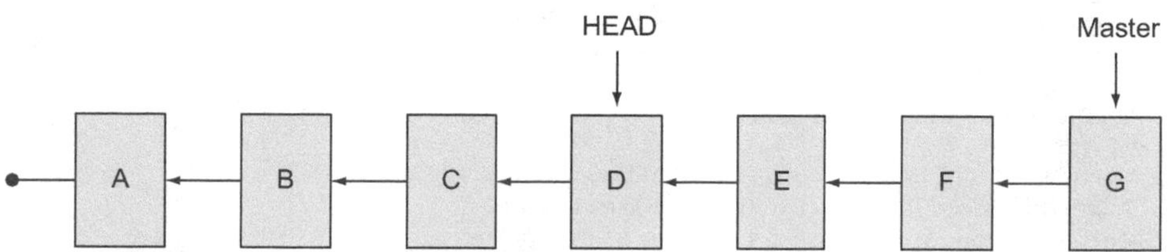

Figure 8.16 Checking out an earlier version

Labeling commits in this fashion is convenient. Now you're ready to learn the command to label your commits: `git tag`.

8.4 Breadcrumbs to previous versions

Referencing commits by their SHA1 IDs can get old quickly, even if you abbreviate them. Git has a mechanism to give names to commits: `git tag`. Let's say that you wanted to go back to the point in time where you added the four files again. Rather than remembering the long SHA1 ID (or even the abbreviated ID, which Git lets us get away with), you can give that particular commit a tag.

TRY IT NOW First, get the SHA1 ID for the `Adding four empty files.` commit. Use the technique from the previous TRY IT NOW.

Now type the following:

```
git tag four_files_galore -m "The commit with four files" YOUR_SHA1ID
```

where `YOUR_SHA1ID` is the SHA1 ID of the commit that has the log message `Adding four empty files.` Notice that you pass in a message (via the `-m` switch) to `git tag`. As with commits, tags can have messages, which can be as detailed as you want. And as with `git commit`, the `-m` switch lets you type the message on the command line, rather than in Git's default editor.

To confirm that you have this tag, type the following:

```
git tag
```

To visualize this tag, type `gitk` in this directory. You should see figure 8.17.

Make sure to exit from gitk to get back to the command prompt.

To look at the tag itself, you use the `git show` command:

```
git show four_files_galore
```

Figure 8.17 A tag in gitk

You should see the following listing.

Listing 8.10 `git show` output

```
tag four_files_galore
Tagger: Rick Umali <rickumali@gmail.com>
Date:   Fri Jun 20 19:42:54 2014 -0400

The commit with four files

commit c2eb6c5f275d18c0432431a66a868565e3078381
Author: Rick Umali <rickumali@gmail.com>
Date:   Mon Jun 16 18:39:39 2014 -0500

    Adding four empty files.
```

```
diff --git a/a b/a
new file mode 100644
index 0000000..e69de29
diff --git a/b b/b
new file mode 100644
index 0000000..e69de29
diff --git a/c b/c
new file mode 100644
index 0000000..e69de29
diff --git a/d b/d
new file mode 100644
index 0000000..e69de29
```

Listing 8.10 contains two Git objects: the tag itself (which is what you passed to the `git show` command), and the commit that the tag points to.

Tags let you give human-readable names to commits. As you work with Git, you'll be looking at and referencing a lot of SHA1 IDs. It's good to know that you can give them names.

Typically, you create a tag as soon as you make a commit. In the workflows that you'll see in chapter 17, you normally use `git tag` immediately after `git commit`. But you can add a tag anytime you want, as you've shown here.

Let's now use the tag you've created.

TRY IT NOW From the gitk window (figure 8.17), you can see that you're at the master's latest commit. To check out the commit containing your four empty files, type the following:

```
git checkout four_files_galore
```

Open the gitk program to confirm that your repository looks like figure 8.18.

Make sure to reset the repository back to master by typing this:

```
git checkout master
```

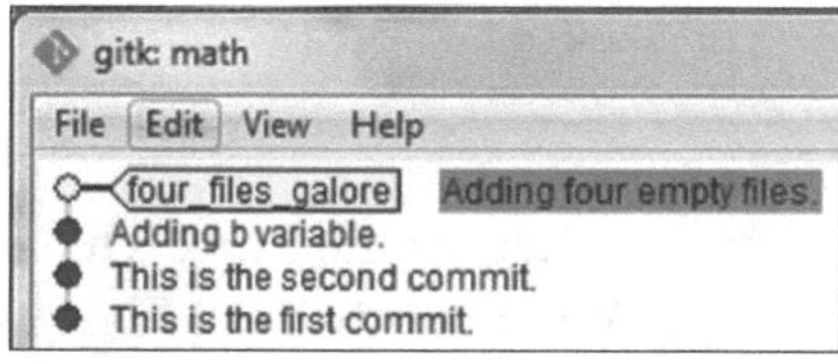

Figure 8.18 Your repository at the "four empty files" commit

8.5 Lab

You've covered a lot of Git commands that help you understand your commit history. You also saw how to travel anywhere in that history via `git checkout`.

8.5.1 Viewing history (part 1)

You spent the first part of this chapter looking at your history via `git log`. This command has a lot of functionality, and these exercises are designed to teach some of these functions:

1 How can you list the history from the first commit to the last?
  What do you think is faster for Git to generate: the default listing or a reverse listing?

2. Is there a way to list just the most recent *N* commits? (Where *N* can be any number?) How?
3. Can you display the date as time relative to the current time (for example, 2 hours ago)?
4. I'm a big fan of `git --oneline`, but it's a shortcut. What is it a shortcut for?

8.5.2 Amending commits

Make any change that you want to any file in your math repository. After you perform `git add` and `git commit`, note the SHA1 ID of this commit. Now type the following:

```
git commit --amend -m "Fixed commit" -m "Second paragraph" -m "Wall of text"
```

Observe the SHA1 ID of this amended commit. Is it different? Notice that your commit now has multiple lines. You can pass in the `-m` switch multiple times to `git commit`, allowing you to form longer messages (though you would probably use Git's default editor or `git citool` to enter a longer message).

8.5.3 Using other names

You examined the `git rev-parse` and `git show` commands. Like `git log`, these commands have much more utility than you looked at during this chapter. To practice these commands some more, type the following listing in your math repository and then answer the following questions.

Listing 8.11 Example `git rev-parse` and `git show` commands

```
git rev-parse master~3
git show master@{3}
git show master^^^
git rev-parse :/"Removed a and b"
```

1. What commits do each of these commands point to?
2. What did the last `git rev-parse` search to find its commit?
3. Does `git rev-parse` work on tag names?
4. Do these symbols (`~3`, `@{3}`, and so forth) work on tag names?

8.5.4 Committing while in detached HEAD mode

When you performed a `git checkout` while on the `four_files_galore` commit, you had a detached HEAD from the master. The note that Git displays indicates that you can make a commit. For this exercise, find out what happens when you make a commit while the HEAD is detached. What does Git do when you run `git checkout master`?

Working in the detached HEAD mode isn't an everyday event. Git has a better mechanism for introducing bug fixes, called *branches*, which you'll learn in the next chapter.

8.5.5 Deleting tags

Tags can be deleted as well! When you no longer need a tag pointing to a SHA1 ID, you can delete this tag. Look up how to delete a tag. For this exercise, create a tag, confirm that it exists via `git show`, and then delete it.

8.5.6 Viewing history (part 2)

Download the script make_lots_of_commits.sh from the book's website (www.manning.com/umali). Run this program from your home directory, in the command-line window, by typing this:

```
cd $HOME
bash make_lots_of_commits.sh
```

The script creates a repository in the directory called lots_of_commits. This working directory will be in the directory that you ran the script from. Go into this directory and then orient yourself by typing `git log`. Now answer the following questions:

1. What does the first commit say? What is the date of the first commit?
2. What is the SHA1 ID of the commit containing the word *ubiquitous* in its message?
3. Which commit was authored by the user with the rgu@freeshell.org address?
4. The dates of the commits have been modified. Display all the commits since yesterday in one `git log` command.
5. Open the repository in gitk and look at the last commit. It has multiple files affected. Use the patch and tree views and select specific files to see how to display changes.

8.6 Further exploration

In section 8.1.1, I used the phrase *cryptographically unique* with regards to the SHA1 IDs produced by the `git commit` command. The SHA1 ID is the foundation of Git's ability to be distributed because the SHA1 IDs are guaranteed to never repeat. Of course, you can never say never. To learn more about SHA1 IDs and computing collisions in hash functions, read the thread at http://marc.info/?l=git&m=111365428717118&w=2.

8.7 Commands in this chapter

Table 8.1 Commands used in this chapter

Command	Description
`git log --parents`	Show the history, displaying the parent commit's SHA1 ID for each commit.
`git log --parents --abbrev-commit`	Same as the preceding command, but shorten the SHA1 ID.
`git log --oneline`	Display history concisely, using one line per each commit.

Table 8.1 Commands used in this chapter *(continued)*

Command	Description
`git log --patch`	Display the history, showing the file differences between each commit.
`git log --stat`	Display the history, showing a summary of the file changes between each commit.
`git log --patch-with-stat`	Display the history, combining patch and stat output.
`git log --oneline file_one`	Display the history for file_one.
`git rev-parse`	Translate a branch name or a tag name to a specific `SHA1 ID`.
`git checkout YOUR_SHA1ID`	Change your working directory to match the version specified in `YOUR_SHA1ID`.
`git tag TAG_NAME -m "MESSAGE" YOUR_SHA1ID`	Create a tag named `TAG_NAME`, pointing to `YOUR_SHA1ID`. The tag will have a short `MESSAGE` associated with it.
`git tag`	List all tags.
`git show TAG_NAME`	Show information about the tag named `TAG_NAME`.

9 Taking a fork in the road

Branches are an important feature in many version control systems, but in Git, they're especially important because they're so easy to create. In this chapter, you'll learn how to make branches in your repository. This is sometimes known as *diverging your code base*. If you want to add a new feature or fix a bug in your repository's code, you'll want to create a branch so you can do that work on a copy of your code.

This chapter also covers how to switch between branches, and how to delete branches by using the `git branch` command. You'll learn how to jump back and forth among branches by using `git checkout`. In the next chapter, you'll bring together (*converge*) your branched code base.

Branching does introduce questions of how best to use branches. You'll tackle these policy questions in chapter 17. This chapter is about the mechanics of creating and using branches. Finally, don't worry if you lose track of the steps as you go through the TRY IT NOW sections; the lab at the end of the chapter introduces a small script that will rebuild your repository to the correct state.

9.1 Introducing branches

In the preceding chapter, you learned that when you create a repository, Git automatically creates a default branch called *master*. In this chapter, all the figures show branches like a tree, growing from bottom to top. Like a real tree, Git's branches grow higher and higher as you make commits.

Figure 9.1 shows the master branch and the last three commits on that branch. This master branch is a line of development, an ordered sequence of the commits you've made since you created the repository.

9.1.1 Creating references

Because each commit points to its parent, you can point to the last commit's SHA1 ID if you want to reference the entire branch. Instead of saying that the branch master consists of a set of commits, you can say that the branch master is the last commit. The last commit on a branch is called the *tip* of the branch, and it's always the most recent commit made to a branch.

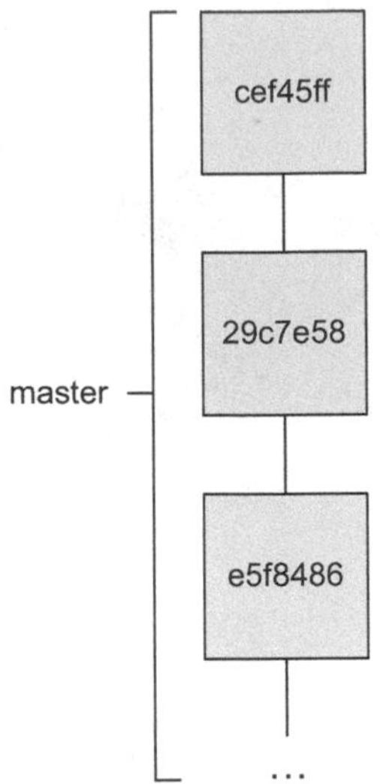

Figure 9.1 The master branch. Each box represents a commit, labeled with its SHA1 ID.

This idea of pointing to a commit is known as a *reference*. In figure 9.2, you label the word *master* as a reference, and you point master to the tip of the branch (the last commit). In the Git software, master contains the SHA1 ID of this last commit.

If you make a new commit, Git changes the reference named master to point to this new commit (the reference moves forward to point at this latest commit). See this in figure 9.3.

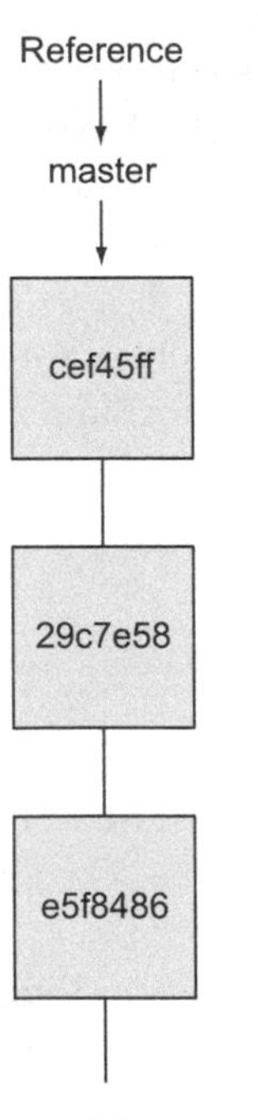

Figure 9.2 master is a reference to the last commit (its SHA1 ID is `cef45ff`).

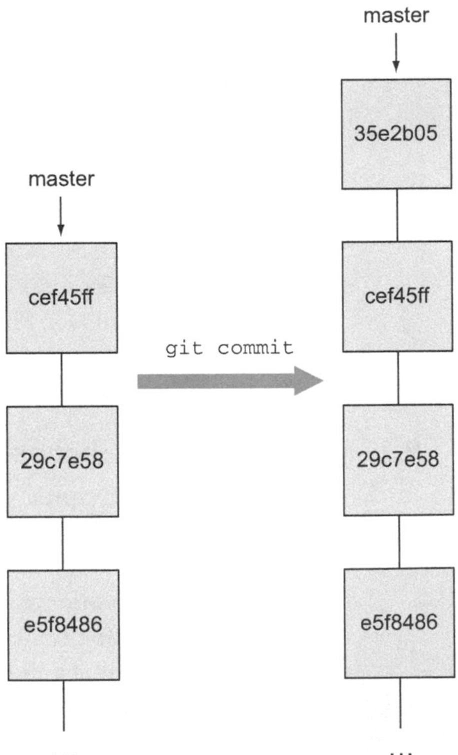

Figure 9.3 After making a new commit, master is changed to point to this new commit (`35e2b05`).

9.1.2 Understanding that master is just a convention

The word *master* has a certain meaning, doesn't it? It suggests that it's somehow essential or important. But if you look up the definition of master in the Git glossary, you'll read this sentence: "In most cases, [master] contains the local development, though that is purely by convention and isn't required." The next example will convince you that the branch master isn't required, as it's an important point.

TRY IT NOW In your home directory, create a new repository:

```
cd
mkdir empty
cd empty
git init
```

Add a file to this repository:

```
touch foo
git add foo
git commit -m "committing the file foo"
```

After you have a commit, Git creates the branch master. To list the branches that you have so far, type the following:

```
git branch
```

Your repository looks like figure 9.4.

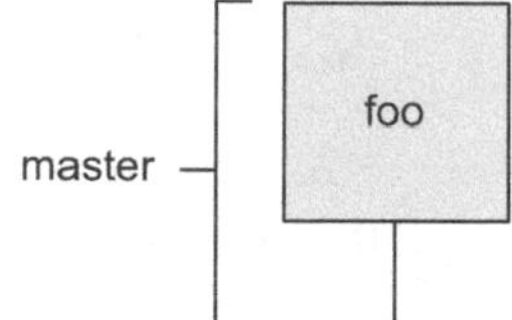

Figure 9.4 Your small repository, with one branch, master

The command should return a single line, `* master`. The * represents the current branch that you're on. You now know that this file belongs to the master branch. But let's create a new branch:

```
git branch dev
```

This creates a new branch called *dev*. Confirm that this branch is available by typing this:

```
git branch
```

You should see two branches: dev and master. The * is still next to master, meaning that you're still in the master branch. Your repository looks like figure 9.5.

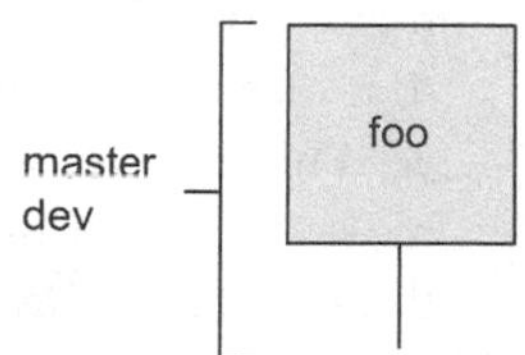

Figure 9.5 Your repository with two branches

Because dev was created as a branch of master, it initially contains the exact same content as master. It may be easier to think of your repository as having

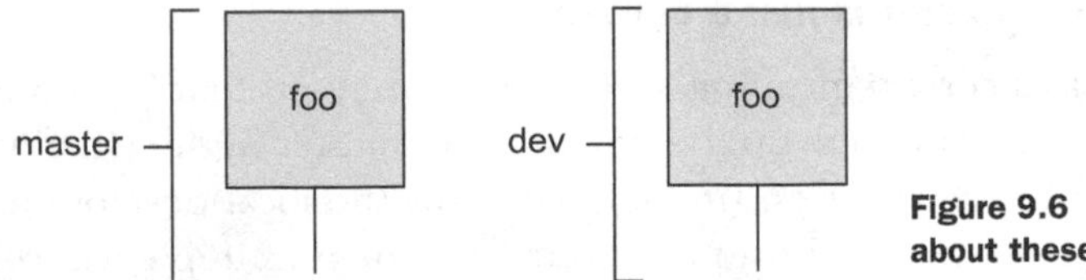

Figure 9.6 Another way of thinking about these two branches

two copies of foo, one belonging to the master branch and one belonging to the dev branch, as in figure 9.6.

Keep in mind that figure 9.6 doesn't show how Git works internally, but it may be clearer to think about branches as having their own copies of files. To access the files of this new branch, you check out the branch by using the `git checkout` command. Type the following:

```
git checkout dev
```

Check out isn't like checking out a book from a library; in the context of Git, check out means changing the working directory to reflect the contents of the branch. (Because dev and master are the same, there's no difference between these two branches.)

Now type the following:

```
git branch
```

You should see two branches, dev and master, and that dev is the current branch. Delete the master branch, using the `-d` switch to `git branch`:

```
git branch -d master
```

To confirm that master is gone, type the following:

```
git branch
git checkout master
```

The first command lists only the dev branch. The second command produces a somewhat cryptic error: `pathspec 'master' did not match any file(s).` Read this to mean that master doesn't exist.

At this point, your empty repository contains only one branch. This is the branch named dev. It does contain a single commit, the one for the file foo. You've proven that master isn't required!

The most important branch to your organization is the one that's designated as the most important branch. This is sometimes the branch named master, but more often than not, it's a branch named v1.0 or dev or even perhaps a code name. You (or your organization) get to pick which branch rules all the other branches. Chapter 17 covers some basic conventions.

9.2 When and how to create branches

In the course of any project, you'll often need additional lines of development. The master branch (or whichever branch your team designates as the master) typically represents the working code-base. The code in master is generally clean, builds properly, can be deployed to the production environment, and so forth. Therefore, you typically don't want to develop on master directly. Instead, you want to create a copy of this master branch and work on that.

Your workplace may have a different approach, but the benefit of having your own copy of the repository means you can do your own work in your own private branches. It's only when you have to collaborate that you must pay strict attention to which branch you're developing on. In this book, you'll treat master as the working code base.

In general, you'll want to create a branch for two common scenarios: introducing new code and fixing existing code. For both scenarios, you want to create a copy of master. In the next section, you'll create new branches of master to cover these scenarios.

9.2.1 Introducing new code with branches

Let's say you're going to add new functionality to your code base. This new work could take multiple days to implement. To keep the master branch in working condition, you'll make a branch (a copy) of master for your new development.

When you create a branch, you introduce a new line of development. This diverges your code base. When you make commits on this new line of development, it's completely separated from the master. You get to make commits and track your work without messing up the build.

In figure 9.7, this new branch grows upward by our convention. Notice that nothing has been committed on this new branch. Instead, Git has made another reference (called *new_feature*). This reference points to the same commit as master, but not for long.

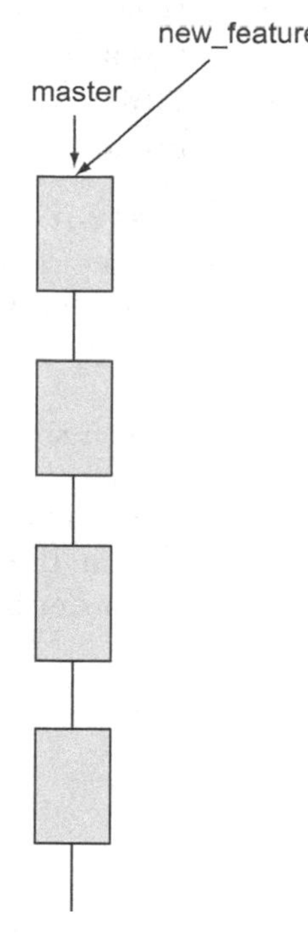

Figure 9.7 The repository, after you type `git branch new_feature`

TRY IT NOW From the command line, go to the math directory that you were working with in the preceding chapter. (All of our TRY IT NOWs in this chapter are run from this directory, unless otherwise stated.)

```
cd $HOME/math
```

List the branches that are in your repository:

```
git branch
```

This should return a single line, `* master`. To create a branch, type the following:

```
git branch new_feature
```

Now type this:

```
git branch
```

You should see two branches. The * is still next to the master branch, as shown in figure 9.8.

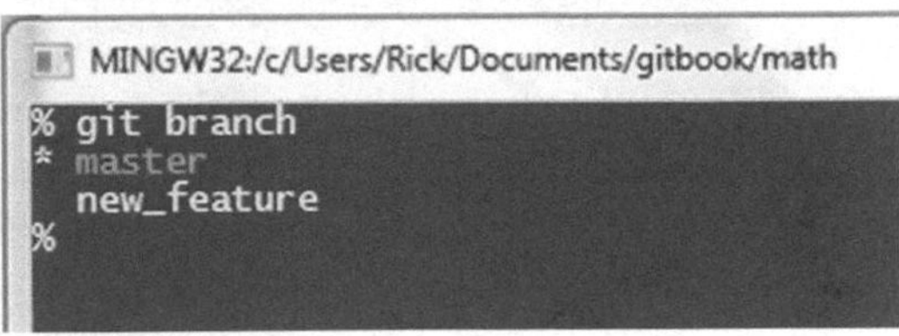

Figure 9.8 Creating a new branch

COMMITTING ON A NEW BRANCH

The previous TRY IT NOW section introduced a new line of development. This is a fork in the road of your master branch. What you can do now is change your working directory to use this new branch, and start making commits on it. After the next TRY IT NOW section, you'll have a repository that looks like figure 9.9.

TRY IT NOW Let's first examine the state of your repository via gitk, which is a useful tool for visualizing branches. In the math directory, type the following:

```
gitk
```

You should see a window that looks like figure 9.10. Notice that the new second branch is immediately next to the master branch. This is how gitk depicts the situation. Also, notice that the master branch is in boldface. This indicates that master is the current branch.

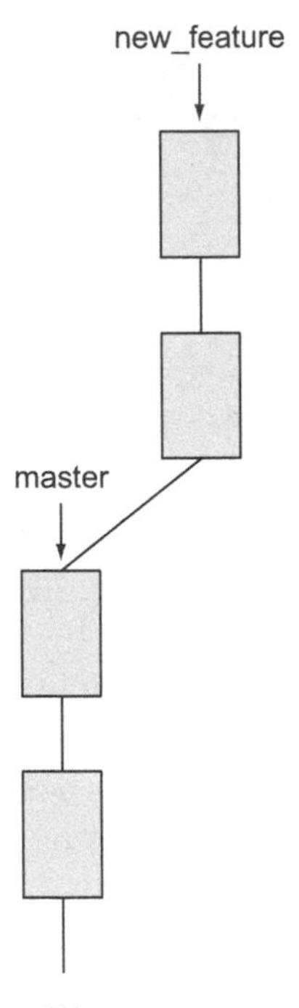

Figure 9.9 Commits in your new line of development. Again, notice that the new commits are growing (and going upward).

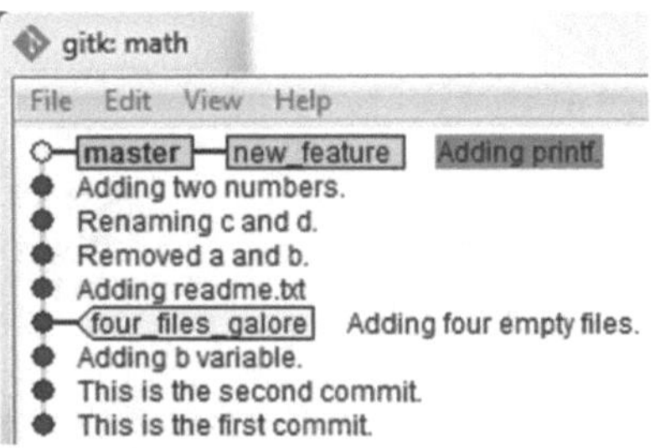

Figure 9.10 gitk's look at a new branch

Exit gitk, and from the command-line window, change into the new_feature branch:

```
git checkout new_feature
```

You should see a message indicating that the working directory has switched to the new_feature branch. At this moment, this new branch is a copy of master, but after you make commits to it, it will be different from master.

To make two new commits in this new line of development, type the following:

```
echo "new file" > newfile.txt
git add newfile.txt
git commit -m "Adding a new file to a new branch"
echo "another new file" > file3.c
git add file3.c
git commit -m "Starting a second new file"
```

Now type `gitk` to see these commits in a UI. It should look like figure 9.11.

You can now exit gitk.

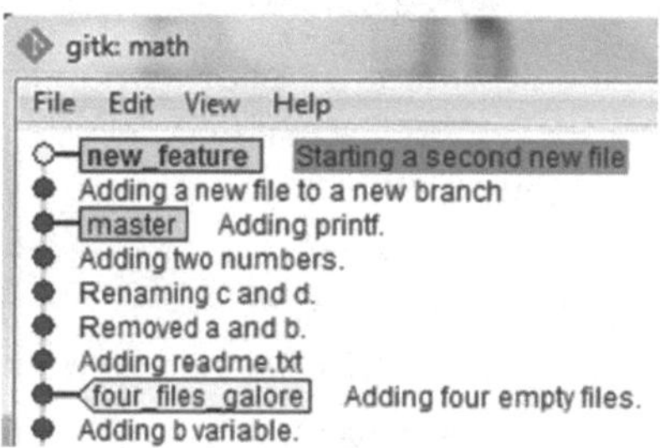

Figure 9.11 Two commits in your new line of work

Compare figure 9.11 with figure 9.9 at the start of this section. It's mostly the same, right? You're starting your development of a new feature in a separate branch (called new_feature), and master is unaffected by any of this development work.

VIEWING BRANCHES VIA GIT LOG

Like `gitk`, `git log` can display branches. The tree display is shown in the terminal by using characters such as `*` and `|`. (You'll see these for yourself in listing 9.3 later in this section.)

TRY IT NOW In the command-line window, ensure that you're in the math directory and then type the following:

```
git log --graph --decorate --pretty=oneline --all --abbrev-commit
```

This produces output like the following listing.

Listing 9.1 Viewing branches via `git log`

```
* 15f75dd (HEAD, new_feature) Starting a second new file
* 0e29c71 Adding a new file to a new branch
* 5bfd3c8 (master) Adding printf.
* 34d51d1 Adding two numbers.
* bd6a2c4 Renaming c and d.
* 3788018 Removed a and b.
* ba8ca57 Adding readme.txt
```

```
* 737f38b (tag: four_files_galore) Adding four empty files.
* 6bb3f6a Adding b variable.
* 5e31795 This is the second commit.
* 56e7d7d This is the first commit.
```

In most terminal windows, the HEAD, the branches, and the tags have different colors, making them easier to see. The command-line switch that enables all the branches to be displayed is `--all`.

If you spend a lot of time on the command line, knowing this is helpful, but as you create more branches, you'll probably need to resort to gitk to manage the entire display.

Above and Beyond

Git has a built-in alias system so that you can shorten a long command line like the one in this section. With an alias, instead of typing `git log --graph --decorate --pretty=oneline --all --abbrev-commit`, you could type `git lol`.

To create an alias, you must use the `git config` command. You did this back in chapter 2 to initialize your username and email address. To create an alias, use this command:

```
git config --global alias.lol "log --graph --decorate
➥ --pretty=oneline --all --abbrev-commit"
```

You have to type this only once. Afterward, you can type `git lol`.

You'll be visiting the `git config` command in later chapters because, in addition to creating aliases, you can modify the default behavior of many of Git's commands. (I use `git lol` because it's a common alias. Search `git lol` on GitHub for this and other snippets.)

NAMING BRANCHES

You might be wondering whether any restrictions or conventions exist for naming branches. Branch names must pass the rules described by the `git check-ref-format` command. These rules can be seen by typing `git check-ref-format --help`. Most of the rules forbid special characters (such as a space, `?`, or `*`) and sequences (such as `..`). As long as you stick to alphanumeric names, you'll be fine.

I've seen the gamut of naming conventions when it comes to branch names. I've seen branch names with camel case (capitalized words that aren't separated by spaces, such as MyBigBranch), bug numbers (for example, BUG14015), and folder-style names (for example, branch/rick/bug1). Git doesn't impose any guidelines on what your branch name should be. It's up to you or your organization to come up with a convention that makes sense.

SWITCHING BETWEEN BRANCHES

When creating a diagram of the repository after the previous section, I tend to draw a diagonal line, to emphasize that these two new commits are on a different branch (see figure 9.12). You may observe from the gitk display that a straight line could be drawn from the tip of the new feature to the tip of master. Even though this is the case, the contents of both branches are different. You can see this for yourself in the next TRY IT NOW.

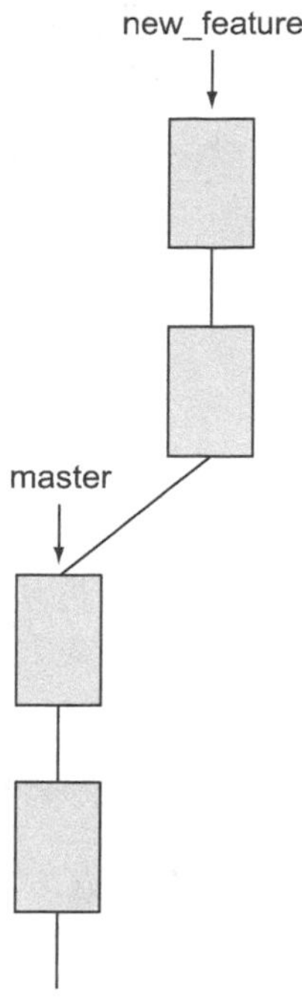

Figure 9.12 Two commits made to the new_feature branch

TRY IT NOW Let's switch between the two branches and observe that the working directory changes appropriately. You should be in the new_feature branch. Confirm that by typing the following:

```
git branch
```

You should see the two branches, master and new_feature. new_feature should have the * next to it, denoting that it's the current branch. Now type this:

```
git branch -v
```

This should show you output like the following listing.

Listing 9.2 `git branch -v` output

```
  master      35e2b05 Adding printf.
* new_feature ebcd35d Starting a second new file
```

The `-v` switch displays the SHA1 ID of the tips of these branches. (If you did the lab in chapter 8, the tip of master will be `Fixed commit`.) Now type the following:

```
ls
```

Your directory should include the two new files, newfile.txt and file3.c. Now type this:

```
git checkout master
ls
```

This time newfile.txt and file3.c aren't present. You should see something close to figure 9.13.

```
% git branch
  master
* new_feature
% ls
another_rename  file3.c  math.sh  newfile.txt  readme.txt  renamed_file
% git checkout master
Switched to branch 'master'
% ls
another_rename  math.sh  readme.txt  renamed_file
%
```

Figure 9.13 Switching between branches changes the working directory

The thing to keep reminding yourself is that the branches are independent of one another. Let's see what happens if you make a commit on master.

> **TRY IT NOW** Go to the math directory and type the following:
>
> ```
> echo "A small update." >> readme.txt
> git commit -a -m "A small update to readme."
> ```
>
> Now the picture of your repository looks like figure 9.14.

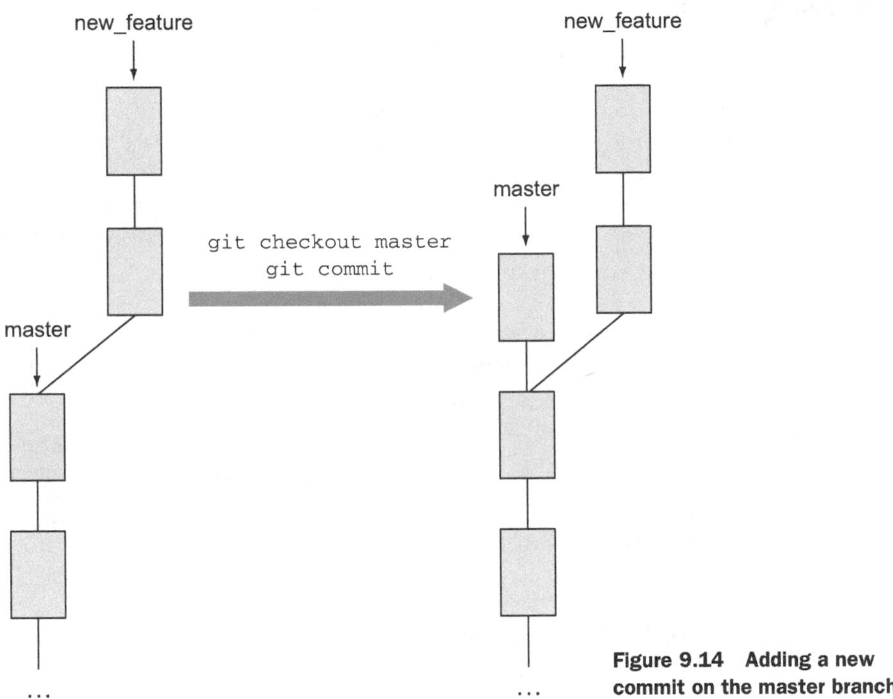

Figure 9.14 Adding a new commit on the master branch

The new commit makes the master branch grow upward by one commit. The new_feature branch isn't affected by this change at all.

ADJUSTING GITK TO VIEW MULTIPLE BRANCHES

It may be more clear after this last commit that both branches are independent. Let's look at this history and learn to switch between branches in gitk.

> **TRY IT NOW** Start gitk. One thing you should notice is that the only branch that appears is master, as in figure 9.15.

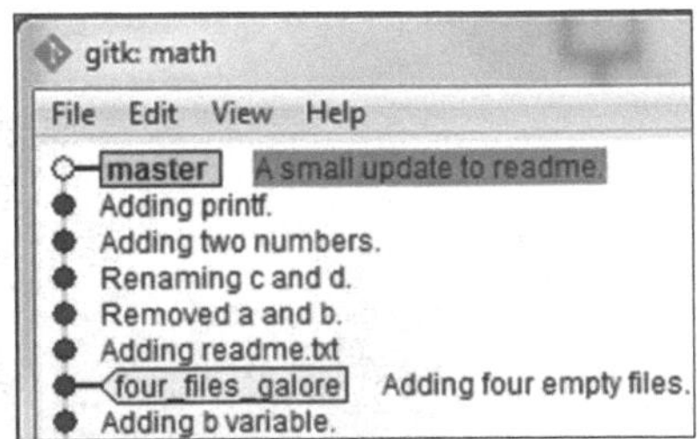

Figure 9.15 Viewing the master branch in gitk

As you move the tips of the branches apart by making commits, they're no longer visible to the other branch.

To configure gitk to show the other branch, click the View menu and choose New View. This brings up a complicated-looking window with the title Gitk View Definition – Criteria for Selecting Revisions, shown in figure 9.16.

Figure 9.16 Configuring the view to show all

The change you'll make is to select the All (Local) Branches option (highlighted in figure 9.16). To make this take effect, click Apply (F5) at the bottom of the window. To make gitk remember this change, click the Remember This View check box at the upper right. (You can give your view a unique name in the View Name field, and the name will appear in the menu.) Then click OK. Your gitk should look like figure 9.17.

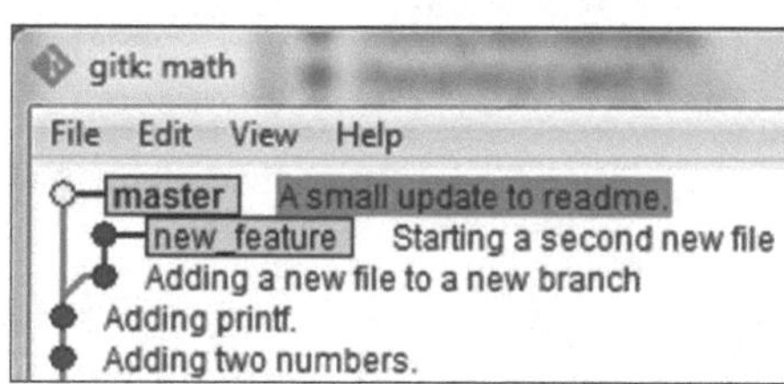

Figure 9.17 Your two branches

You can now exit gitk.

Compare figure 9.17 with figure 9.14, and you should see the similarities between the two graphs. You've seen that `git log` can mimic this view as well.

TRY IT NOW In the repository's directory, type the following:

```
git log --graph --decorate --pretty=oneline --all --abbrev-commit
```

(The Above and Beyond sidebar earlier in this chapter showed how to create an alias for this long command. If you created that alias, give it a try here.) This should give you a listing like the following.

Listing 9.3 git log --graph --decorate --pretty=oneline --all --abbrev-commit

```
* e150c19 (master) A small update to readme.
| * b1641b2 (new_feature) Starting a second new file
| * eafc3ce Adding a new file to a new branch
|/
* f48c719 Adding printf.
* 58ee0fc Adding two numbers.
* d3ae3ea (HEAD, another_fix_branch) Renaming c and d.
* dd87c91 Removed a and b.
* 11a90b4 Adding readme.txt
* 12a7b37 (tag: four_files_galore) Adding four empty files.
* 907b870 Adding b variable.
* 56d7919 This is the second commit.
* c57cd5c This is the first commit.
```

The key thing to notice is that listing 9.3 splits at `f48c719`. This is how Git depicts the branch in figure 9.17 using ASCII. (Your commit messages may be different, depending on how much of the earlier labs you've completed, but the branch should be visible.) Now let's switch between branches in gitk.

TRY IT NOW Restart the gitk program. Make sure you're using the same view as in the previous TRY IT NOW.

In the branch window pane, hover over the new_feature branch, and then bring up the context menu. In Windows or Unix/Linux, context menus are raised by clicking the right mouse button, but on the Mac, you have to click the mousepad with two fingers. You should see a menu pop up, as in figure 9.18.

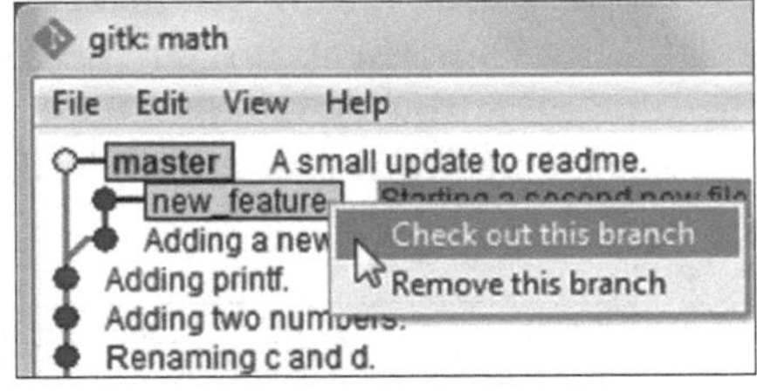

Figure 9.18 The context menu to check out branches in gitk

Click the Check Out This Branch item. Exit gitk, and confirm on the command line that you've changed to the new_feature branch by typing the following:

```
git branch
```

The output should show new_feature with the asterisk in front of it.

9.2.2 Introducing fixes with branches

Branches provide a way to isolate changes from the rest of the code base. In the previous section, you created a new branch right from the tip of master. This branch could represent a new feature that you're developing. But perhaps more common is the need to create a branch to develop a fix. Fixes tend to be made on an earlier part of your repository's history.

In figure 9.19, a bugfix branch is created at commit `29c7e58`, which is two commits behind master. With Git, you can specify any commit point as the starting point for a new branch.

master

35e2b05

cef45ff

bugfix

29c7e58

e5f8486

...

Figure 9.19 Branching for a bug fix

TRY IT NOW On the command line, in the math directory, type this:

```
git checkout master
git log --oneline
```

The first command makes sure you're on the master branch. The second command lists the history with the corresponding SHA1 IDs. Next, identify the SHA1 ID for the commit labeled `Renaming c and d`. Write down or copy this SHA1 ID. Now, on the command line, type the following:

```
git branch fixing_readme YOUR_SHA1ID
git checkout fixing_readme
```

Replace YOUR_SHA1ID in the preceding step with the SHA1 ID for the `Renaming c and d` commit.

At this point, your repository should look like figure 9.20. Confirm that by typing `gitk`.

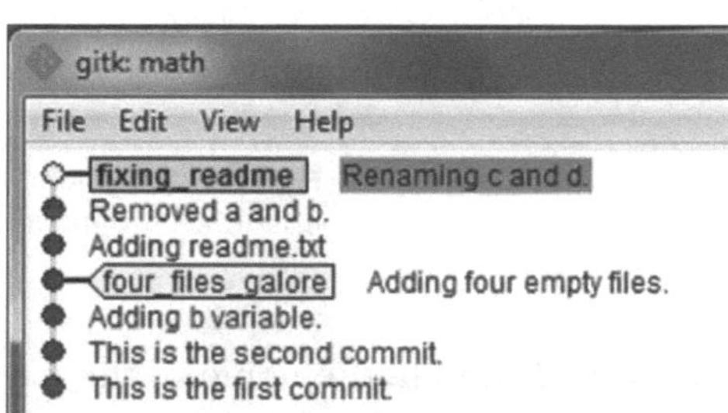

Figure 9.20 Making a branch from an earlier start time

9.3 Performing other branch operations

You've created two branches so far and made commits to one of them. You have a handful of other branch operations to learn.

9.3.1 Branching faster

Branches are incredibly fast to create. Unlike other version-control systems, Git requires no server to talk to and no copying of files. This is due to Git's architecture (commits point back to their parents). Because of this, the `git branch`/`git checkout` operations are fast. You can use your branches just as quickly as you can create them.

A useful shortcut can help you more fully embrace branching. Using `git checkout`'s `-b` switch, you can create a branch and check it out in one step.

TRY IT NOW On the command line, in the math directory, type the following:

```
git checkout -b another_fix_branch fixing_readme
```

Now type this:

```
gitk
```

Your branch list should look like figure 9.21.

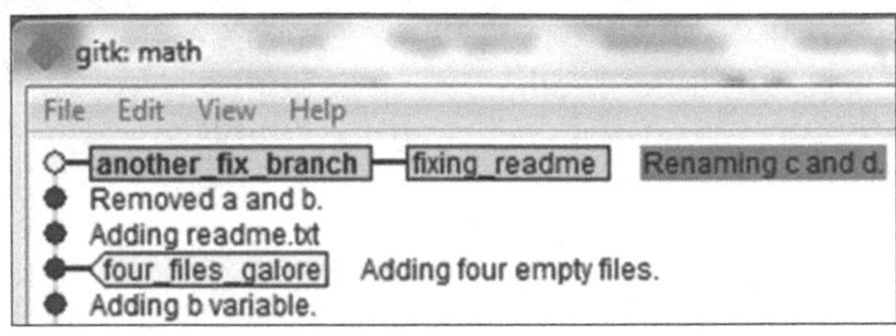

Figure 9.21 Making another branch

You can now exit gitk.

Look carefully at this command in figure 9.22, because it does two things at once.

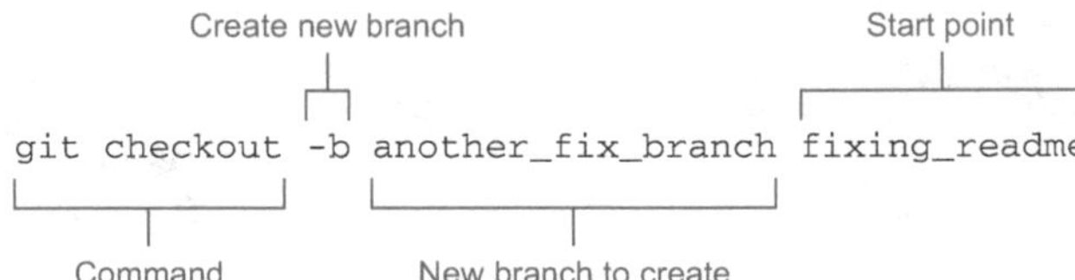

Figure 9.22 The `git checkout -b` command

This command takes three arguments: the `-b` switch, the name of the new branch to create, and the starting point on which to create this new branch.

The starting point that you used in the TRY IT NOW section was the branch you created before. There's nothing wrong with using a current branch as a starting point! The starting point can also be a commit SHA1 ID or a tag.

The new branch name is the name you'd normally pass into the `git branch` command. The `-b` switch is the signal to `checkout` that you want to create the branch and then enter that branch immediately.

Memorize this form of `git checkout`, because when you create a branch, you almost always want to use it immediately.

9.3.2 *Deleting branches*

Sometimes you need to delete branches. Remember that branches are forks in the road. Some branches you'll walk down only a short way. You create branches because you want to try something, but sometimes your experiment doesn't succeed.

You may inadvertently create a branch by accident. As you saw in the previous section, branches are easy to create. Deleting is just as easy. To delete a branch, use the `-d` switch of `git branch`. Remember that you deleted a branch already in the first TRY IT NOW of this chapter.

TRY IT NOW In the command-line window, in the math directory, type this:

```
git checkout master
git branch -d fixing_readme
```

To see what branches you have, type this:

```
git branch
```

You immediately deleted the fixing_readme branch that you created moments ago. Hopefully, this encourages you to create branches anytime you want to try an experiment with your code base. Branches isolate your development from the rest of your work and from the master branch. Anytime you think, "I wish I had a copy of my repository," you should immediately think, "I'll just make a new branch!"

You should be careful with `git branch -d`. This operation has no fail-safes, and Git does delete the branch. Git protects you from the obvious blunder of deleting the branch you're currently on, but if you delete a branch that you meant to keep, you can easily re-create it with the delete message that Git provides.

TRY IT NOW Make sure you're in the master branch and then type this:

```
git branch -d another_fix_branch
```

Notice the message that Git provides after you perform a deletion. It should read something like the following listing.

Listing 9.4 Output from `git branch -d`

```
% git branch -d another_fix_branch
Deleted branch another_fix_branch (was d6cc762).
```

The SHA1 ID in the delete message is the starting point of the branch you just deleted. At this point, if you realize that this delete was the wrong thing to do, you can re-create the branch by immediately typing the following:

```
git checkout -b another_fix_branch d6cc762
```

Notice that you specify the same SHA1 ID from the delete message.

Confirm that the branch is now back by typing this `git log` command from earlier:

```
git log --graph --decorate --pretty=oneline --all --abbrev-commit
```

If for some reason you don't have the SHA1 ID, you can resort to `git reflog`. This command shows a record of all the times that you've changed branches.

Type the following:

```
git reflog
```

You should see output like the following listing.

Listing 9.5 `git reflog` output

```
158b7ef HEAD@{0}: checkout: moving from master to another_fix_branch
2bd20cb HEAD@{1}: checkout: moving from another_fix_branch to master
158b7ef HEAD@{2}: checkout: moving from fixing_readme to another_fix_branch
158b7ef HEAD@{3}: checkout: moving from master to fixing_readme
```

Locate the line `moving from master to another_fix_branch`. The SHA1 ID at the start of this line (in listing 9.5, it's `158b7ef`) is the SHA1 ID of another_fix_branch. You can now perform the `git checkout` command from this section by using this SHA1 ID.

9.4 Switching branches safely

Multitasking in your code base becomes easy, because creating and switching branches is so easy. But if you're in the middle of some work (you have uncommitted changes in your working directory), Git won't allow you to check out another branch. You must put this work aside properly.

9.4.1 Stashing away your work

Git does have a facility for putting aside your work temporarily: the `git stash` command. You can use this to save all your work temporarily, leaving you with a clean working directory. Think of it as a temporary commit! Let's explore its most common use case.

TRY IT NOW In the math repository, confirm that you're in the another_fix_branch branch. Type the following:

```
git status
```

If you aren't in this branch, type this:

```
git checkout another_fix_branch
```

Now that you're in this branch, make a change to math.sh by using your favorite editor. Add this line at the end:

```
c = 0
```

Confirm that your repository is in the middle of some work:

```
git status
```

math.sh should be marked as modified. Now imagine that your manager has asked you to look at something important on the master branch. Try to switch to it by typing this:

```
git checkout master
```

You should get an error message like the following listing.

Listing 9.6 An error from the `git checkout` command

```
error: Your local changes to the following files would be overwritten
➥ by checkout:
        math.sh
Please, commit your changes or stash them before you can switch branches.
Aborting
```

This error is self-evident: changing branches at this time would erase your current work. You could commit your change now (using `git commit -a`), but if your work isn't finished, you may not want to commit it. You could reset the code (using the instructions from the `git status` command), but what you want to do is temporarily save your changes without doing a formal commit. This is what `git stash` does.

TRY IT NOW In the math repository, type the following:

```
git stash
```

You should get a message like the following listing.

Listing 9.7 `git stash` output

```
Saved working directory and index state WIP on
➥ another_fix_branch: 29c7e58 Renaming c and d.
HEAD is now at 29c7e58 Renaming c and d.
```

This message says Git is saving your WIP, short for *work in progress*. Under the covers, Git stores the work in a commit, but it's reachable only by the `git stash` command. Now type this:

```
git status
```

At this point, you can do a `git checkout` of the master branch:

```
git checkout master
```

9.4.2 Popping the stash

Let's continue the scenario: you and your manager are no longer looking at master. You want to go back to whatever work you were doing in another_fix_branch.

TRY IT NOW Go back to another_fix_branch:

```
git checkout another_fix_branch
```

Now let's take a look at what you've stashed away. For this, you'll use `git stash list`:

```
git stash list
```

This shows you one item, the work that you stashed away in the previous section. To reapply this work to your current branch, type this:

```
git stash pop
```

You should see output like the following listing.

Listing 9.8 `git stash pop` output

```
On branch another_fix_branch
Changes not staged for commit:
  (use "git add <file>..." to update what will be committed)
  (use "git checkout -- <file>..." to discard changes in working directory)

        modified:   math.sh

no changes added to commit (use "git add" and/or "git commit -a")
Dropped refs/stash@{0} (48569a9917d430bad9aaa856e6cc1a05be1701da)
```

After you pop the stash, it's removed from the stash list, and the work that was stashed is added back to your working directory. After `git stash pop`, the working directory is in the same state as it was before the first `git stash`. You can think of the stash as a sticky note that you leave for yourself. If you get interrupted, stash your work so you can safely switch context (both in your brain and in your repository).

9.5 Lab

The following lab exercises use the math repository from the TRY IT NOW sections. If you've been able to follow along without problems, feel free to jump into the lab exercises now.

However, if you've found yourself in a mixed-up state with regards to the math repository, or just want to work on a math repository that's in the right state, download the zip file LearnGitMoL_SourceCode.zip from the book's website. It contains a script named make_math_repo.sh. Run this script on the command line:

```
bash make_math_repo.sh
```

This creates a new math repository in a directory named math. You'll have to delete your existing math directory. The script creates a new repository following all of the steps and exercises up to section 9.4.2. The script leaves the repo in another_fix_branch branch, with one edited file that needs to be committed (via `git add` and `git commit`). Be sure to look at the code, as it documents the steps.

9.5.1 Using the GUI for branch work

1. Using Git GUI, try a `git checkout` of the new_feature branch.
2. Use gitk to add another branch starting at the same place as the another_fix_branch branch.

3 Create a branch off the tag from the previous chapter. Try this operation with both Git GUI and gitk. (You'll need to give these two branches different names.) Which is easier to you?
4 Delete these three branches by using both Git GUI and gitk.

9.5.2 Warm-up questions

1 If you create a branch in error, could you rename the branch instead of deleting it?
2 In section 9.2.2, you had to search for the SHA1 ID of the commit containing the string `Renaming c and d`. How would you identify this SHA1 ID when using the command `git rev-parse`?
3 Section 9.2.1 introduced a lengthy `git log` command. Look up what all the switches do! (Try running the command and leaving some of the switches out.)
4 What happens to the commits of a branch if you delete that branch?

9.5.3 Working on another_fix_branch

When you last left another_fix_branch, you had changed math.sh but hadn't committed it yet. Instead of committing it, remove this impending change by typing the following:

```
git checkout -- math.sh
```

1 What form of the `git checkout` command are you using?
Look up this answer via `git checkout --help`.
2 Add the following line to the math.sh file (still in another_fix_branch):

```
c=1
```

Commit this change.

Does your repository look like figure 9.23?

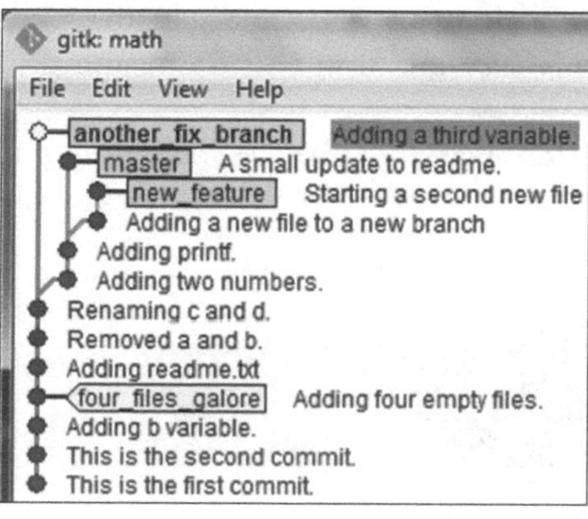

Figure 9.23 Adding a commit to another_fix_branch

9.5.4 *Viewing branches*

The same zip file mentioned at the beginning of this section contains a script, make_lots_of_branches.sh. Run this script in your command-line window:

```
bash make_lots_of_branches.sh
```

The script creates a repository in the directory called lots_of_branches. It may run for a long time, but it will eventually finish, after creating 40 branches in this repository. Answer and perform the following tasks:

1. What commit do all the example branches start from?
2. How many commits are in branch_30 from this start point?
3. Three branches are tagged random_prize_1, random_prize_2, and random_prize_3, respectively. What are these branches? Confirm by looking at the file answers.txt in branch_40.
4. Locate the tag labeled random_tag_on_file. Which branch contains this tag? (Use the `git branch` command to find this answer.)
5. Type `git log --oneline --decorate --simplify-by-decoration --all`. What is this command telling you? Add the `--graph` switch. What do you see then?

9.6 *Further exploration*

In figure 9.24, you see a small bit of the command line. The command that was typed was `ls`, but all the text preceding it is a detailed command prompt.

You may be familiar with the standard prompt of `$` or `>` (or # if the user is a super user). On Windows, entering the command line will give you a prompt that contains the drive letter, as in `C:\>`. On the Mac, the prompt is `$`, usually preceded by the hostname.

The default prompt on most command-line systems can be modified to be more descriptive. Git BASH for Windows comes with a customized prompt that always announces the branch name for the current directory. If you're using the BASH environment (and you definitely are using BASH with Git BASH, the Mac, and with most Unix/Linux servers), explore this customization by reading the prompt customization code at the following site:

https://github.com/git/git/blob/master/contrib/completion/git-prompt.sh

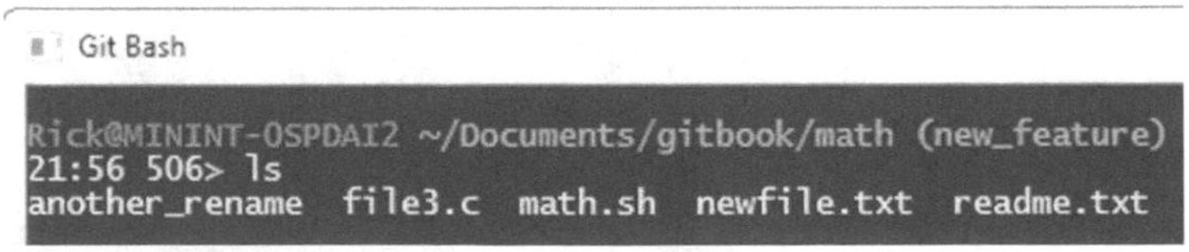

Git Bash

```
Rick@MININT-OSPDAI2 ~/Documents/gitbook/math (new_feature)
21:56 506> ls
another_rename  file3.c  math.sh  newfile.txt  readme.txt
```

Figure 9.24 A fancy command-line prompt

9.7 Commands in this chapter

Table 9.1 Commands used in this chapter

Command	Description
`git branch`	List all branches.
`git branch dev`	Create a new branch named dev. (This branch points to the same commit as HEAD.)
`git checkout dev`	Change your working directory to the branch named dev.
`git branch -d master`	Delete the branch named master.
`git log --graph --decorate --pretty=oneline --all --abbrev-commit`	View history of the repository across all branches (see section 9.2.3).
`git config --global alias.lol "log --graph --decorate --pretty=oneline --all --abbrev-commit"`	Make an alias named `lol` for the `git log` command in the previous row (see the Above and Beyond sidebar).
`git branch -v`	List all branches with SHA1 ID information.
`git branch fixing_readme YOUR_SHA1ID`	Make a branch using YOUR_SHA1ID as the starting point.
`git checkout -b another_fix_branch fixing_readme`	Make a branch named another_fix_branch using branch fixing_readme as the starting point.
`git reflog`	Show a record of all the times you changed branches (via `git checkout`).
`git stash`	Set the current work in progress (WIP) to a stash (holding area), so you can perform a `git checkout`.
`git stash list`	List works in progress that you've stashed away.
`git stash pop`	Apply the most recently saved stash to the current working directory; remove it from the stash.

10 Merging branches

In the preceding chapter, you created multiple branches, diverging your code base. You learned that when you work on a branch, you're working on a separate line of development. If you want to incorporate the work from your separate branch back into your main line of development, you need to use `git merge`, as shown in figure 10.1.

Branching diverges code bases, and merging converges code bases. In figure 10.1, we diverge the code base at commit B, making two branches: master and new_feature. We then make some commits on both branches. Next, using `git merge`, we converge new_feature and master back together in commit

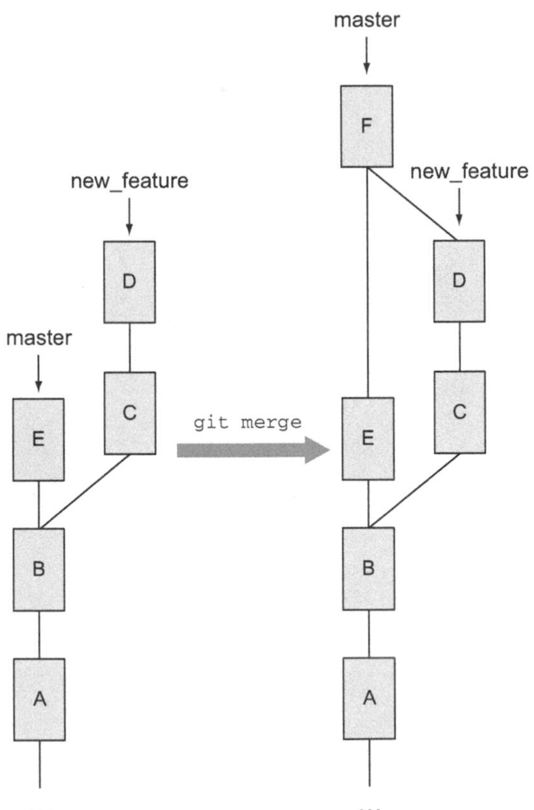

Figure 10.1 A typical merge. The commits are the boxes with letters.

F. git merge is the focus of this chapter, as well as some of the graphical tools accessed by git mergetool (you'll see a few of them). Because producing branches is so easy, the git merge command is an important tool.

10.1 Considering point of view: Traffic merges into us

When you run the git merge command, you merge branches into whatever branch you have currently checked out, as shown in figure 10.2.

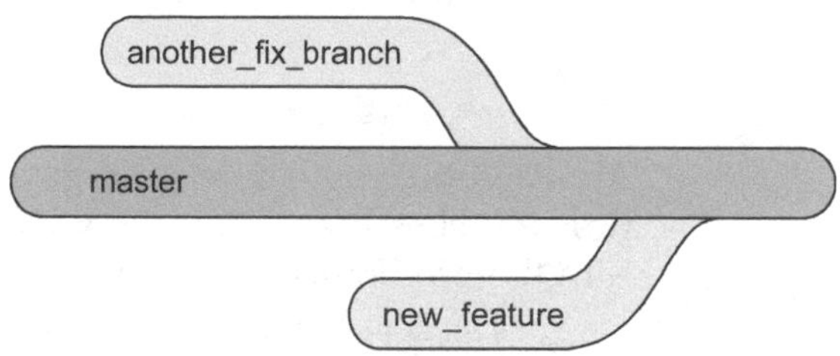

Figure 10.2 Branches merge into whatever branch you're on.

As an example, suppose that the master branch from your math repository has been checked out, using git checkout master. This repo has two other branches, another_fix_branch and new_feature, that could merge into master. Your working directory contains the master branch, and you can run git merge to bring in either another_fix_branch or new_feature.

Now imagine traveling down a highway. The highway you're on is the master branch, and on-ramps join the highway. Those on-ramps are the git merge commands, bringing in traffic from another branch. In Git, you decide when those on-ramps appear.

You can check out any branch, and merge any branch into it. But in practice, one branch is usually designated as the branch that accepts all merges. Typically, this is master, but as you saw in the previous chapter, it can be any branch.

10.2 Performing a merge

A merge results in a commit that has two (or even more) parent commits. In figure 10.1, the merge commit F has two parent commits: E and D. Performing a merge is easy: you call git merge. In these next sections and TRY IT NOWs, you'll review your understanding of branches and then practice using git merge.

10.2.1 Starting with at least two branches

To try merging, you must create at least two branches. In this chapter, you'll work on a bugfix branch whose commit point is earlier than the master branch. This repository looks like figure 10.3, and you'll create the bugfix branch by using the steps in chapter 9, in section 9.2.2.

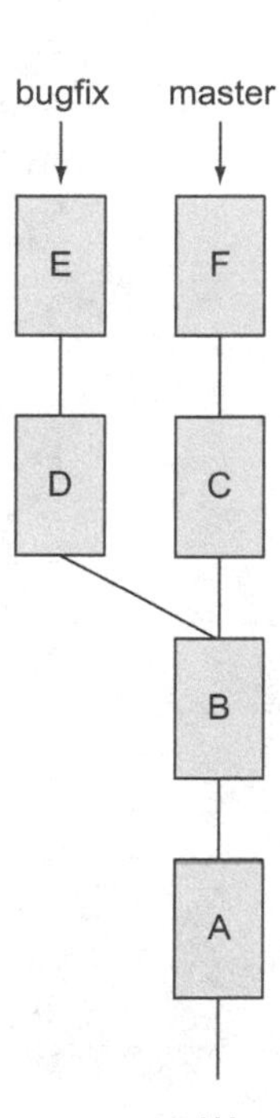

Figure 10.3 The bugfix branch

In this repository, after the code bases have diverged, the master branch has added commits C and F, and the bugfix branch has added commits E and D.

You'll now create this repository and perform a merge between master and bugfix.

TRY IT NOW Download the script make_merge_repos.sh from the book's website, and run that in the command window. Do something like the following:

```
cd $HOME
bash make_merge_repos.sh
```

At this point, you'll have a mergesample directory, containing a repository of two branches that should look like figure 10.3. As before, get comfortable with the two branches. Type this:

```
cd mergesample
git checkout master
```

This puts the contents of the master branch into your working directory. This branch has four files: README.txt, bar, baz, and foo. The file baz is a small script that you can run by typing this:

```
bash baz
```

The script has a bug in it that causes a division-by-zero error message. Take a look at the contents of the file (type `cat baz`), and it should be apparent why this error occurs. Now type this:

```
git checkout bugfix
bash baz
```

This branch contains a fix for the program, and running it should output the number 1 at the command line. Now type this:

```
git log --graph --decorate --pretty=oneline --all --abbrev-commit
```

This should produce the following listing.

Listing 10.1 Two branches that you'll eventually merge

```
* 115df4c (HEAD, bugfix) Ugh, I was dividing by zero!
* 6e0c5d3 Adding echo to check error.
| * f771da4 (master) Committing bar.
| * 1d4640c Committing foo.
|/
* b47c153 (tag: bug_here) Committing baz.
* a3c8e23 Committing the README.
```

You should also open this repository in gitk. After you enable the viewing of all (local) branches in gitk's view configuration, you should see the tree in figure 10.4.

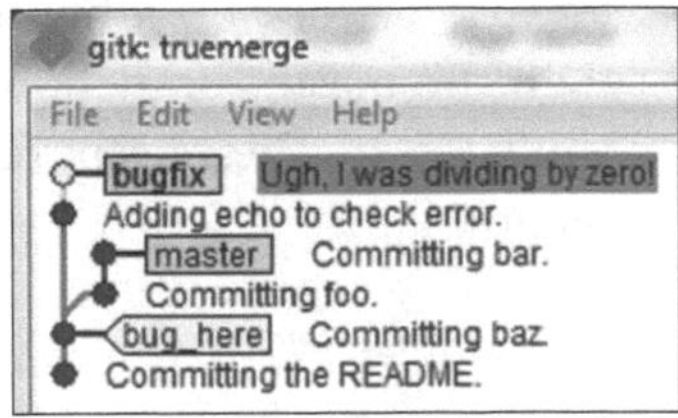

Figure 10.4 Two branches that you'll merge

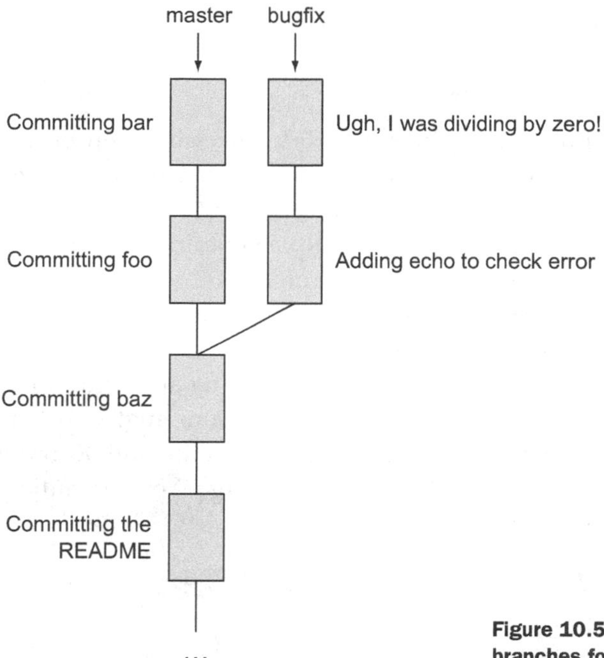

Figure 10.5 Drawing out the commits and branches for the last TRY IT NOW section

Hopefully, you can see that it represents the scenario in figure 10.5. In this figure, the commit messages are shown next to each box. In the next section, you'll learn how to figure out the differences between these two branches.

10.2.2 Checking the difference between two branches

The `git diff` command has an interesting syntax you can use to determine the differences between two branches.

TRY IT NOW In the mergesample directory from the preceding section, type this:

```
git diff master...bugfix
```

Notice the three periods between the two branches. This produces the following listing.

Listing 10.2 `git diff` between two branches

```
diff --git a/baz b/baz
index 56d6546..1c52108 100644
--- a/baz
+++ b/baz
@@ -1,3 +1,4 @@
 a=1
```

```
-b=0
+b=1
 let c=$a/$b
+echo $c
```

The output of this form of `git diff` shows the differences between bugfix and the master branch, relative to when they first became different. This is a preview of what the merge will do. The order of branches to the `git diff` command is significant: master is listed first, and bugfix is listed second. Remembering the highway analogy, bugfix (the on-ramp) will merge onto master (the highway).

In listing 10.2, the string `a/baz` represents the baz file as it exists on master, and the string `b/baz` represents the baz file as it exists on bugfix.

The `diff` output (refresh your memory of this by looking at chapter 6) shows how to turn the file baz from master (`a/baz`) to the file baz from bugfix (`b/baz`). This is why you list master first, and then bugfix in the `git diff` command. Knowing the difference between branches is helpful before you do a merge. You can anticipate what the merge of the file baz will look like.

> **Above and Beyond**
>
> Listing 10.2 gives some insight into how Git can perform the merge.
>
> The variable `b` is set to 0 (`b=0`) in the master branch, but in the bugfix branch, `b` is set to 1 (`b=1`). Git can tell that these changes happen in a certain order, and that only the number has changed. Git will then merge this by changing 0 to 1.
>
> The `echo` statement is a new line, added by the bugfix branch. master didn't have this before. Git will merge this by adding this line.

One other variation of `git diff` is helpful in analyzing branch differences.

TRY IT NOW In the mergesample directory, type the following:

```
git diff --name-status master...bugfix
```

This displays the following listing.

Listing 10.3 `git diff --name-status` listing

```
M       baz
```

Here, the `M` means baz will be merged into master. For our case, this command is overkill, but in larger repositories, multiple files may be merged, and this command provides a useful summary of those files.

Let's complete this merge, now that you know baz is the file that will be affected.

10.2.3 Performing the merge

To incorporate the changes from bugfix into master, you'll use `git merge`.

TRY IT NOW Type the following in the mergesample directory:

```
git checkout master
git merge bugfix
```

Depending on your version of Git, this may put you in Git's default editor so you can type in a new message. A default message already exists, indicating `Merge branch 'bugfix'`. Type this in the editor:

```
:wq
```

This saves the message, at which point Git will merge the two branches.

You'll see output like the following listing.

Listing 10.4 `git merge` output

```
Merge made by the 'recursive' strategy.
 baz | 4 +++-
 1 file changed, 3 insertions(+), 1 deletion(-)
```

Now check the `git log` output to see the merge. Type the following:

```
git log --graph --decorate --pretty=oneline --all --abbrev-commit
```

You saw how to abbreviate this long command into an alias called `git lol` in chapter 9. The log output looks like the following listing.

Listing 10.5 Detailed `git log` output

```
*   71a0b88 (HEAD, master) Merge branch 'bugfix'
|\
| * 115df4c (bugfix) Ugh, I was dividing by zero!
| * 6e0c5d3 Adding echo to check error.
* | f771da4 Committing bar.
* | 1d4640c Committing foo.
|/
* b47c153 (tag: bug_here) Committing baz.
* a3c8e23 Committing the README.
```

Now look at the output of the gitk screen. It should look like figure 10.6. Make sure you can see how the gitk picture matches listing 10.5.

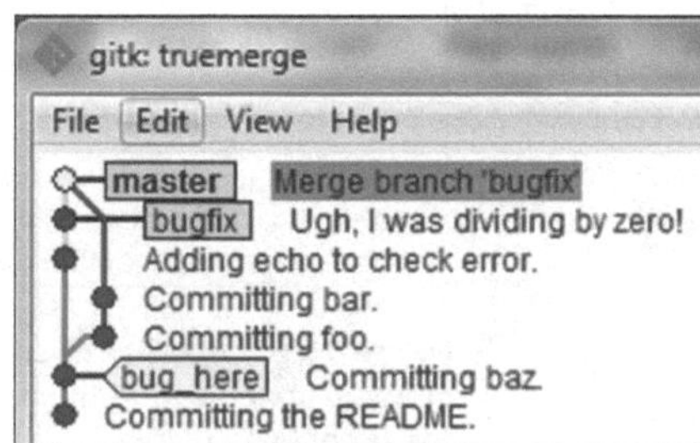

Figure 10.6 A merge in gitk

10.2.4 Working with a merge commit's parents

A merge results in a new commit that represents the merge. In our example from the previous TRY IT NOW, this merge commit has two parents: the latest commit from the master branch and the latest commit from the bugfix branch.

TRY IT NOW In the mergesample directory, type the following:

```
git log -1
```

You should see output like the following listing.

Listing 10.6 `git log` output

```
commit 65f538a53a0d530ce0ca2e06069b8f13f7385e8b
Merge: 8d13856 3c00c46
Author: Rick Umali <rickumali@gmail.com>
Date:   Sun Jul 13 19:29:51 2014 -0400

    Merge branch 'bugfix'
```

Notice the line that starts with `Merge:`. This line lists the two commits that produced this merge commit. This is easier to see in gitk, as shown in figure 10.7.

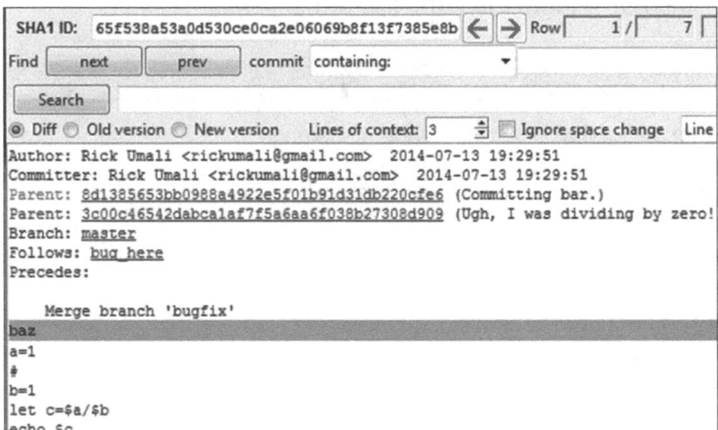

Figure 10.7 The merge commit in gitk

gitk shows that the commit has two parents. On a color screen, you'd see that one parent is in red, and one is in blue. These colors correspond to the branch colors in the tree window of gitk (see figure 10.8).

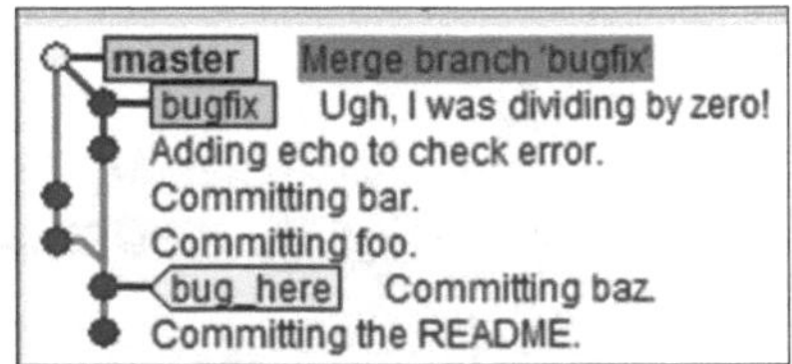

Figure 10.8 Another tree from mergesample. Notice the colors of the branches.

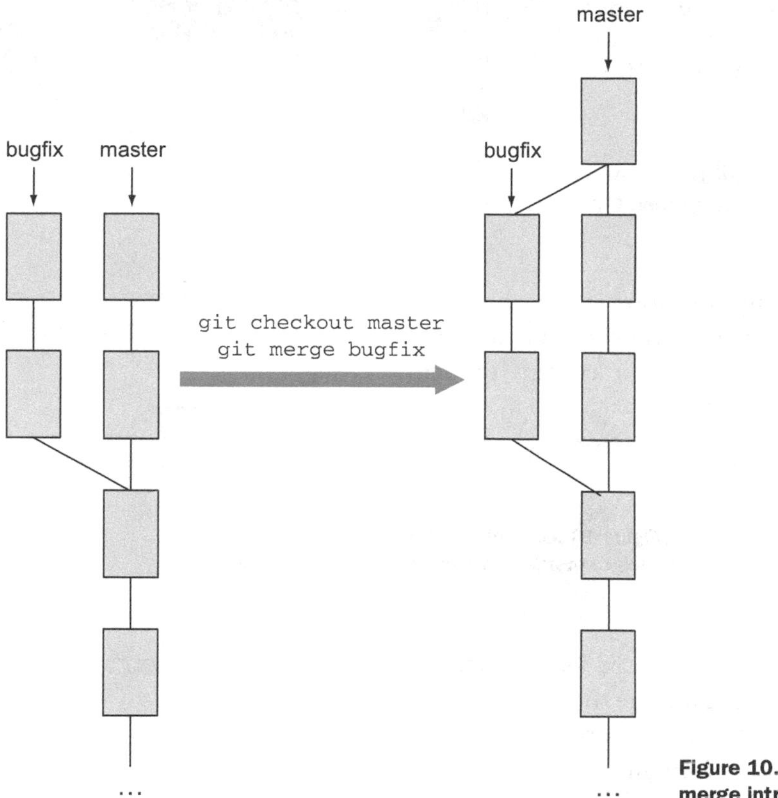

Figure 10.9 Notice how the merge introduces a new commit.

Figure 10.9 depicts this merge in a diagram.

10.2.5 Performing merges in Git GUI

The `git gui` command has the ability to select branches that you can merge into your current branch.

> **TRY IT NOW** Let's re-create mergesample to get to a known starting point. To do this, you'll delete the mergesample directory and then re-create it from the make_merge_repos.sh script. Type the following:
>
> ```
> cd $HOME
> rm -rf mergesample
> bash ./make_merge_repos.sh
> ```
>
> Now you'll start Git GUI in the mergesample directory and make sure you're in the correct branch:
>
> ```
> cd mergesample
> git checkout master
> git gui
> ```

When the GUI appears, choose Merge > Local Merge, as in figure 10.10.

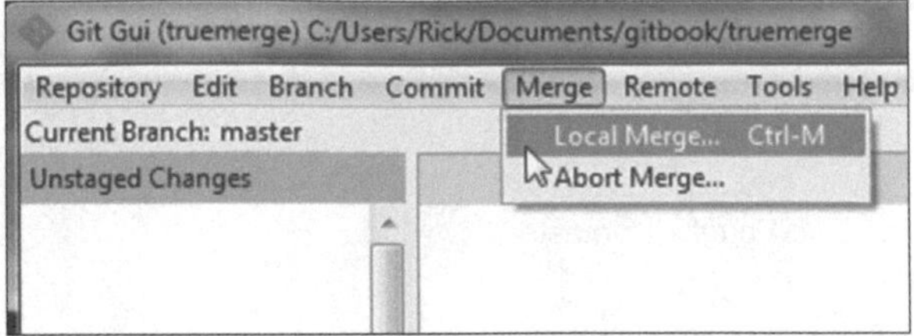

Figure 10.10 Doing a merge from Git GUI

This brings up a window, shown in figure 10.11, that lists the branches that are eligible to merge with your current branch, master.

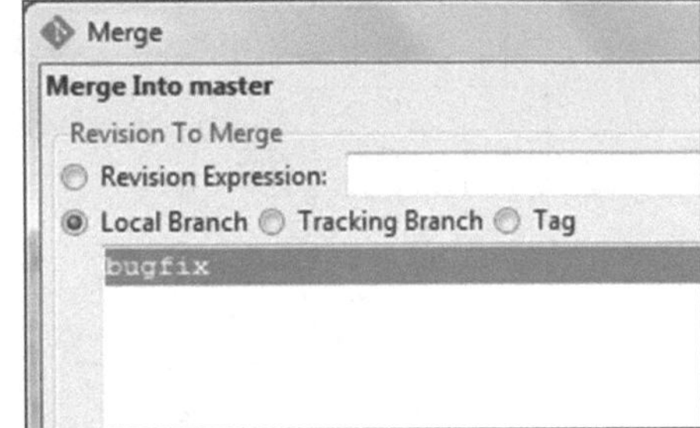

Figure 10.11 Selecting a branch to merge into master

Because this merge can be done automatically, the output window that appears will show success, as you can see in figure 10.12.

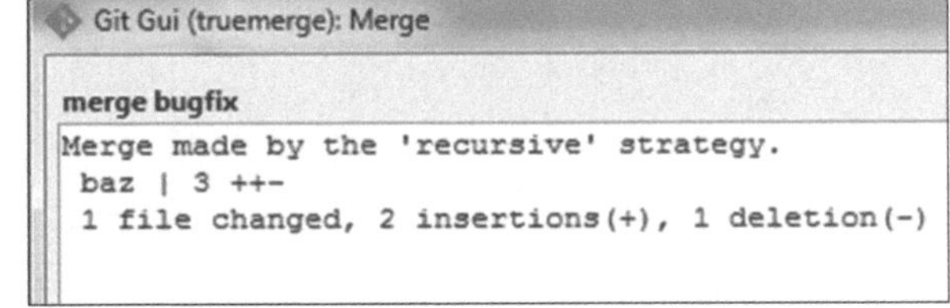

Figure 10.12 Merging the bugfix successfully via Git GUI

10.3 Handling merge conflicts

In the preceding section, one file, baz, was modified on two branches. The merge of these two branches produced a clean merge. Git determined how to merge the two changes to baz from the two branches automatically, but this won't always be the case.

10.3.1 Understanding differences that Git can't handle

Figure 10.13 depicts the two changes made to the file baz from the previous section. In the bugfix branch, you changed the value of the `b` variable from 0 to 1 (fixing the divide-by-zero error), and you added an `echo` statement.

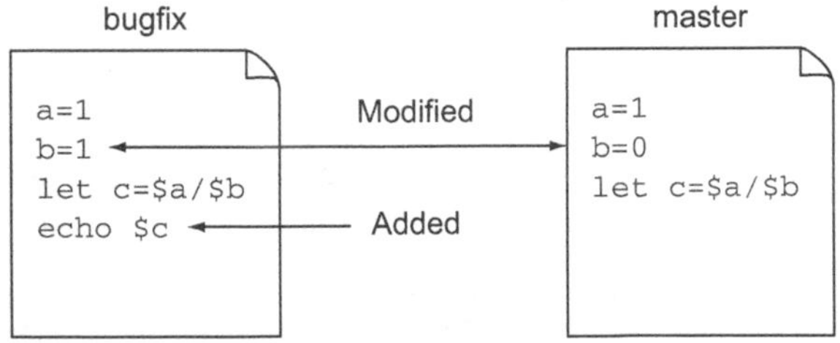

Figure 10.13 Diffs between bugfix and master

The underlying algorithms that Git uses to determine how to perform a merge consider these kinds of changes to be simple. But most changes aren't simple, and those kinds of changes result in a conflict.

TRY IT NOW Let's re-create the mergesample directory to get to a known starting point, as you did in the previous TRY IT NOW. Remember, to do this, you'll delete the mergesample directory and then rerun the make_merge_repos.sh script. Type the following:

```
cd $HOME
rm -rf mergesample
bash make_merge_repos.sh
```

To create a state that will cause a conflict, you'll edit the master branch's baz file to have a `printf` statement. Type the following:

```
cd mergesample
git checkout master
```

Using your favorite editor, add the following as a new line at the end of the baz file:

```
printf "The answer is %d" $c
```

Alternatively, if you don't want to open an editor, carefully type the following line on the command line:

```
echo 'printf "The answer is %d" $c' >> baz
```

The difference between the two versions of baz is shown in figure 10.14.

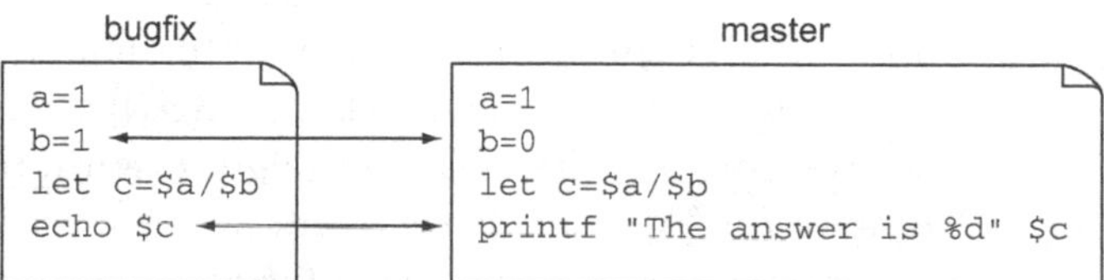

Figure 10.14 A diff that results in a conflict (in the last line)

Commit this change:

```
git commit -a -m "Adding printf"
```

Now try to merge in the bugfix branch:

```
git merge bugfix
```

It should produce an error message like the following listing.

Listing 10.7 A merge conflict

```
Auto-merging baz
CONFLICT (content): Merge conflict in baz
Automatic merge failed; fix conflicts and then commit the result.
```

The previous TRY IT NOW presented a typical scenario when developing with other people: two or more people will make changes to a single file. In our case, someone made a change to the baz file in the bugfix branch, and someone made a change to the same file in the master branch. In some cases, as in the previous section, Git can figure out how to do the merge. But in other cases, Git can't figure out how to do the merge, and therefore you must handle it.

10.3.2 Merging files by directly editing conflicting hunks

When Git generates the message in listing 10.7, it has already modified the conflicted file. It merges the lines that it can calculate, but it leaves special markers that indicate where it needs help. If you open the file, you'll see the area delimited by lines that start with <<<, >>>, and ===. baz will contain the lines in the following listing.

Listing 10.8 A merged file with a conflicted hunk

```
a=1
b=1
let c=$a/$b
<<<<<<< HEAD                       ❶ Start of HEAD changes
printf "The answer is %d" $c
=======                            ❷ Separator between conflicting changes
echo $c
>>>>>>> bugfix                     ❸ End of bugfix changes
```

The lines between ❶ and ❷ represent the code that is in the HEAD commit. Remember that HEAD is always on the current branch, and in this situation you're on the master branch (you just did a `git checkout` of the master branch). The lines between ❷ and ❸ represent the code that is in the bugfix branch. The change between lines ❶ and ❷ is a local change, and the change between lines ❷ and ❸ is a remote change.

Merging any file in Git is straightforward: pick the correct hunk and remove the other hunk. Also, you have to remove the markers. For example, if you decide you like the master branch, a corrected file would look like the following listing.

Listing 10.9 A possible fixed file

```
a=1
b=1
let c=$a/$b
printf "The answer is %d" $c
```

Here you picked the HEAD (or master, or local) change over the bugfix (remote) change. If you decide you like the `echo` in bugfix, your corrected file would look like the following listing.

Listing 10.10 Another possible fixed file

```
a=1
b=1
let c=$a/$b
echo $c
```

Once you make the file look like either listing 10.9 or listing 10.10, using perhaps your favorite editor, you can perform `git add`/`git commit` on this file as usual. This completes the merge. (But don't perform these steps yet. In the next section, you'll explore how to fix the conflicts by using another tool.)

10.3.3 Merging files by using a merge tool

If you have multiple conflicted hunks or want to more clearly see the conflicted hunks, you should consider using a merge tool. This graphical tool displays the details of the three-way merge. Three-way merge? Consider figure 10.15.

When you edited the file baz in your various branches, you produced three versions of baz. The first two may be obvious by now: the versions on your two branches. You have one version of baz on the bugfix branch, and the one on master (HEAD). The third isn't so obvious: it's the original version of baz from which you produce the master and bugfix versions.

Merging assigns names to these three versions, which are labeled in figure 10.15. Base is this original version. Git refers to this as the *common ancestor*. Local is the version of your current branch, which is master. Remote is the version of the branch you're trying to merge, which in this example is bugfix.

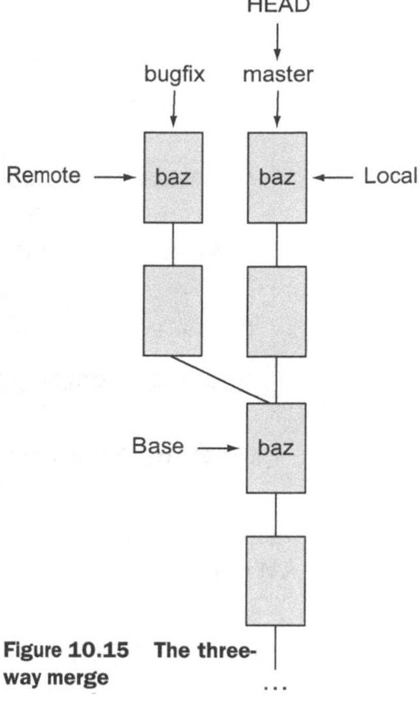

Figure 10.15 The three-way merge

In the next TRY IT NOW, you'll look at a few merge tools and how they display the three-way merge presented by the conflict.

TRY IT NOW In the mergesample directory, type the following:

```
git mergetool
```

(Remember, this TRY IT NOW assumes you haven't fixed the conflicts. If you have, redo the steps in the TRY IT NOW in section 10.3.1.)

The command window prompts you with something that looks like the following listing (depending on your platform).

Listing 10.11 Prompt for `git mergetool`

```
Merging:
baz

Normal merge conflict for 'baz':
  {local}: modified file
  {remote}: modified file
Hit return to start merge resolution tool (kdiff3):
```

When you press Return/Enter, you'll see a window appear. This window is your merge tool. The screen capture in figure 10.16 is a merge tool in Unix/Linux (gvimdiff). This particular merge tool can also be configured for Windows.

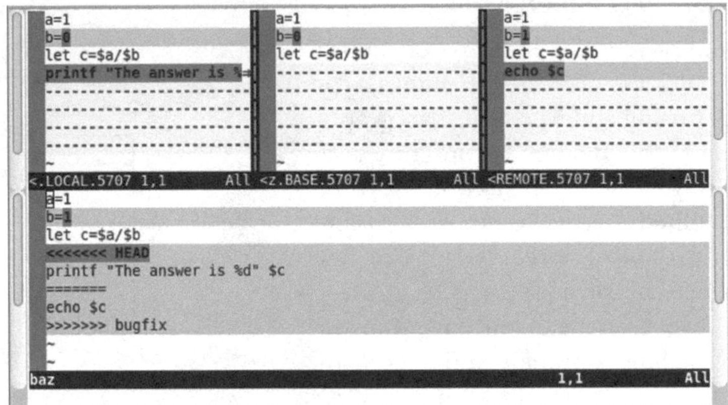

Figure 10.16 The merge tool gvimdiff (for Linux/Unix and Windows)

Figure 10.17 is a screen capture for KDiff3, a merge tool that you can use on Windows.

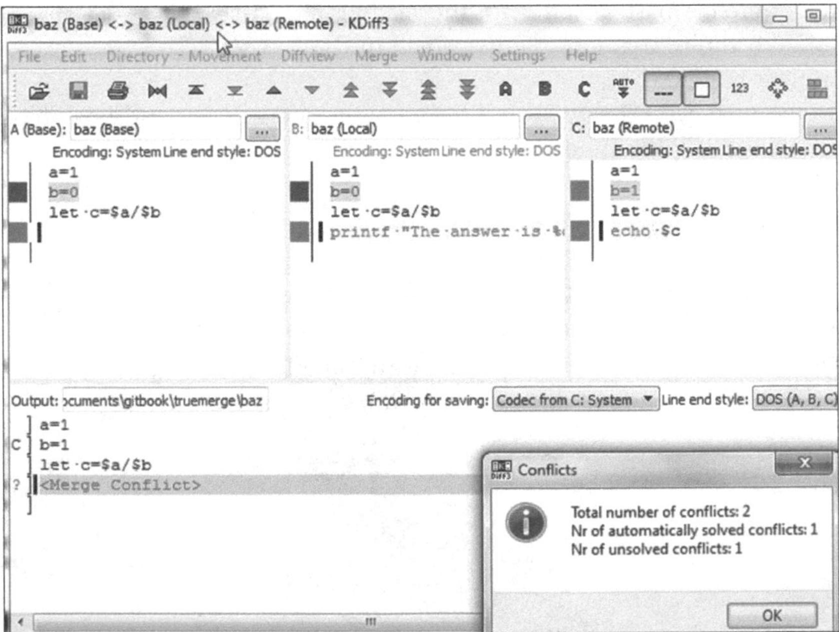

Figure 10.17 The merge tool KDiff3 (for Windows)

Figure 10.18 is a merge tool for Mac, called opendiff.

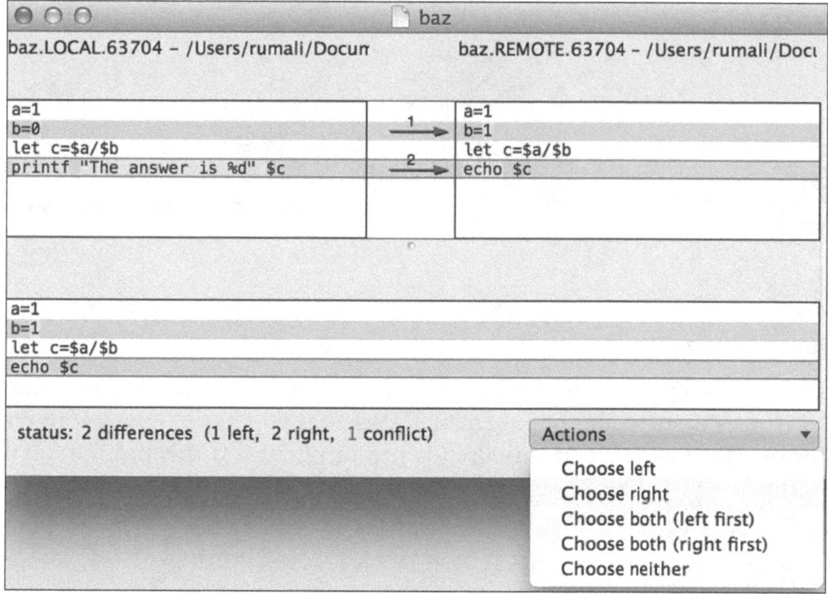

Figure 10.18 The opendiff merge tool (Mac)

Configuring these tools can be challenging, because they aren't supplied by the Git distribution. Online resources describe how to set up these tools. Please visit this book's forum (from the book's website) for additional help or pointers.

After the tool is running, it will let you select a conflicted line. For each conflict, the tool presents a choice of selecting the corresponding line from the current branch or the branch that you're merging. In figures 10.16, 10.17, and 10.18, the line with the `?` is the conflicted line.

The screenshot in figure 10.19 shows how to resolve this conflict with KDiff3 (Windows). You first position the cursor on the conflicted line. Then you click B from the toolbar to pick the change from the local branch.

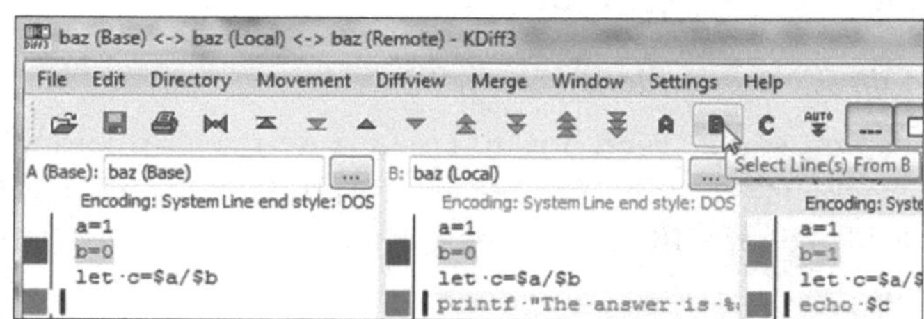

Figure 10.19 Selecting the B (or local) version for the conflict

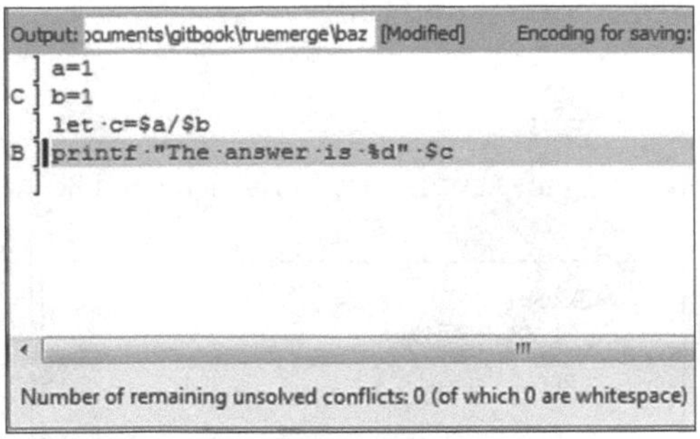

Figure 10.20 All conflicts resolved

After you click B, the main window that represents the merged file no longer displays the ? (figure 10.20). At this point, you can save the file.

The `git mergetool` command detects whether the merge was completed, and in this case, typing `git status` shows that you have a change to commit. This means that `git add` was already run implicitly. At this point, type the following:

```
git commit
```

10.3.4 Aborting a merge

At times you have to abandon or abort a merge. This is usually because you've selected the wrong branch to merge into, or you forgot to check out the correct branch to use as the starting point. One way to avoid this is to use Git GUI to at least figure out the list of branches that are eligible to merge into your current branch!

If you're in the middle of a merge, you should be able to type `git diff` to see the conflicted hunks. This will help you determine whether you should abandon the merge. To stop a merge, type `git merge --abort`.

If you've already performed a merge and need to revert, or roll back, to a previous version, you'll have to use more advanced techniques. Chapter 16 covers this procedure.

10.4 Performing fast-forward merges

One special case of merging in Git is the fast-forward merge. This special case takes effect when the target branch is a descendant of the branch that it will merge with. This section will help you understand what it means to be a descendant, and then you'll perform a fast-forward merge of your own.

10.4.1 Understanding the direct-descendant concept

In the preceding chapter, you learned how to create a new_feature branch on top of master, using `git checkout -b new_feature` while in the master branch. The resulting repository looks like figure 10.21. The new_feature branch is available, and it's waiting for a commit.

After making a few commits to the new_feature branch, your repository looks like figure 10.22. Notice that master hasn't made any commits yet!

Each box is a commit, and it's easy to see that there's a path from the latest commit of the new_feature branch to the latest commit of the master branch. From the figure, the commits made to the new_feature branch are descendants of master's last commit.

To be more precise, a commit is a descendant of a target branch if you can follow its parents all the way to the target branch. Remember that each commit points back to its parent. In figure 10.23, you can see by the arrows that commits in new_feature point all the way back to master.

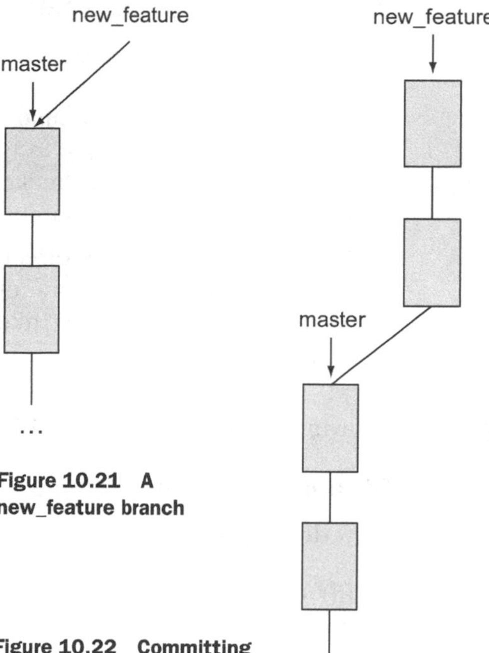

Figure 10.21 A new_feature branch

Figure 10.22 Committing on the new_feature branch

If you perform a merge of two branches that are connected to one another in this manner, Git will perform a fast-forward merge. You can try this for yourself by producing a repository that has two branches in this fashion.

TRY IT NOW You can either type the following to create a new repository in a new directory (called ff), or use the script make_merge_ff.sh from the zip file of code from the book's website, and run that (using `bash make_merge_ff.sh`):

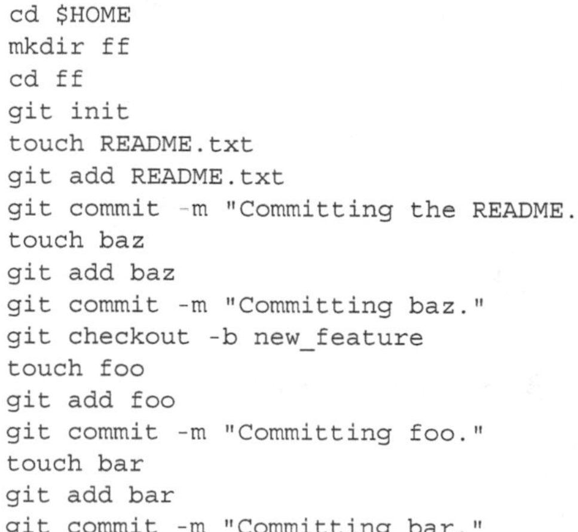

```
cd $HOME
mkdir ff
cd ff
git init
touch README.txt
git add README.txt
git commit -m "Committing the README."
touch baz
git add baz
git commit -m "Committing baz."
git checkout -b new_feature
touch foo
git add foo
git commit -m "Committing foo."
touch bar
git add bar
git commit -m "Committing bar."
```

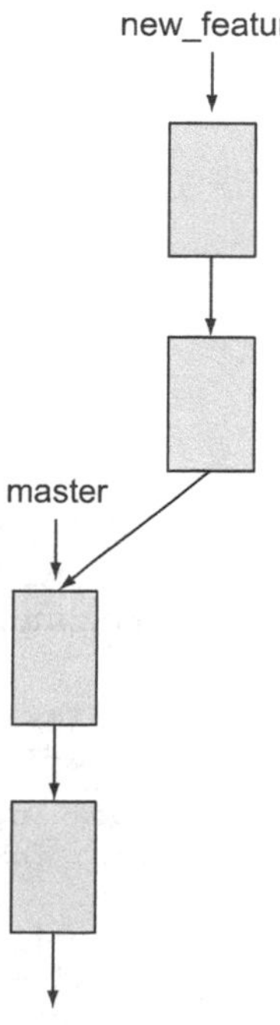

Figure 10.23 If you can follow a commit's parents all the way to another commit, they're descendants.

Now type the following:

```
git log --graph --decorate --pretty=oneline --all --abbrev-commit
```

Remember that this command can be aliased so you don't have to type all of this. The command to make this an alias is in the previous chapter.

Also, open gitk in this directory, and make sure to view all branches. You should see something like figure 10.24.

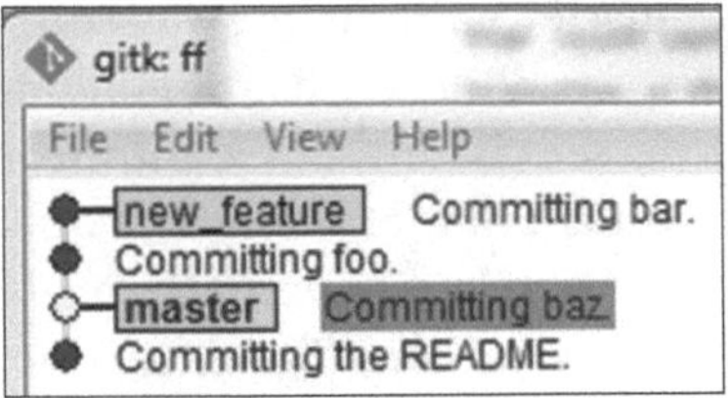

Figure 10.24 The result of the TRY IT NOW section, in gitk

Convince yourself that figure 10.24 is the equivalent of figure 10.23.

10.4.2 Making a fast-forward merge

Now that you have the right configuration, let's perform the fast-forward merge.

> **TRY IT NOW** In the ff directory that you created in the previous section, type the following:
>
> ```
> git checkout master
> git merge new_feature
> ```
>
> This performs the fast-forward merge. You don't have to do anything special to invoke this. You'll now perform the same checks from the previous TRY IT NOW section. Type the following:
>
> ```
> git log --graph --decorate --pretty=oneline --all --abbrev-commit
> ```
>
> Also, open gitk in this repository. It should look like figure 10.25.
>
> 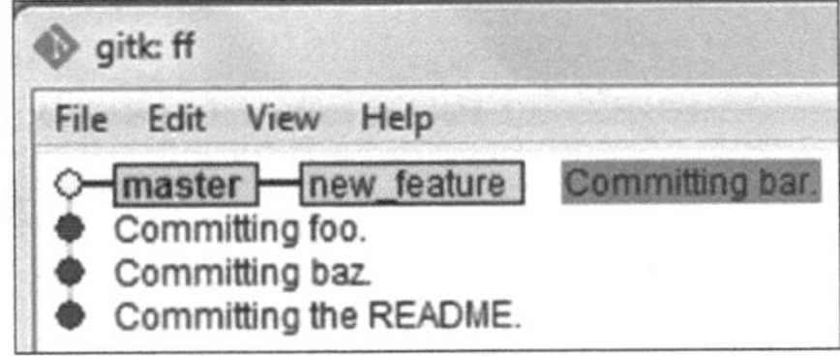
>
>
> **Figure 10.25 Branches now merged**

The merge command produces output like the following listing.

Listing 10.12 `git merge` output (from a fast-forward)

```
Updating 9d7b1b8..3e7c402
Fast-forward
 bar | 0
 foo | 0
 2 files changed, 0 insertions(+), 0 deletions(-)
 create mode 100644 bar
 create mode 100644 foo
```

1. SHAI IDs of our branches

In listing 10.12, ❶ shows the SHA1 IDs of our two branches. `9d7b1b8` is the SHA1 ID of master, and `3e7c402` is the SHA1 ID of new_feature.

Once Git detects that the branch being merged is a direct descendant of the current branch, it moves the local branch (master) up to the remote branch (new_feature). This is what is meant by *fast-forward*. Take a look at figure 10.26. Our repo before the merge has the branches at two locations, but after the merge (designated by the arrow), master is now at the same place as the new_feature branch. Master was fast-forwarded.

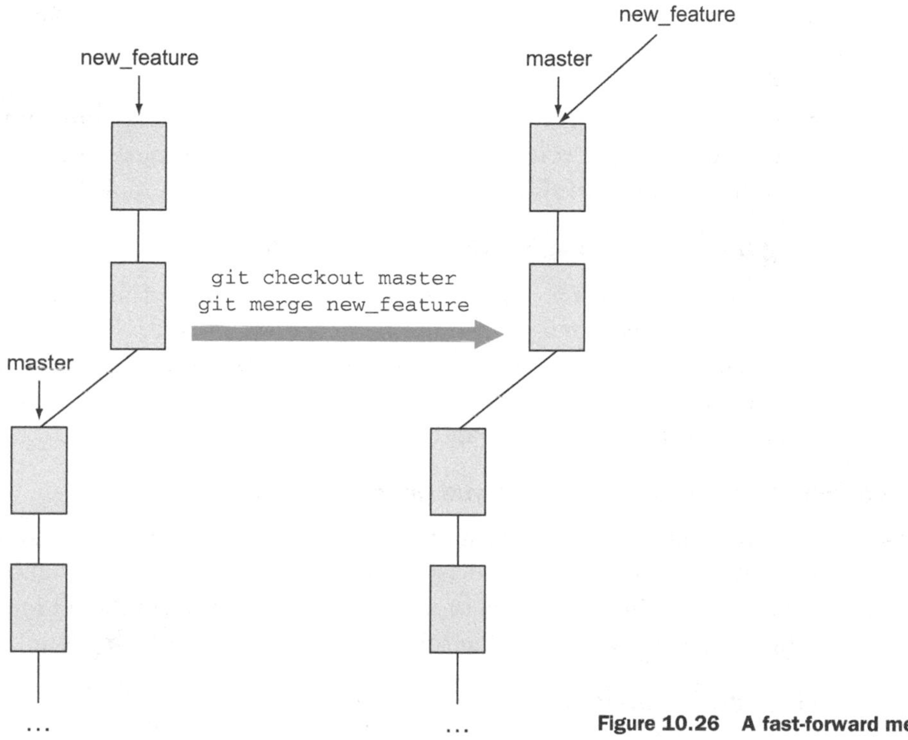

Figure 10.26 A fast-forward merge

One thing to point out from figure 10.26 is that after `git merge`, the new_feature branch isn't deleted. This branch remains in the repository. You can add and commit files to new_feature, causing master and new_feature to diverge again.

10.5 Lab

Git forces you to do more merges by making it easy for you to create branches. The key thing to remember is that merges bring into the current branch (the HEAD) changes from another branch. Complete the following questions and tasks:

1. Read the HOW TO RESOLVE CONFLICTS section of the `git merge` command.
2. Create a branch from master and then try to merge master into it. What happens?

3. In section 10.2.2, you typed `git diff master...bugfix` before performing a merge. What happens if you type it now? Is there another word you can substitute for either `master` or `bugfix`? What happens to the diff if you try a different order?
4. Delete the mergesample directory. Then re-create it via the make_merge_repos.sh script. Add a new file in the bugfix branch. What is the output of `git diff --name-status master...bugfix`?
5. The fast-forward merge from section 10.4 doesn't produce a commit after the merge is completed. Look up the switch to `git merge` to add a commit even though it's a fast-forward merge, and retry the TRY IT NOW using this switch.

10.6 Further exploration

Merging branches is a deep computer problem. Git's underlying architecture allows it to calculate both clean and conflicted merges quickly. Git has plenty of tooling, configuration options, and controls that facilitate its handling of merges.

10.6.1 Calculating the base of a merge with git merge-base

One of the key steps in performing a merge is the calculation of the common ancestor. This base is displayed when you use the `mergetool` command. Git has a command-line tool that determines the commit of this base: `git merge-base`. Reset the mergesample directory (as you did in step 4 of the lab), and use `git merge-base` to display the SHA1 ID of the base between master and bugfix.

10.6.2 Changing how conflicts are displayed (merge.conflictstyle)

Git has a configuration setting that subtly changes the way conflicting hunks are displayed in a file. Look at merge.conflictstyle in the `git merge` documentation. Enable this configuration (using `git config`) and examine how the conflicted hunk is presented differently when you do the merge of master and bugfix from section 10.4.

10.6.3 Performing octopus merges

An *octopus merge* is a merge that consists of more than two parents. All of the merges you've considered in this chapter have two parents: the master branch, and the branch that you're merging into master.

Git has the ability to merge multiple branches into the branch you're working on (the HEAD). Project maintainers may use this to bring in the work of multiple branches into the current branch. To see this for yourself, reset the mergesample directory (as you did in step 4 of the lab), and create another branch. Add a file in that branch, and commit it. Now check out the master branch, and try to merge both the bugfix branch and this new branch that you created.

10.7 Commands in this chapter

Table 10.1 Commands used in this chapter

Command	Description
`git diff BRANCH1...BRANCH2`	Indicate the difference between BRANCH1 and BRANCH2 relative to when they first became different.
`git diff --name-status BRANCH1...BRANCH2`	Summarize the difference between BRANCH1 and BRANCH2, by listing each file and its status.
`git merge BRANCH2`	Merge BRANCH2 into the current branch that you're on.
`git log -1`	A shorthand for `git log -n 1` (show only the most recent commit).
`git mergetool`	Open a tool to help perform a merge between two conflicted branches.
`git merge --abort`	Abandon a merge between two conflicted branches.
`git merge-base BRANCH1 BRANCH2`	Show the base commit between BRANCH1 and BRANCH2.

11 Cloning

Cloning, the act of making a physical copy of a Git repository, is the first step in collaborating with others. You'll be covering collaboration over this and the next three chapters. When you clone a repository, you make an exact replica of that repository. The clone has a special reference to the original repository. This reference lets your clone push (send) and pull (receive) changes to and from the original repository. You'll study this special reference (called a remote) in the next chapter, and then read about push and pull in chapters 13 and 14.

In this chapter, you'll make a copy of your repository by using `git clone`. You'll examine your clone and figure out how to confirm that it's an exact copy of the original, where it stores your branches, and how it knows about the original repository. All this will help you become oriented after you run `git clone` on an existing repository. Finally, you'll cover a special type of clone called the *bare directory* that lets you set up collaboration, as in figure 11.1. This is a technique that this book uses to teach collaboration on a single machine, and it's the basis for server-based Git systems like GitHub.

It won't take you long to learn these operations, which are the foundation of Git collaboration.

11.1 Cloning: making copies locally

What if you wanted someone else to work on your math program, the one you've been developing the past few chapters? You have two options for sharing your repository. In the first option, you can make a copy of your entire working directory by using the standard operating system Copy command. You can then give it to anyone you want. That person will have an exact duplicate of your repository.

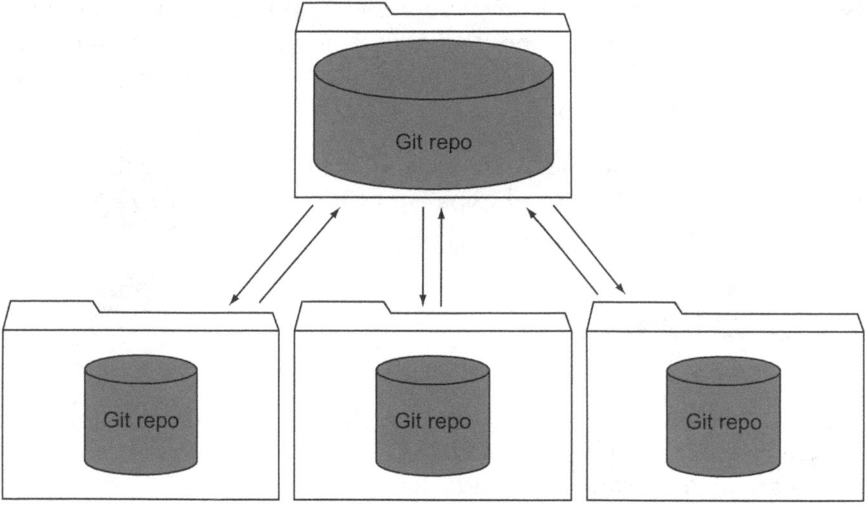

Figure 11.1 A set of repositories collaborating with one another

The second option is to make a special repository that someone else can clone from. That person would use the `git clone` command to make a copy of your repository. This second option differs from the first in an important way: the copy created with `git clone` is linked to the original and can send and receive changes back to the original. No such capability is available in the first option.

This capability is so important that `git clone` is the universally accepted mechanism to make a copy of a Git repository. Another key advantage in using `git clone` is that you can clone over the Internet, something you can't do with the first option.

11.1.1 Using git clone

The `git clone` command copies a Git repository from a source location to a local directory on your machine. It also sets up the special linking between the clone and the source repository, which you'll fully explore in the next chapter. The syntax of the `git clone` command is shown in figure 11.2.

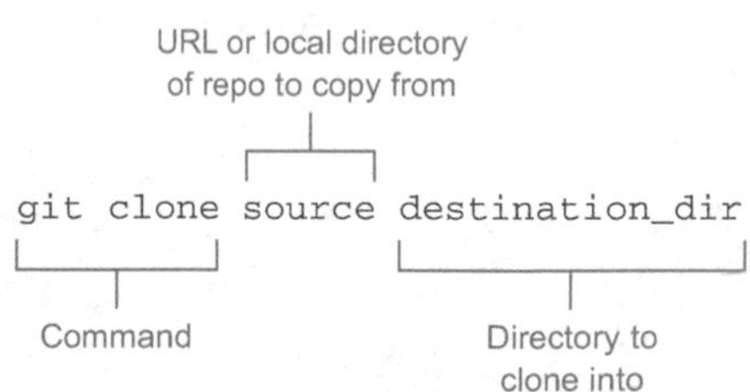

Figure 11.2 The form of the `git clone` command

Only two arguments are needed: the source repository and the directory to copy it into. If you omit the name of destination directory, `git clone` will make one up based on the source repository. The destination directory is also local to your machine.

Reading the `git clone` documentation makes you realize that the source can be local to your machine, or a repository that is at another remote location (such as GitHub or Bitbucket). You'll practice on remote locations in chapter 18, but in this chapter you'll consider the local case.

> **TRY IT NOW** In this exercise, you'll make two clones, one using the command-line technique, and the second using Git GUI.
>
> First, you'll make a clone using the command line:
>
> ```
> cd $HOME/math
> git checkout master
> cd $HOME
> git clone math math.clone1
> ```
>
> Note that you check out the master branch so it's the active branch in this repository. After you type the preceding `git clone` command, you'll have a clone of math in the directory math.clone1.
>
> Now you'll make a clone using Git GUI. First, start Git GUI via the menu or by typing the following:
>
> ```
> git gui
> ```
>
> Click the Clone Existing Repository option. In the window that appears, click the Browse button next to the Source Location field. In the file browser that appears, select the math repository directory. Then click the Browse button next to the Target Directory field, and enter the directory corresponding to $HOME/math.clone2 (see figure 11.3). Remember, the target directory doesn't exist yet, so you'll have to type it out. If you see a Clone Type prompt, select Standard.

At this point, you have two copies of the math repository. The first one is math.clone1, which you created with the `git clone` command, and the second one is math.clone2, which you created via the Git GUI's Clone Existing Repository feature.

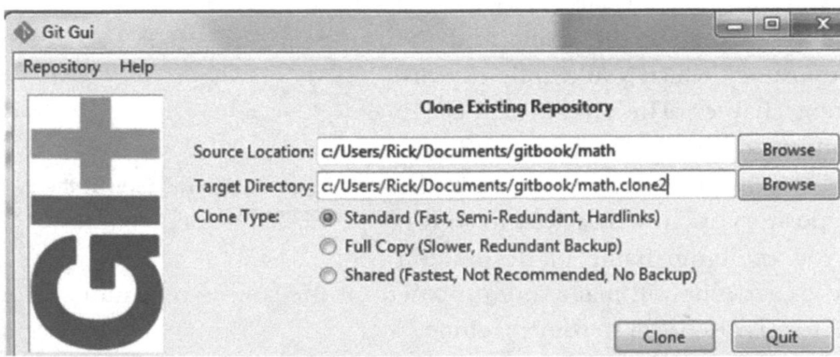

Figure 11.3 Cloning via Git GUI

You can visit either of these repositories, and confirm that they contain the current history (by typing `git log --oneline --all`). Each of these repositories is linked back to the original repository, and you can begin to see that link by examining how your new repositories handle branches.

11.1.2 Viewing branches in your clone

In this section, you'll use the `git branch` command in your clones to see the link back to the original repository. Remember that the original math repository contains three branches: master, new_feature, and another_fix_branch. Let's confirm this.

TRY IT NOW To confirm the branches that are in the original math repository, type the following:

```
cd $HOME/math
git branch
```

The `git branch` command should show the three branches. The output should look like the following listing.

Listing 11.1 `git branch` output

```
  another_fix_branch
* master
  new_feature
```

You could also see the branches by looking at gitk, as shown in figure 11.4. (You should have at least the three listed here, but don't worry if you see more.)

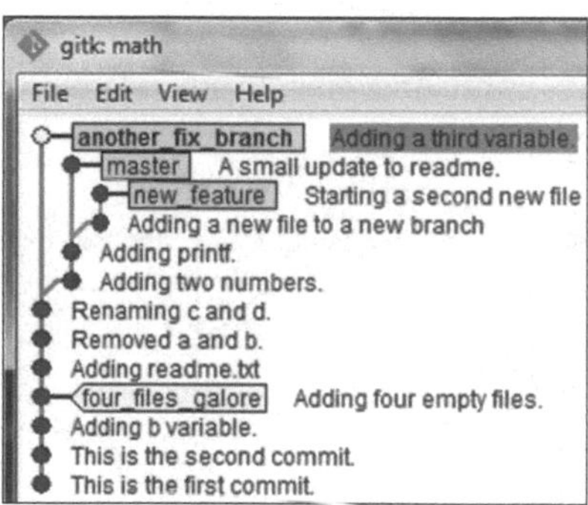

Figure 11.4 The branches of the math repository

Recall that to get the listing of all the branches, your gitk view must be edited to show All (Local) Branches (see section 9.2.1). To simplify this display, you can also edit the gitk view to use a simple history (under Miscellaneous options in the gitk view configuration). The gitk window looks like figure 11.5.

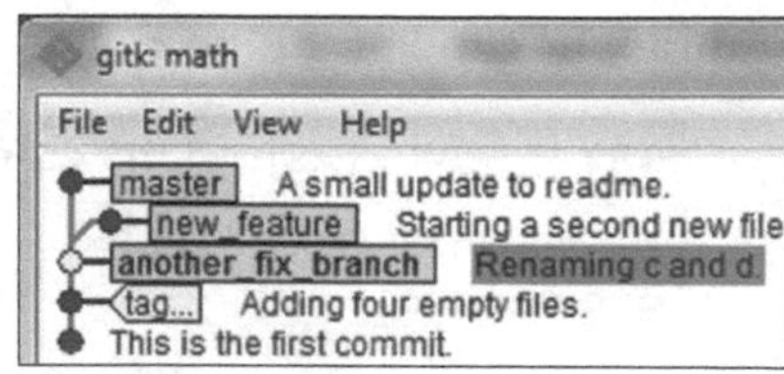

Figure 11.5 The gitk output with simple history

Knowing the branches available in a repository is important. Here's the command-line equivalent to get a simplified list of branches, as in figure 11.5:

```
git log --simplify-by-decoration --decorate --all --oneline
```

This displays output like the following listing. (Again, don't be worried if your listing is different. You should, however, see the master, new_feature, and another_fix_branch branches.)

Listing 11.2 `git log` listing to show branches

```
2f84c2a (master) A small update to readme.
835ad57 (new_feature) Starting a second new file
547a17b (HEAD, another_fix_branch) Renaming c and d.
1c18222 (tag: four_files_galore) Adding four empty files.
0231899 This is the first commit.
```

This `git log` output contains more information than the `git branch` output. It shows the SHA1 ID and the first part of the commit message for each branch.

A repository contains branches, so figure 11.6 is another helpful way to think about a Git repository and its branches. Your math repository contains three branches.

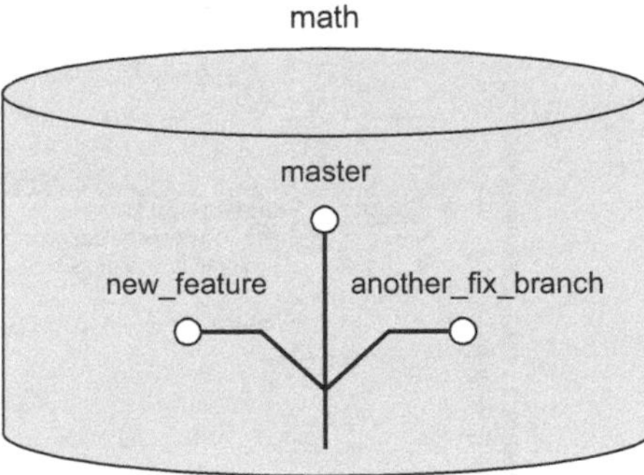

Figure 11.6 The math repository and its branches

Now that you've confirmed the branches in the original repository, let's see how these branches appear in the clone.

> **TRY IT NOW** Go into the cloned directory and list the branches:
>
> ```
> cd $HOME/math.clone1
> git branch
> ```
>
> You'll see just one branch (depending on which branch was active in the math directory).

When you clone a repository, the only branch that appears in the clone is the active branch (the one HEAD points to) from the original repository. In this case, it should

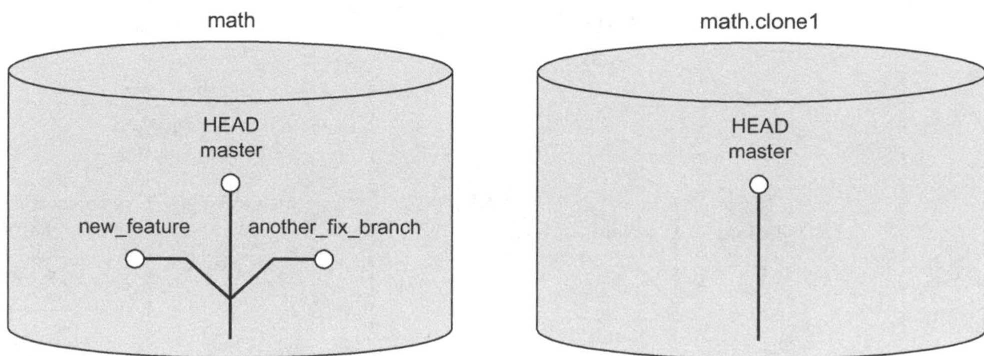

Figure 11.7 The clone of the math repository has only one branch checked out.

be master. In figure 11.7, I placed HEAD next to master in the original math repository. Your clone in math.clone1 has only this one branch, and the HEAD of math.clone1 points to it.

Where did the other two branches go? When you make a clone, only the active branch (the one that HEAD points to in the original repository) is checked out. To see the other branches in your clone, you must use the `git branch` command's `--all` switch.

> **TRY IT NOW** To list all the branches in your clone's repository, use the `git branch`'s `–all` switch:
>
> ```
> cd $HOME/math.clone1
> git branch --all
> ```
>
> This produces output like the following.

Listing 11.3 Annotated `git branch --all` output

```
* master
  remotes/origin/HEAD -> origin/master
  remotes/origin/another_fix_branch
  remotes/origin/master
  remotes/origin/new_feature
```

❶ **Available branches on the remote**

Listing 11.3 shows that three branches ❶ are available on the remote, including another_fix_branch and new_feature, but they have the string `remotes/origin/` prepended to their names. This indicates that these branches are tracked from the remote that is named *origin*. The remote is the link from the clone to the original repository. The output ❶ reads `remotes` because Git allows you to have multiple remotes. You'll study remotes more closely in the next chapter. Figure 11.8 illustrates these new terms.

In figure 11.8, the original repository (math) is labeled a remote that is named origin. Think of *remote* as an address for another Git repository. When you perform a

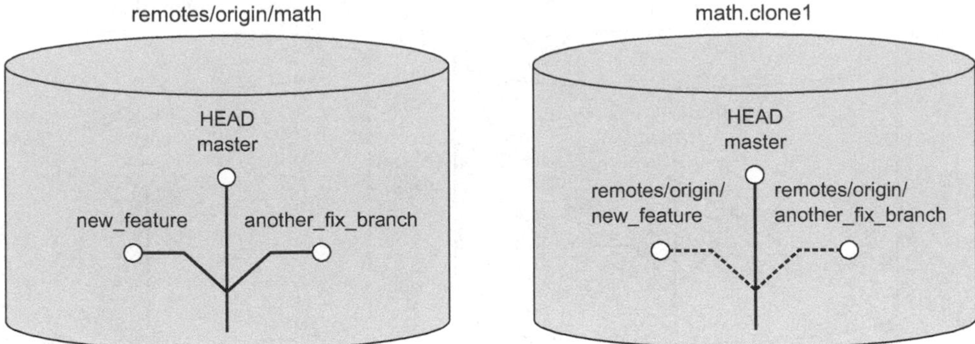

Figure 11.8 Remotes, origins, and tracking branches

clone of this repository to the directory math.clone1, Git checks out only the master branch in the math.clone1 directory, but because the clone copies in the entire repository, it can record, or track, the other branches from the original repository.

The dashed lines in figure 11.8 indicate these remote-tracking branches. Their complicated names (remotes/origin/new_feature and remotes/origin/another_fix_branch) are another indication that these branches also exist on the remote that is named origin. In the next section, you'll learn to how access these branches in your cloned repository.

11.1.3 Checking out branches

Remember: `git clone` always copies the entire repository. As a result, your clone has the files and the history it needs to re-create any branch that existed in the original repository. How? Using `git checkout`.

TRY IT NOW Let's check out another_fix_branch, which existed in the original repository. In the math.clone1 directory, type the following:

```
git checkout another_fix_branch
```

This should produce the output in the following listing.

Listing 11.4 Checking out a tracking branch

```
Branch another_fix_branch set up to track remote branch another_fix_branch
➥ from origin.
Switched to a new branch 'another_fix_branch'
```

Now the situation looks like figure 11.9.

Now there are two references to another_fix_branch labeled *another* in figure 11.9: the local branch named another_fix_branch, and the original remote-tracking branch. The line is solid, indicating that you've checked it out, and HEAD is next to another_fix_branch.

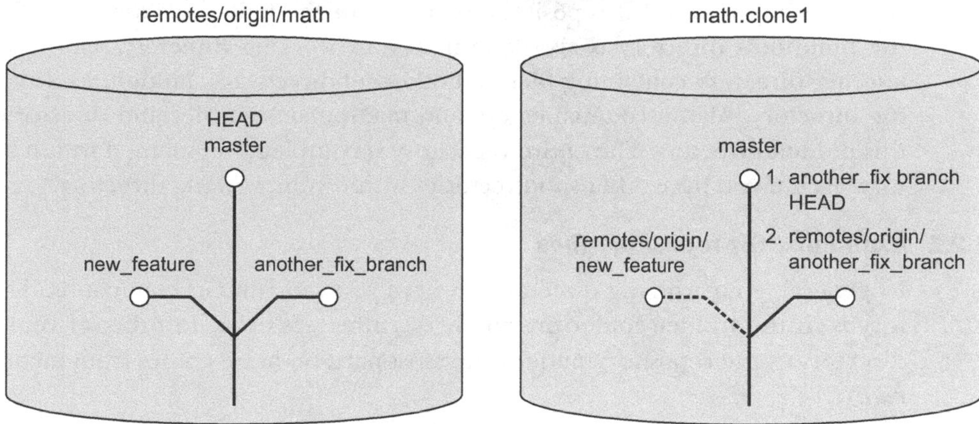

Figure 11.9 Checking out a remote-tracking branch

The command `git checkout another_fix_branch` is a shortcut for this longer form of `git checkout`:

```
git checkout -b another_fix_branch remotes/origin/another_fix_branch
```

You saw this command in section 8.3.2. The local branch named another_fix_branch set its starting point as the remote-tracking branch named remotes/origin/another_fix_branch.

It's important to understand this section before proceeding to the next chapter. If you use your imagination, you might be able to see how the origin might exist on another server, and further, you might be able to envision how multiple people might access a common remote location. As I said: cloning is an essential step to collaborating.

11.2 Working with the bare directory

Because we're now talking about cloning the repository, it's a good time to more thoroughly discuss what the Git repository is and where it exists.

In chapter 4, you spent time with the `git init` command. This command initializes a Git repository in whatever directory you're in. Figure 11.10 shows what happens.

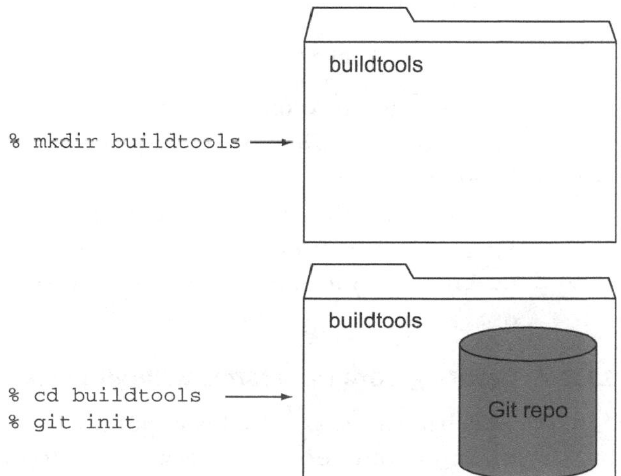

Figure 11.10 `git init` creates the repository.

In this example, the Git repository exists within the buildtools directory. Recall that the buildtools directory is also known as your working directory. The repository is another directory containing files and other subdirectories, hidden inside your working directory. All Git commands use and manipulate the files and directories within this hidden directory. The entire repository is completely contained within this directory, even if you have other subdirectories within your working directory.

11.2.1 Examining Git repository files

In general, your working directory is where you can run Git commands. The repository is in the hidden folder previously described. It's easy to think of your working directory as the repository, and for the most part, no harm comes from mixing up the two.

TRY IT NOW In the math.clone1 directory, let's examine the repository files by using the `ls` command. Type the following:

```
ls -a
cd .git
ls -F
```

The `ls -a` command shows a directory listing with all hidden directories revealed. The `ls -F` command shows a directory listing with all the folders marked with a slash (/). The preceding commands should give you the following output.

Listing 11.5 Getting to and examining the repository directory

```
$ ls -a
./  ../  .git/  another_rename  math.sh  readme.txt  renamed_file

$ cd .git

$ ls -F
HEAD    description  index  logs/     packed-refs
config  hooks/       info/  objects/  refs/
```

You need to know only that this .git directory exists, and that it's called by a particular name: the *bare directory*. When you clone repositories, this bare directory is manipulated and copied around.

The bare directory's contents include the objects that you're tracking and references (for branches, which you already learned about in chapter 9). It's not important to know about these internals, but they're described in the Git help page for `gitrepository-layout`.

11.2.2 Creating bare directories with git clone

This bare directory itself is important because when you later push commits that you make in your repository back to the original repository (which you'll do in chapter

13), the original repository should ideally also be a bare directory. The bare directory is therefore crucial for collaboration!

There's a way to create a Git repository that consists of just this bare directory: `git clone --bare`.

TRY IT NOW You'll create a bare Git directory from the math repository:

```
cd $HOME
git clone --bare math math.git
cd math.git
ls -F
```

The `-F` switch to `ls` reveals the directories. The last command gives you a listing like the following.

Listing 11.6 math.git directory contents

```
$ ls -F
HEAD  config  description  hooks/  info/  objects/  packed-refs  refs/
```

Notice that it's mostly the same as listing 11.5. (The only exception is that the original repository has a logs directory, indicating that it's active.) Figure 11.11 illustrates what you've just done. You start at the top and then run the `git clone` command. The arrow shows the result.

You started with one repository and then made a clone of that repository into another directory named math.git. Because you used the `--bare` switch, the math.git

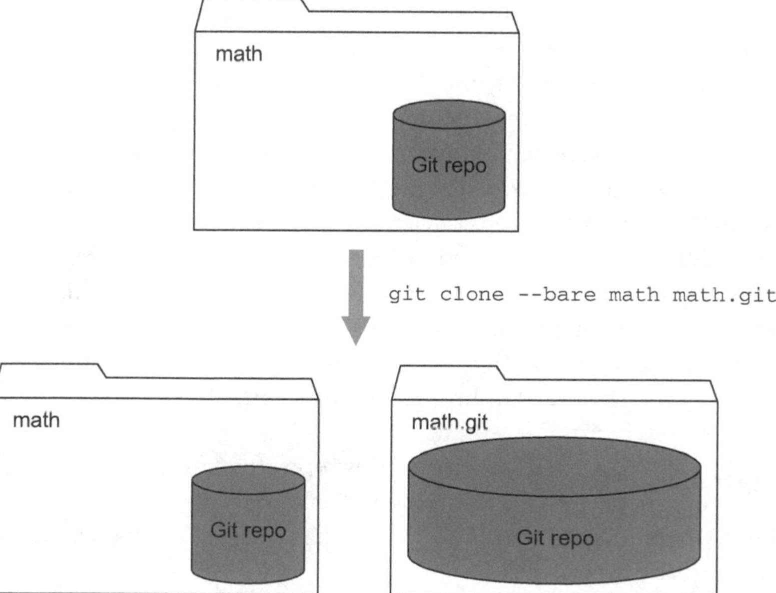

Figure 11.11 `git clone --bare`

directory is known as the bare directory. In figure 11.11, the repo fills up the entire math.git directory, which is how this book depicts a bare directory. This drawing indicates that there's no room for a working directory in the bare directory.

The math.git directory is just the repository files (the bare directory). You can't perform any Git operations within the math.git directory, because there's no working directory. But you can clone this math.git directory and push commits to it.

11.2.3 Cloning from bare directories

Another important aspect of a bare directory is that it has no reference to the original repository. Unlike a clone, which has a reference to its originating repository, the bare directory is a completely standalone repository. Because of this, and the fact that it has no working directory, bare directories are often the official copy of a repository. The only way to update it is to push to it, and the only way to retrieve its contents is to clone, or pull, from it.

TRY IT NOW Let's make a clone from the repository (bare) directory:

```
cd $HOME
git clone math.git math.clone3
```

To confirm that you have a working repository, type the following:

```
cd math.clone3
git log --oneline --all
```

It should give you a listing like math.clone1.

Figure 11.12 illustrates what you've done with the last two TRY IT NOWs.

You've created two new directories, each a copy of the math repository. One of the copies is a bare repository. (Remember in this book, bare directories are drawn in figures as a directory containing only the repository.)

You've run `git clone` to make copies of your math repository. In three of the directories (math, math.clone2, and math.clone3), you could perform Git work such as committing new changes, and making and merging new branches. One of the directories (math.git) is just the repository, and, because it has no working directory, it might be considered the official version of your code.

If our environment supported sharing directories and multiple users, you might decide to declare the math.git repository as the official version, and have people clone from and push commits to it. (By now you might be wondering when you'll push across the Internet to another Git hosting site like GitHub. You'll get to that in chapters 13 and 18.)

In figure 11.13, you've made a clone of this repository, using `git clone math.git`. (You use the convention your/math to represent your directory, and the directory math within it.) Notice again that the your/math is a directory that contains the working directory and the repository.

math
Git repo

git clone --bare math math.git

math
Git repo
math.git
Git repo

git clone math.git math.clone3

math
Git repo
math.git
Git repo
math.clone3
Git repo

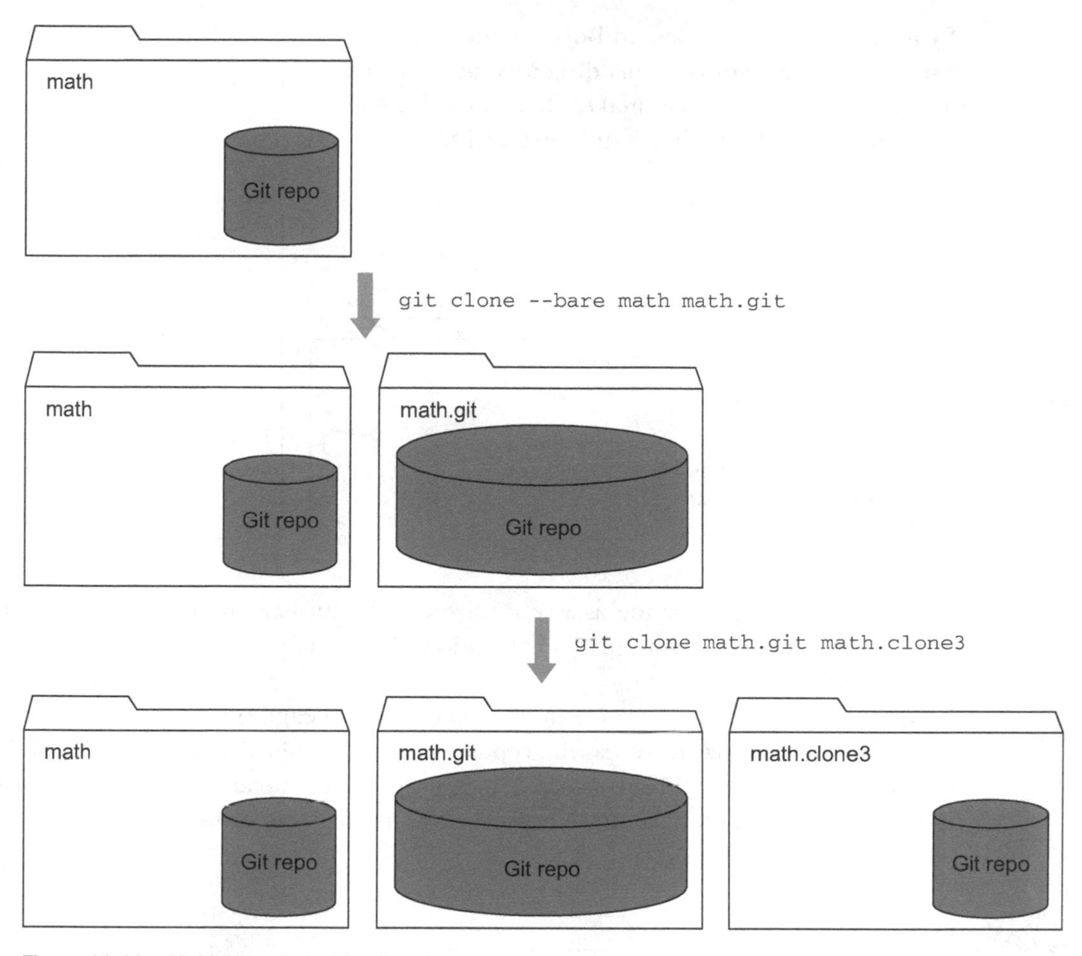

Figure 11.12 Making copies with `git clone`

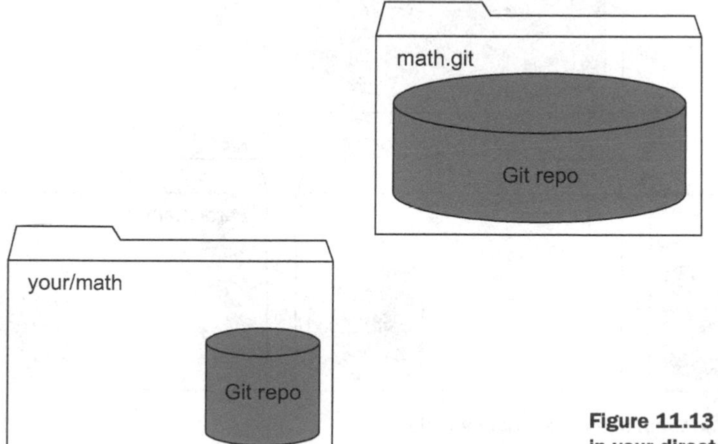

Figure 11.13 Making an initial clone in your directory from math.git

If you had a colleague named Bob, he might make a clone of this repository in his directory, again using `git clone math.git`. He makes this math clone in his directory (bob/math), as in figure 11.14.

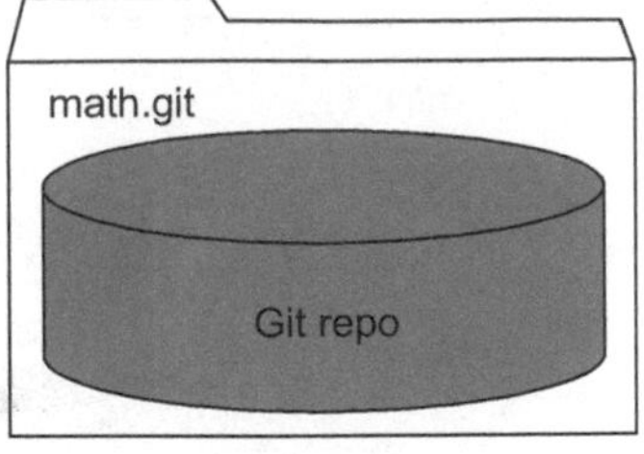

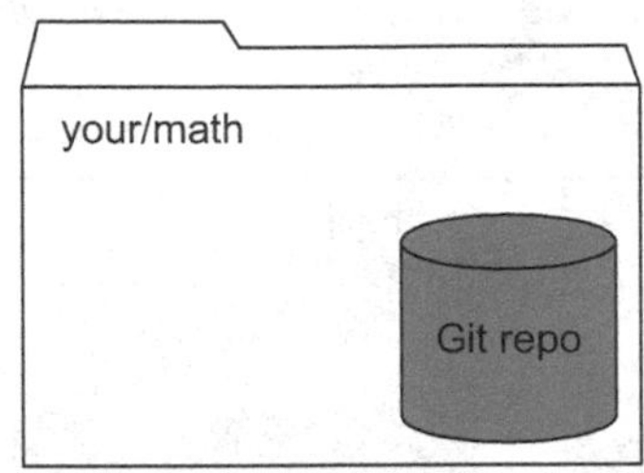

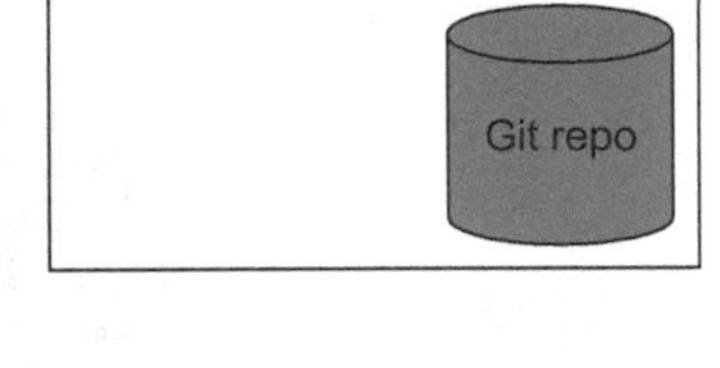

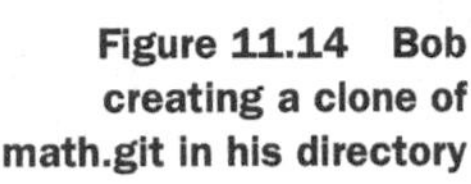
Figure 11.14 Bob creating a clone of math.git in his directory

Notice how math.git is acting as a centralized copy. But because Git supports cloning from any source, Bob could just as easily clone from the your/math directory, if Bob could access it.

To extend the example once more, imagine a colleague named Carol. She joins your company and has to access the repository. You tell her that from her carol directory, she can type `git clone math.git`. That produces a carol/math directory (see figure 11.15). You now have three copies of the math.git repository!

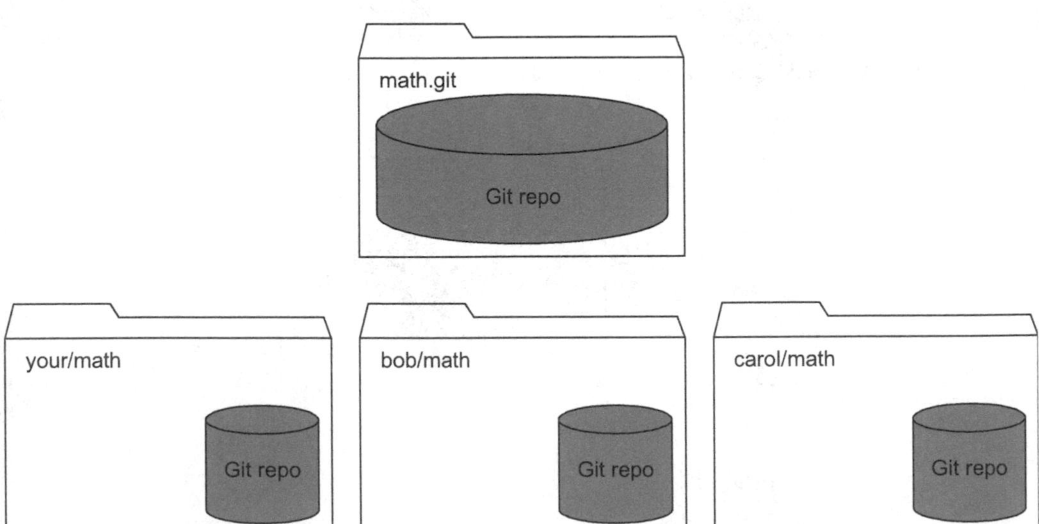

Figure 11.15 Carol joins the team and makes her own clone.

I'll leave it as an exercise for you to create this scenario on your PC. Spend time thinking about this situation, and how the bare directory serves as an official version, but only because as an organization we decided that to be the case. Nothing stops Carol from cloning Bob's copy of the repository. And nothing stops us from deleting math.git and declaring bob/math to be the new official version. This is an important feature of a distributed version-control system.

Making a bare directory on your local computer is a way for you to learn about collaboration without resorting to using an external server. Bare repositories are what are hosted on GitHub or any Git server, and over the course of these chapters on collaboration, you'll treat the math.git bare directory as a simple, practice GitHub.

11.3 Listing files in the repo by using git ls-tree

In the previous sections, you convinced yourself that you had the entire repository by using the command `git log --oneline --all`. But this shows you only that you have the entire history (the entire list of commits). Can you convince yourself that you have all the files? Yes, you can. The Git command `git ls-tree` lists all the files in a tree.

Remember, every commit in Git contains a tree of files. You saw this in the gitk tool, and how selecting a commit showed its corresponding tree. In figure 11.16, if you highlight the commit `Adding four empty files`, gitk shows its SHA1 ID to be `bcaa6e` (see the arrow in figure 11.16).

This pane contains the files a, b, c, d, and math.sh (see the highlighted box in figure 11.16).

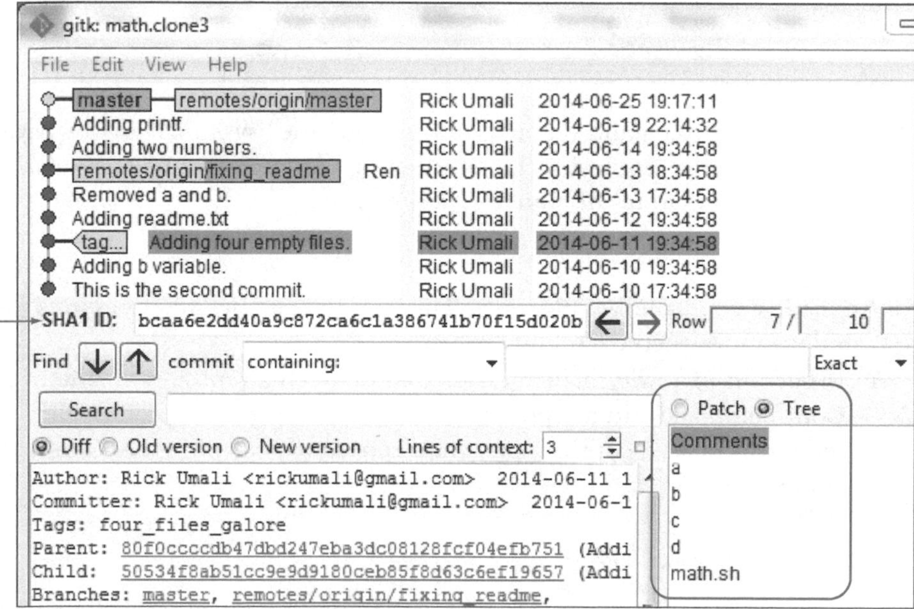

Figure 11.16 Each commit contains a tree of files.

TRY IT NOW You'll now go to a math.clone3 directory and try the `git ls-tree` command. This command takes one argument: a SHA1 ID (a commit ID), or a branch or tag. Note that these point to commits. Type the following:

```
cd $HOME/math.clone3
git checkout master
git ls-tree HEAD
```

This should give you the following output.

Listing 11.7 `git ls-tree HEAD`

```
100644 blob e69de29bb2d1d6434b8b29ae775ad8c2e48c5391    another_rename
100644 blob 41c57fac1f6c7eab44a0c2c181f934eb3b0040e0    math.sh
100644 blob 26f994161380366e6fed57f80203c0af2dfb9fe8    readme.txt
100644 blob e69de29bb2d1d6434b8b29ae775ad8c2e48c5391    renamed_file
```

This shows you the files at the HEAD, also known as the current branch. You can view the files that have been tagged by the `git tag` command (from section 8.4). You made a four_files_galore tag. To see its files, type the following:

```
git ls-tree four_files_galore
```

This should display the output in the following listing.

Listing 11.8 `git ls-tree four_files_galore` **output**

```
100644 blob e69de29bb2d1d6434b8b29ae775ad8c2e48c5391    a
100644 blob e69de29bb2d1d6434b8b29ae775ad8c2e48c5391    b
100644 blob e69de29bb2d1d6434b8b29ae775ad8c2e48c5391    c
100644 blob e69de29bb2d1d6434b8b29ae775ad8c2e48c5391    d
100644 blob 5bb7f6370f458be09d74514bab11178bf39fe4d8    math.sh
```

The `git ls-tree` command is a helpful way to get the list of files for any part of your Git history. When you work with clones, using `git ls-tree` can help you confirm that you have all the files from the source repository.

11.4 Lab

You've made a good number of clones. These short exercises and questions will check your understanding of the cloning process.

1. In section 11.1, you made a distinction between a repository made by a simple Copy command and a repository made by `git clone`. Type the following commands to make a copy of your math repository, using the command-line `copy` command:

   ```
   cd $HOME
   cp -r math math.copy
   ```

   Now type `git log --oneline --all` in both the math and the math.copy directories. You should confirm that these yield the same output. Compare that this copy is different from the clone by using `git branch --all`. The copy has no remote-tracking branches.

2 Make a clone with the current active branch in the original repository set to something else besides master.

   Go to the math directory and use `git checkout new_feature`. Now make a clone of the math repository. Confirm that the initial branch of the new clone is the fixing_readme branch.

3 What happens when you try `git checkout` on one of the remote-tracking branches? Are you able to?

   Remember that the remote-tracking branches are the ones that have the word *remotes* in the name when you do a `git branch --all`.

4 Use the longer form of `git checkout` (discussed in section 11.1.3) to make a whole new local branch with a different name from the remote-tracking branch.

5 Is there a limit to the number of branches that use the starting point?

6 Use the `--origin` switch of the `git clone` command to specify another name instead of *origin*. Confirm that the `git branch --all` command in your clone doesn't contain the string `origin` anymore.

   The word *origin* can be replaced by any name you want.

11.5 Further exploration

With `git clone`, you have the ability to clone only the most recent parts of a repository. To do this, you must use the `--depth` switch to the `git clone` command. For this section, use the `--depth` switch with varying arguments (`--depth 1`, then `--depth 2`, and so on), and convince yourself that you're getting a smaller repository.

To use the `--depth` switch, you must use a more formal manner of specifying the local directory. Your command looked like this: `git clone math math.clone1`. In this command, `math` references the source, the local directory math. But if you pass in the `--depth` switch to this command, you'll encounter this error: `--depth is ignored in local clones; use file:// instead.`

To get past this error, specify the source with the file:// URL. Learn about this syntax in the GIT URLs section of the `git clone` documentation. For my machine, the math directory is specified with file:///home/rick/math.

Another `git clone` switch to explore is `--no-single-branch`. Using this switch combined with `--depth 1` enables you to produce a repository that consists of only one commit for all the branches in your repository. In gitk, this repository looks like figure 11.17.

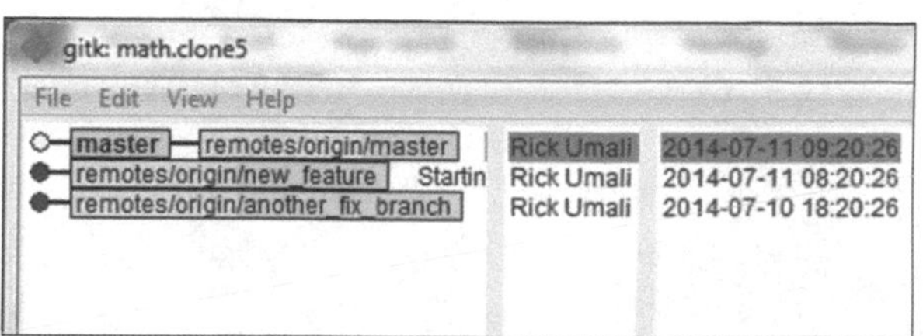

Figure 11.17 gitk showing a clone of math with `--depth 1` and `--no-single-branch` passed to `git clone`

11.6 Commands in this chapter

Table 11.1 Commands used in this chapter

Command	Description
`git clone source destination_dir`	Clone the Git repository at source to the destination_dir.
`git log --oneline --all`	Display all commit log entries from all branches. (Normally, `git log` displays only entries from the current branch.)
`git log --simplify-by-decoration --decorate --all --oneline`	Display the history in a simplified form.
`git branch --all`	Show remote-tracking branches in addition to local branches.
`git clone --bare source destination_dir`	Clone the bare directory of the source repository into the destination_dir. By convention, destination_dir should end with .git.
`git ls-tree HEAD`	Display all the files for HEAD (the current branch).

12 Collaborating with remotes

This is the second chapter (of four) on Git collaboration. In chapter 11, you learned about cloning. In this one, you'll learn about remotes and the command that manipulates them, `git remote`. Each clone that you create contains a reference to where it came from. This reference is a *remote*. Remotes serve as pointers back to a location, either on your computer or on the Internet. It's the basis for collaboration, along with cloning.

You'll learn how `git remote` allows you to examine and update the remote. You'll also learn about the `git ls-remote` command, which lets you know if your local repository and your original (remote) repository are out of sync. This chapter will help you understand Git's collaboration model, and will make the next two chapters on `git push` and `git pull` clearer.

12.1 Remotes are distant places

In figure 12.1, you have three repositories: math.clone, math.bob, and math.carol. All of these repositories are created from the math.git repository, which is a bare directory.

In your collection of repositories, each clone in the second row (math.clone, math.bob, and math.carol) is created by `git clone`, using math.git as the source. Each clone contains a reference to math.git. This reference, called a *remote*, indicates the location of the original repository.

> **TRY IT NOW** If you performed all the TRY IT NOW sections in the preceding chapter, you already have parts of this exercise completed. This section repeats all the steps, so you might be able to skip some. The only prerequisite is to have a known math repository.

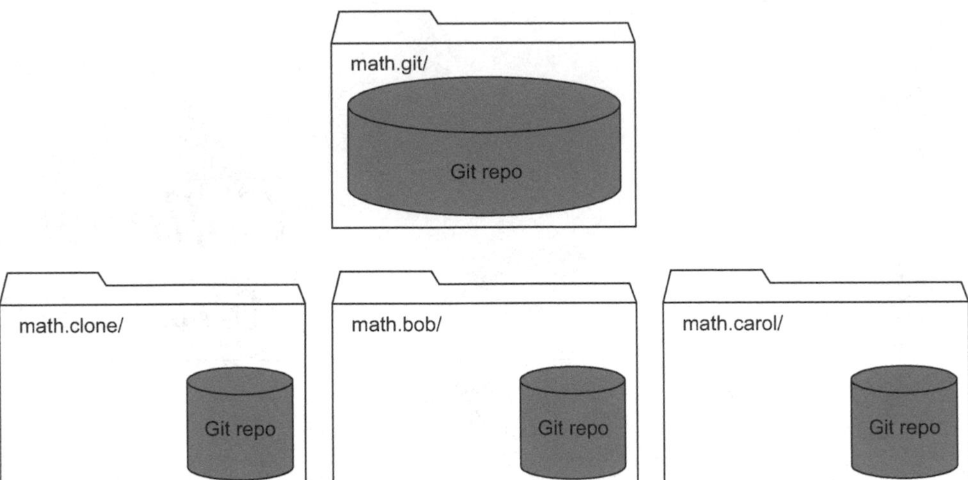

Figure 12.1 A small collection of repositories

If you don't have a working math repository, download the LearnGitMoL _SourceCode.zip file from the book's website, and obtain the make_math _repo.sh script from it (this is the script from the chapter 9 exercises). Run that script, using the following:

```
cd $HOME
bash make_math_repo.sh
```

This makes a math directory containing your starting point. As described in chapter 9, the script leaves the repo in the state described at the end of section 9.4.2. To get out of that state, perform the following steps:

```
cd math
git checkout -f master
```

This changes the branch to master, without saving any of the edits. Now to make the clone, type this:

```
cd ..
git clone --bare math math.git
```

If you get an error saying that math.git already exists, you already have the math.git repository. The preceding `git clone` command makes a bare repository that will serve as the source of your future collaboration (cloning, pushing, pulling).

To make the rest of the clones in your collection, using this bare repository as the source URL, type the following:

```
git clone math.git math.clone
git clone math.git math.bob
git clone math.git math.carol
```

This makes the three clones that are below math.git in figure 12.1.

Collaboration on a software project using Git's distributed architecture starts with `git clone`. As you learned in the preceding chapter, when you make a clone, you make a copy of a repository. Remember, each clone contains a reference to where it came from, and this reference is a remote.

12.1.1 Analyzing a clone's origin (git remote)

Each clone that you created in the previous section knows where it came from. It knows its origin, thanks to the remote. You can confirm this by using the `git remote` command. This command provides multiple ways to manipulate references to a clone's original repository. The `git remote` command has many forms, and you'll now try out a few.

> **TRY IT NOW** Let's go into your math.clone repository and try `git remote`. Type the following:
>
> ```
> cd $HOME
> cd math.clone
> git remote
> ```
>
> The simplest form of the `git remote` command shows only the name of the remote. Each remote has a name, and the preceding `git remote` command shows that you have one remote, called origin. Now type this:
>
> ```
> git remote -v show
> ```
>
> The `git remote -v show` command shows the remote URL. Running these two `git remote` commands should produce the following output.

Listing 12.1 `git remote` output

```
$ git remote
origin

$ git remote -v show
origin  c:/Users/Rick/Documents/gitbook/math.git (fetch)
origin  c:/Users/Rick/Documents/gitbook/math.git (push)
```

The last output (from `git remote -v show`) shows where the origin exists for the fetch and push operations. Fetch is an operation that downloads (receives) files from a remote, and for purposes of the `git remote` command, you can replace `fetch` with `pull`. Push is an operation that uploads (sends) files to a remote. In both cases, your remote is a separate directory on your local machine, but soon in this chapter, it'll be a server on the Internet somewhere. I have more to say about `push` and `fetch` in the following chapters.

It's important to note that the source URL that you gave to the `git clone` command is the URL that the remote is set to. Git expands math.git to the full path to the math.git directory.

Figure 12.2 depicts that *origin* is the name (local to your repository) of the remote that points back to the math.git directory/repository. (The arrow could be thought of as the remote—a reference, or pointer, back to the source.)

To be absolutely precise, `git clone` creates remote-tracking branches in the new repository. These branches, discussed in the preceding chapter, are given a distinct name to indicate that they are tracking branches that originally existed in the source repository.

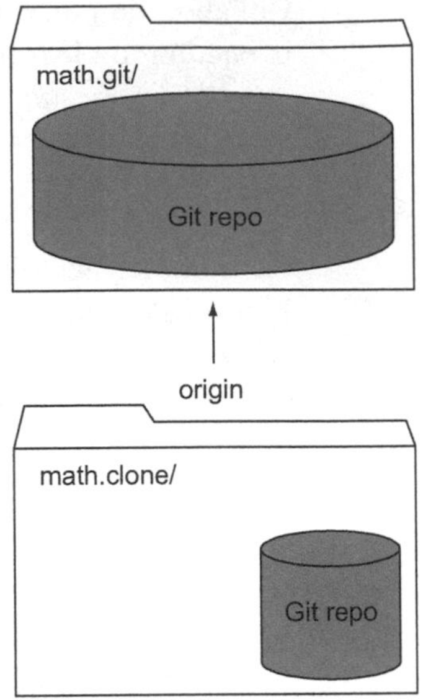

Figure 12.2 origin is a name that exists locally.

In figure 12.3, the math.clone repository has three remote-tracking branches:

- remotes/origin/master
- remotes/origin/new_feature
- remotes/origin/another_fix_branch

These three branches existed in the original repository, and these remote-tracking branches serve as bookmarks to those branches in your local repository.

When you cloned a repository, you brought over every commit and every directory and file that make up those commits. The remote-tracking branches, like regular branches, point to the last commit of that line of development. Because every commit points to its parent, you can see how you have the entire history.

12.1.2 Renaming a remote

The word *origin* is the name that a remote is given by default by the `git clone` command. But you can rename this string if you want to.

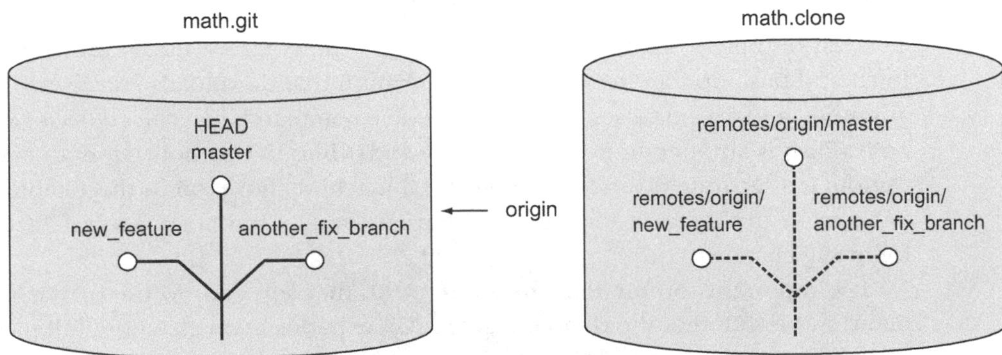

Figure 12.3 Remote-tracking branches

TRY IT NOW Make sure you are in the math.clone directory and type the following:

```
git branch --all
git remote -v show
```

This should give you the following output.

Listing 12.2 `git branch --all` and `git remote` output

```
$ git branch --all
* master
  remotes/origin/HEAD -> origin/master
  remotes/origin/another_fix_branch
  remotes/origin/master
  remotes/origin/new_feature

$ git remote -v show
origin  c:/Users/Rick/Documents/gitbook/math.git (fetch)
origin  c:/Users/Rick/Documents/gitbook/math.git (push)
```

Now use `git remote` to rename *origin*:

```
git remote rename origin beginning
```

This variation of `git remote` renames the remote from *origin* to *beginning*. Notice that this is all happening locally. Now confirm that the rename took place by typing the following:

```
git branch --all
git remote -v show
```

This should give you the following output.

Listing 12.3 Looking at the branches and remotes

```
$ git branch --all
* master
  remotes/beginning/HEAD -> beginning/master
  remotes/beginning/another_fix_branch
  remotes/beginning/master
  remotes/beginning/new_feature

$ git remote -v show
beginning       c:/Users/Rick/Documents/gitbook/math.git (fetch)
beginning       c:/Users/Rick/Documents/gitbook/math.git (push)
```

In this example, *origin* is just a name for a remote. It can be renamed, as you see here. The only person affected by this change is you. Figure 12.4 depicts your situation.

Compare figure 12.4 with figure 12.2, and listing 12.3 with listing 12.2. Everything is the same, except for the name of your remote.

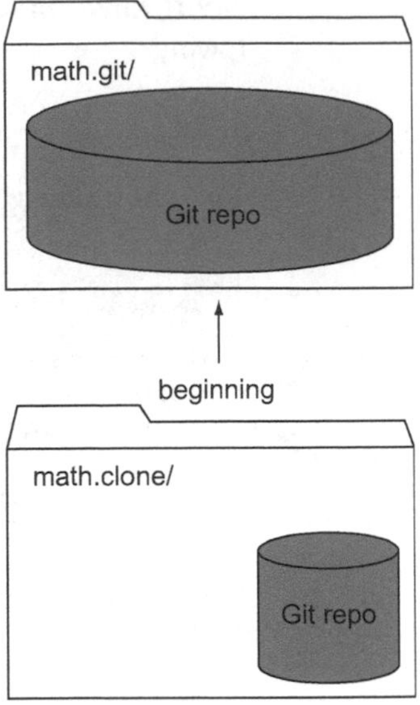

Figure 12.4 Changing the name of the remote

12.1.3 Adding a remote

You can even add a remote. That's right. Even after you make a clone, you can add another remote, which represents another repository that you want to track. If you collaborate on a repository where contributors are actively developing on their own repositories, it may be useful to add their repositories, in addition to the remote that was created when you first did the clone. Let's consider Bob's and Carol's repositories. They look like figure 12.5.

Notice that math.bob and math.carol both have a remote named *origin*, pointing back to math.git. If Carol and Bob wanted to collaborate, Carol could create a remote that points to Bob's repository (and vice versa). Let's try this.

TRY IT NOW In the math.carol repository, to point to Bob's repo, you use the `git remote add` command. Type the following:

```
cd $HOME/math.carol
git remote add bob ../math.bob
```

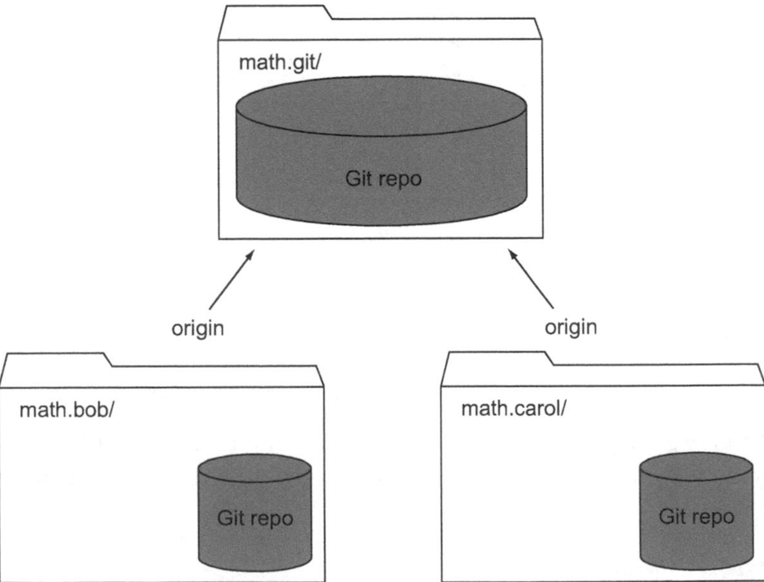

Figure 12.5 Bob and Carol's repositories, both cloned from math.git

This creates a new remote called *bob*. You can view it by typing this:

```
git remote
git remote -v show
```

The last two `git remote` commands produce the output in the following listing.

Listing 12.4 Output from `git remote` commands

```
$ git remote
bob
origin

$ git remote -v show
bob     ../math.bob (fetch)
bob     ../math.bob (push)
origin  c:/Users/Rick/Documents/gitbook/math.clone (fetch)
origin  c:/Users/Rick/Documents/gitbook/math.clone (push)
```

After the preceding TRY IT NOW, your environment looks like figure 12.6.

By now, I hope it's clear that you can make a remote point to any repository. But what's the point of creating these remotes? The point of remotes is to collaborate. The remotes are the other repositories with which you can collaborate.

Because Carol's repository is cloned from math.git, she can bring in any changes from that repository into her repository. Because Carol added a remote to Bob's repository, she can also bring in any changes from Bob's repo into hers. And she

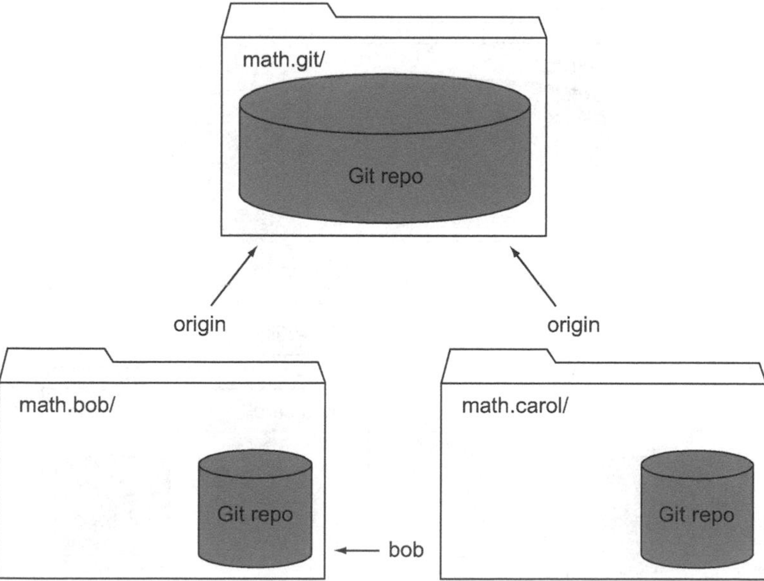

Figure 12.6 Carol has a remote named *bob* that points to Bob's repository.

could go the other way too: she can push any changes she makes on her repository to either Bob or the official repository at math.git.

In the real world, both Carol and Bob would most likely just point to math.git, and math.git would be stored on GitHub, which you'll read about in chapter 18. But Git supports this kind of cross-repository sharing, and the `git remote` command establishes this.

12.2 Interrogating a remote

As you saw from the `git remote show` command, Git enables a repository to push to, and fetch from, another repository. Listing 12.4 shows the output of `git remote show -v` from the math.carol repository. For each repo, she can push or fetch. You can see this in figure 12.7.

In the next few chapters, you'll be pushing and fetching (a.k.a. pulling), but let's take the baby step of interrogating (querying) the remote repository by exercising the `git ls-remote` command. This command returns a list of the SHA1 IDs of each branch and tag (each reference) on the remote repository.

TRY IT NOW In the math.carol directory, type the following:

```
git ls-remote
git ls-remote origin
git ls-remote bob
```

The `git ls-remote` command can be typed by itself or with an argument. The argument must be a remote that is the name of the original repository

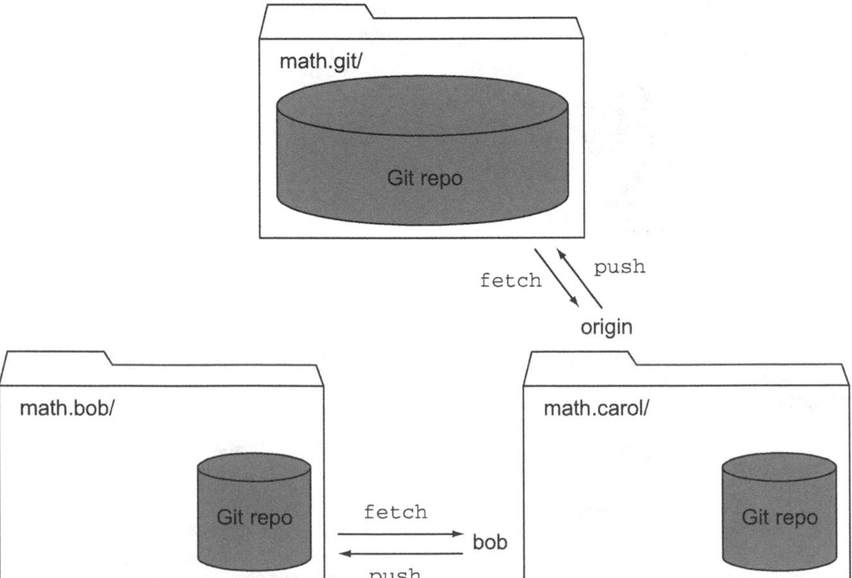

Figure 12.7 From math.carol, she can fetch from or push to bob or origin.

(for example, *origin*) or the name of another remote that you added (for example, *bob*). All three outputs should look like the following listing.

Listing 12.5 `git ls-remote` output

```
From c:/Users/Rick/Documents/gitbook/math.git
4465c540dc79718076bcf66951d27fb65152a895 HEAD
23d30770e5b8b0e42bc5927a0a348a6912963aff refs/heads/another_fix_branch
4465c540dc79718076bcf66951d27fb65152a895 refs/heads/master
dc6f60f417c011bafe6284d362a06e39f9f3cb69 refs/heads/new_feature
f4b5a261dfdcdc5d9081b2ecc252a62f198b01c3 refs/tags/four_files_galore
ef47d3fd293bc13321270e88af284f63d6f85f84 refs/tags/four_files_galore^{}
```

These are the SHA1 IDs of the references on the remote named *origin*. When you cloned the repository, you also cloned these SHA1 IDs. You can compare this to SHA1 IDs of the current local repo (math.carol) by typing the following:

```
git ls-remote .
```

In this command, the period (`.`) represents the current local repository. This should give you the same listing!

Look at the first line of listing 12.5. This line shows which remote you're connecting to. The command doesn't show the remote's name, however. Instead, it shows the remote path or URL. Notice that the second time you used `git ls-remote`, you specified the remote on the command line (`git ls-remote origin`). This makes it easier to remember what remote you're listing.

After the first line, each line consists of a SHA1 ID and a reference name. The references should be somewhat familiar: they're the names of your branches (prefixed by `refs/heads`) and the names of your one tag (prefixed by `refs/tags`).

The SHA1 IDs should be familiar as well. They're the SHA1 IDs of your own branches and tags. You can see this by using `git ls-remote`. The period indicates the current local repository. The SHA1 IDs are the same because math.carol is a clone of math.git! Figure 12.8 illustrates your situation: two repositories, one a clone of the other.

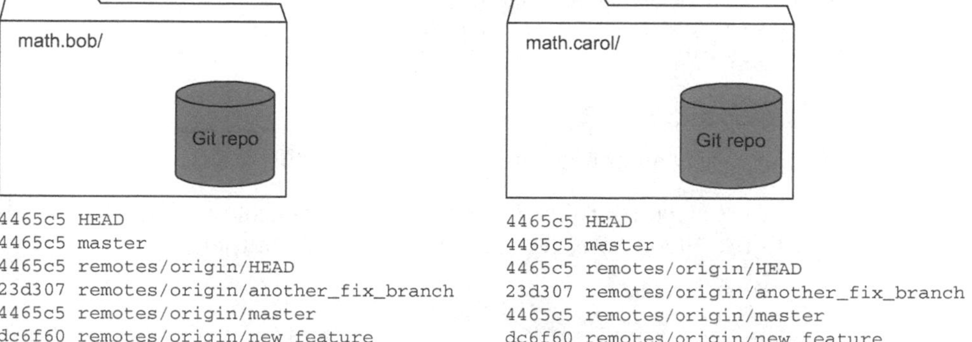

Figure 12.8 Each repo has a list of references. Because these are clones, they're exactly alike initially.

Notice that each of the SHA1 IDs is the same. The other key thing to realize is that `git ls-remote` makes a network connection to the remote repository, and displays its list of references. Now, let's make a change to the math.bob repository.

TRY IT NOW Type the following to commit a small change to the math.bob repository:

```
cd $HOME/math.bob
echo "Small change to file" >> another_rename
git commit -a -m "Updating this file."
```

Now let's visit the math.carol repository and interrogate the remotes:

```
cd $HOME/math.carol
git ls-remote
git ls-remote origin
git ls-remote bob
git ls-remote .
```

If you look at all four outputs, the one from `git ls-remote bob` is different from all the others. The SHA1 ID of its HEAD is different!

The `git ls-remote bob` output should show the SHA1 ID of the latest commit that you made to the math.bob repo. HEAD (and refs/heads/master) should now have this new SHA1 ID. Let's confirm this.

TRY IT NOW Let's obtain the last commit to the math.bob repository (on its current branch, which is master). Type the following to do this:

```
cd $HOME/math.bob
git branch
git log -1
```

Remember that `git log -1` shows only one commit—the most recent one. The listing should look similar to the following.

Listing 12.6 Committing a change in the math.bob directory

```
commit db106c748e5b6aa90cc63de3d25cb5dcbebbcfc6
Author: Rick Umali <rickumali@gmail.com>
Date:   Sat Aug 9 19:59:46 2014 -0400

    Updating this file
```

(Don't worry if your commit message is slightly different.)

Now go to the math.carol repository, and confirm that this SHA1 ID is associated with the HEAD from the `git ls-remote bob` output:

```
cd $HOME/math.carol
git ls-remote bob
```

That listing should be something like the following listing.

Listing 12.7 `git ls-remote bob` output

```
db106c748e5b6aa90cc63de3d25cb5dcbebbcfc6 HEAD
db106c748e5b6aa90cc63de3d25cb5dcbebbcfc6 refs/heads/master
4465c540dc79718076bcf66951d27fb65152a895 refs/remotes/origin/HEAD
23d30770e5b8b0e42bc5927a0a348a6912963aff refs/remotes/origin/another_fix_
branch
4465c540dc79718076bcf66951d27fb65152a895 refs/remotes/origin/master
dc6f60f417c011bafe6284d362a06e39f9f3cb69 refs/remotes/origin/new_feature
f4b5a261dfdcdc5d9081b2ecc252a62f198b01c3 refs/tags/four_files_galore
ef47d3fd293bc13321270e88af284f63d6f85f84 refs/tags/four_files_galore^{}
```

Notice how the HEAD (and refs/heads/master) is the same as the new commit. Notice, too, that refs/remotes/origin/HEAD is the old commit. Why? You made a change to the math.bob repository that you haven't pushed to the math.git repository (see figure 12.9). You'll see how to push changes in the next chapter.

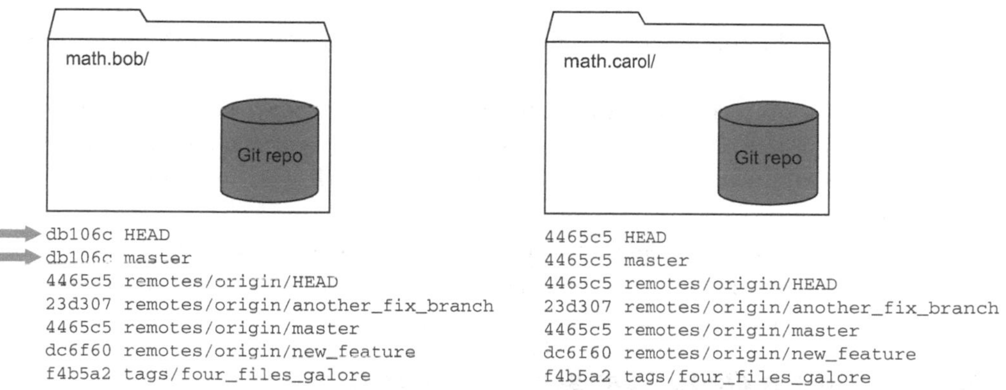

Figure 12.9 math.bob has a new commit. Its list of references is now different from math.carol.

Figure 12.9 shows that math.bob's master branch and its HEAD have changed. Let's make one more change to the math.bob repository: let's add a branch to this repo.

TRY IT NOW To add a branch to the bob repository, type the following:

```
cd $HOME/math.bob
git checkout -b a_new_branch
```

As you may recall, this not only creates a new branch (a_new_branch), but also checks out that branch (changing HEAD). Now confirm that `git ls-remote` can see this new branch from math.carol. Type the following:

```
cd $HOME/math.carol
git ls-remote bob
```

The last output shows a new line for the new branch:

```
db106c748e5b6aa90cc63de3d25cb5dcbebbcfc6        refs/heads/a_new_branch
```

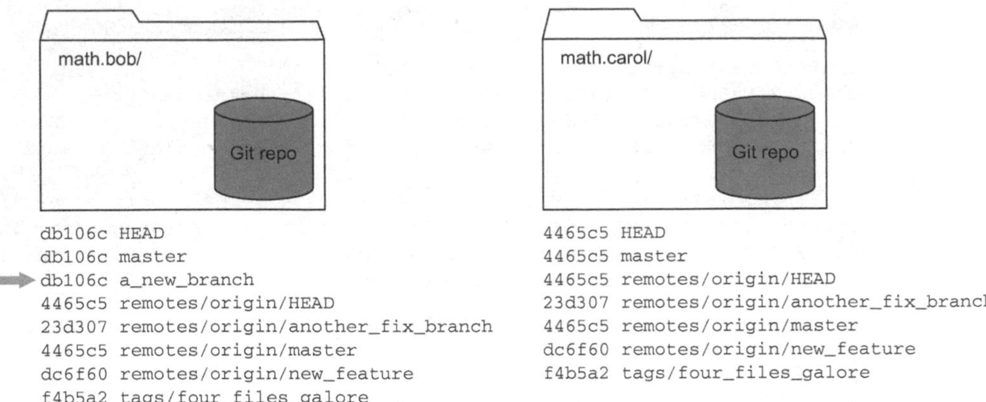

Figure 12.10 math.bob has a new branch now. Notice this is missing from math.carol.

In figure 12.10, this new branch (a_new_branch) appears in the `git ls-remote bob` output.

You're interrogating the remote and getting its up-to-date information. It's important to note again that no server is running to tell you that a new branch has been created. Only when you ask the other repository do you find out that something has changed.

Interrogating a remote is the basis for the interactions you'll look at in the upcoming chapters. Please take the time to understand this section, as it's the foundation for the next two chapters!

12.3 Getting a clone from somewhere remote

Let's now finally interact with a repository that is remote in the literal sense of the word: a repository that exists on the Internet. You already did this at the end of chapter 2. In that chapter, you went to a GitHub page, obtained its clone URL, and used that as the source URL in the `git clone` command. This time, you'll clone a known URL.

TRY IT NOW On the command line, you're going to retrieve a version of the math repository that I stored on GitHub. Type the following:

```
cd $HOME
git clone https://github.com/rickumali/math.git math.github
```

In this command, you use https://github.com/rickumali/math.git as the source URL. Your destination URL is math.github. The output of this command should look like the following listing.

Listing 12.8 `git clone` of the math repo from GitHub

```
Cloning into 'math.github'...
remote: Counting objects: 35, done.
```

```
remote: Compressing objects: 100% (17/17), done.
remote: Total 35 (delta 7), reused 34 (delta 6)
Unpacking objects: 100% (35/35), done.
Checking connectivity... done.
```

The only difference between this `git clone` command and the other ones that you typed earlier in this chapter is that this `git clone` specifies a source location that's on the Internet. Let's examine the remote by using the `git remote` command.

TRY IT NOW Go into the math.github directory, and type this:

```
git remote -v show
```

This yields the following listing.

Listing 12.9 `git remote -v show` output

```
origin      https://github.com/rickumali/math.git (fetch)
origin      https://github.com/rickumali/math.git (push)
```

Compared to your earlier clones, this remote is a URL that's not on your local machine. It's a true remote location, but observe that the only time it interacted with GitHub was to download the repository to your directory. After the repository is downloaded, no further interaction occurs with the remote.

12.4 Lab

Remotes serve as pointers back to a location, either on your computer (as in our math clone directories) or on the Internet (as in the previous section). Remotes are the basis for collaboration, along with cloning.

12.4.1 Exploring your math.github clone

In the following steps, you'll confirm that SHA1 IDs are constant between clones. This is the basis for distributed version control.

1. In the math.github directory, type `git log --oneline --decorate`.
2. What is the SHA1 ID of the remote-tracking branch another_fix_branch?
3. Is there a tag or branch for the SHA1 ID `ef47d3f`?

   One important aspect of this exercise is to recognize that the SHA1 IDs will always be the same for a clone of this repository. Convince yourself of this by creating a clone of this same repository on another computer.
4. Are the SHA1 IDs of math.github the same as math.bob or math.clone? Why not?

12.4.2 Making remotes manually

In the math.bob repository, create a remote named *carol* that points to the math.carol repository. Now repeat the exercise of making a change in Carol's repository, and use `git ls-remote` to see the change.

12.4.3 Using other git remote subcommands

In this chapter you used `git remote` to add, rename, and show remotes. The `git remote` command can also delete remotes and update the URL of a remote. Explore the `remove` and `set-url` subcommands by looking at the help for `git remote`. Experiment with these commands on a clone of the math repository.

12.4.4 Creating clones with Git GUI

Create a clone by using Git GUI. When you start Git GUI, you'll be greeted with a screen like figure 12.11.

From here, click Clone Existing Repository. Specify this GitHub URL: https://github.com/rickumali/math.git. Now specify a target directory. In figure 12.12, I specify the same target directory as in the preceding TRY IT NOW. Will that work?

Confirm that the end result is the same as performing a `git clone` on the command line.

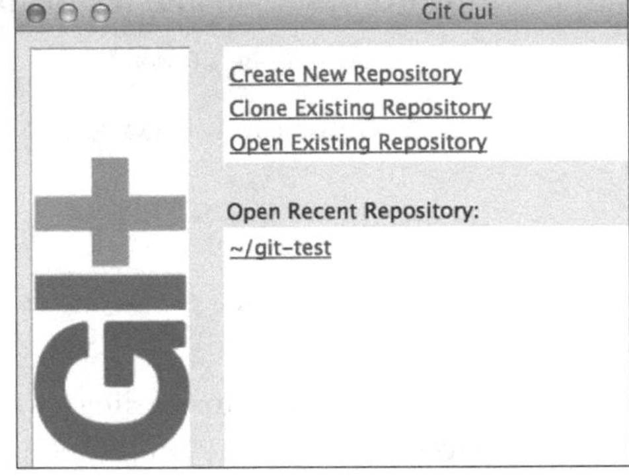

Figure 12.11 The initial Git GUI screen

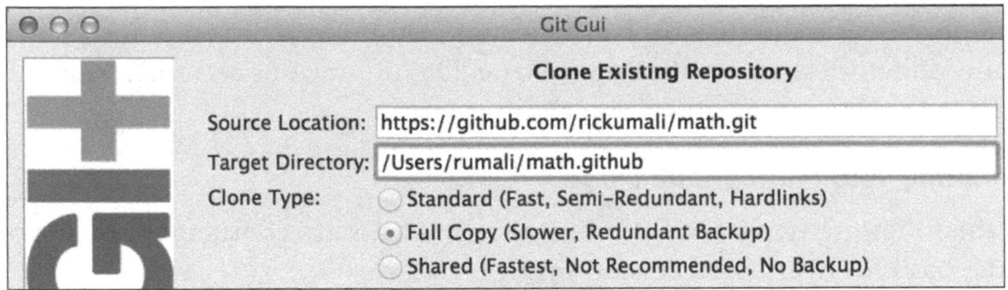

Figure 12.12 Obtaining a clone by using Git GUI

12.4.5 Accessing another remote URL

I've created the math repository on another code-sharing website called GitLab. Visit the URL: https://gitlab.com/rickumali/math. Create a clone from this repository (you'll need to use the URL https://gitlab.com/rickumali/math.git as the source to `git clone`). Are the SHA1 IDs the same?

12.5 Further exploration

On the command line, you run the `git ls-remote` command with environment variables that enable you to trace the network activity. On the command line in either the math.github or any of the math clones, type `GIT_TRACE_PACKET=1 git ls-remote`.

This should produce the following output.

Listing 12.10 Tracing the network activity from `git ls-remote`

```
packet:          git< # service=git-upload-pack
packet:          git< 0000
packet:          git< 4465c540dc79718076bcf66951d27fb65152a895
➥ HEAD\0multi_ack thin-pack side-band side-band-64k ofs-delta
➥ shallow no-progress include-tag multi_ack_detailed
➥ no-done symref=HEAD:refs/heads/master agent=git/2.0.3
packet:          git< 23d30770e5b8b0e42bc5927a0a348a6912963aff
➥ refs/heads/another_fix_branch
packet:          git< 4465c540dc79718076bcf66951d27fb65152a895
➥ refs/heads/master
packet:          git< dc6f60f417c011bafe6284d362a06e39f9f3cb69
➥ refs/heads/new_feature
packet:          git< f4b5a261dfdcdc5d9081b2ecc252a62f198b01c3
➥ refs/tags/four_files_galore
packet:          git< ef47d3fd293bc13321270e88af284f63d6f85f84
➥ refs/tags/four_files_galore^{}
packet:          git< 0000
From https://github.com/rickumali/math.git
4465c540dc79718076bcf66951d27fb65152a895  HEAD
23d30770e5b8b0e42bc5927a0a348a6912963aff  refs/heads/another_fix_branch
4465c540dc79718076bcf66951d27fb65152a895  refs/heads/master
dc6f60f417c011bafe6284d362a06e39f9f3cb69  refs/heads/new_feature
f4b5a261dfdcdc5d9081b2ecc252a62f198b01c3  refs/tags/four_files_galore
ef47d3fd293bc13321270e88af284f63d6f85f84  refs/tags/four_files_galore^{}
```

12.6 Commands in this chapter

Table 12.1 Commands used in this chapter

Command	Description
`git checkout -f master`	Check out the master branch, throwing away any changes in your current branch.
`git remote`	Display the name of the remote(s) in the current repository.
`git remote -v show`	Display the names of the remotes along with the corresponding remote URL.
`git remote add bob ../math.bob`	Add a remote named bob that points to the local repository in ../math.bob.
`git ls-remote REMOTE`	Display the references of a remote repository (use . as the REMOTE when you want the current local repository).
`GIT_TRACE_PACKET git ls-remote REMOTE`	Display the underlying network interaction.

13 Pushing your changes

You're now halfway through your study of how to collaborate with Git. You've learned how to clone repositories (`git clone`) and how to work with remotes (`git remote`). In the preceding chapter, you created the environment shown in figure 13.1 on your local computer. Cloning and making remotes establishes the groundwork for the next two commands: `git push` and `git pull`.

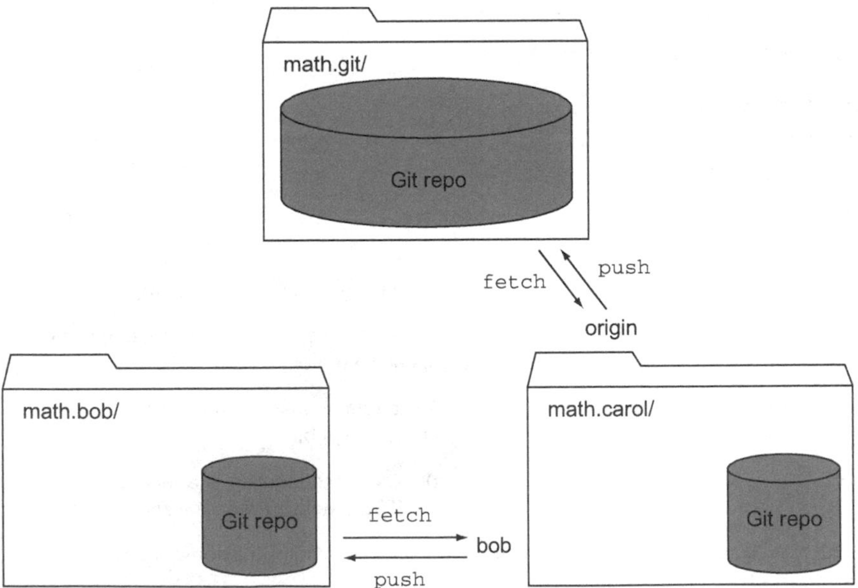

Figure 13.1 Carol can push files.

In this chapter, you'll learn about the `git push` command, the first half of the Git collaboration process (`git pull` is the second half, covered in the next chapter). `git push` is important because it's how you share (publish) your changes. You'll see that `git push` merges those changes with the repository you're pushing to. This may cause conflicts, and you'll see how to recover from those. `git push` operates on your changes in branches, and you'll see how it does double-duty creating and deleting branches and tags on remotes.

13.1 Pushing sends changes to a remote

Everything that you've done up to now has affected only your local repository. In chapter 12, you learned about the `git ls-remote` command, which enables you to list the contents of a remote repository. The `git push` command is the first command that will directly affect another repository besides your own.

13.1.1 Permissions are required

Because `git push` makes changes to another repository, you must have the right permissions to perform `git push` to this remote. Let's try `git push` with the math clone that you made from GitHub.

> **TRY IT NOW** Make sure you have a directory math.github, which is a clone of the math repository from https://github.com/rickumali/math. This directory was created in the previous chapter. To do a push, type the following:
>
> ```
> cd $HOME/math.github
> git push origin master
> ```
>
> You may be prompted for a GitHub username and password. If you have these, enter them, and you should see an error that looks like the following listing. If you don't have them, you can press Ctrl-C to get past the login prompt.

Listing 13.1 A `git push` error due to permissions

```
remote: Permission to rickumali/math.git denied to ff-rumali.
fatal: unable to access 'https://github.com/rickumali/math.git/': The
requested URL returned error: 403
```

The error (after you enter GitHub credentials) happens because you don't have permission to do a push to this repository on GitHub. Before this chapter, you had free reign over your entire repository. But when you venture into the realm of Git collaboration, you must have the appropriate permissions. These permissions aren't as complicated as those needed in other version-control systems, but they do exist nonetheless. In this specific case, the repository owner (me) must grant you permission to push to this repository. You'll spend a lot of time on this in chapters 17 and 18.

Because the repositories you created in chapter 12 are all in your control on your machine, you can freely push to each of them. Let's make `git push` work between your local math repositories.

13.1.2 Pushing requires a branch and a remote

To see `git push` work, you'll use the math repositories that you created in chapter 12. Specifically, you'll focus on the bob and carol repositories, and the math.git repository, which you've designated as the official repository on your local machine. Figure 13.2 shows these repositories.

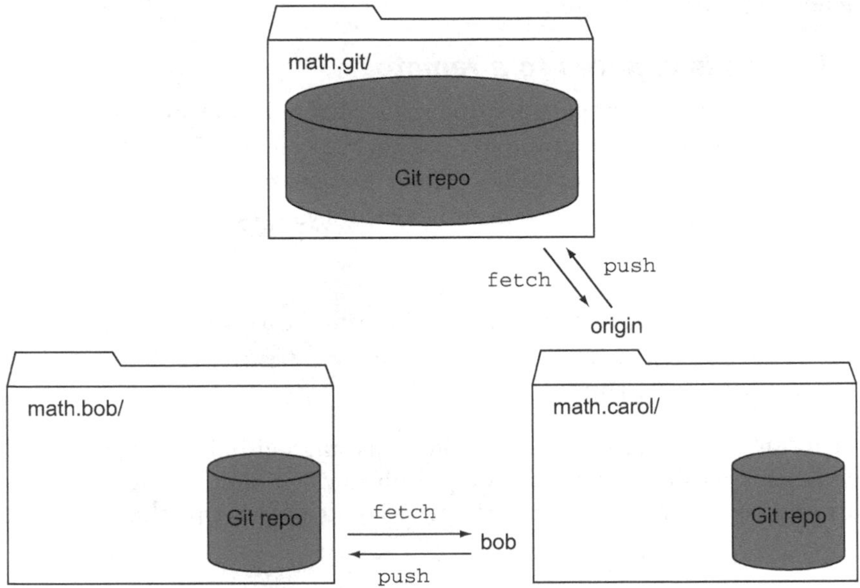

Figure 13.2 You'll push changes to math.git.

> **TRY IT NOW** Let's go to the math.carol repository and do a push to the math.git repository:
>
> ```
> cd $HOME/math.carol
> git push origin master
> ```
>
> You should see the message `Everything up-to-date.`

I know that was anticlimactic, but believe me when I say that Git did a lot of work; it compared the local repository with the remote repository. If there were changes in the local repository, Git would have pushed them over. Because you haven't made any changes, this `git push` doesn't do anything.

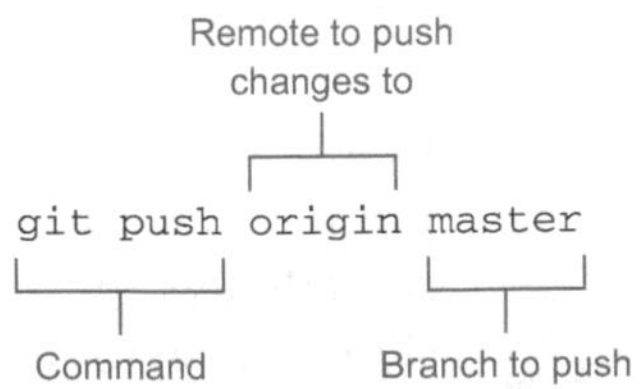

Figure 13.3 Anatomy of the `git push` command

Let's break down this command a bit more. Take a look at figure 13.3.

This form of the `git push` command specifies two arguments: the remote to which you want to send your changes, and the branch that contains the changes you want to push. This is the safest form of git push, in that everything is spelled out. Later in this chapter, you'll see how git push works when you don't pass it any arguments. (You'll also see another breakdown of figure 13.3 in section 13.4.)

Finally, let's push a change from carol to the main repository.

TRY IT NOW In the math.carol directory, let's edit one of the files. Type the following:

```
cd $HOME/math.carol
git checkout master
echo "Added a line here." >> renamed_file
git commit -a -m "Updated renamed_file"
```

You first make sure you're in the master branch. When you do the commit, Git announces the new SHA1 ID (in an abbreviated form).

At this point, Git knows that you've made a change that the remote doesn't have. Type this:

```
git status
```

You should see a message like the following listing.

Listing 13.2 Output of git status (that is ahead of its remote-tracking branch)

```
On branch master
Your branch is ahead of 'origin/master' by 1 commit.
  (use "git push" to publish your local commits)

nothing to commit, working directory clean
```

This message contains a lot of information, which is mostly self-explanatory. The key line says that your local branch is ahead of the origin/master by 1 commit. This new commit is the change you just made to renamed_file. origin/master is another way of saying the branch master on the remote origin.

Instead of doing git push as the instructions say, let's use a more careful command and type the following:

```
git push origin master
```

This syntax does suggest that you can push any local branch, including one that you're not already on.

Our git push command should report its activity and status, as shown in the following listing.

Listing 13.3 Output from git push command

```
Counting objects: 7, done.
Delta compression using up to 4 threads.
Compressing objects: 100% (2/2), done.
Writing objects: 100% (3/3), 287 bytes | 0 bytes/s, done.
Total 3 (delta 1), reused 0 (delta 0)
To /Users/rumali/math.git
   4465c54..b9af80a  master -> master
```

Here Git does a lot of work compressing and writing objects. This is just debugging output. The key line in the output is the one at the end:

```
4465c54..b9af80a  master -> master
```

This is saying that on the remote repository (the directory math.git), the master branch's SHA1 ID was changed from `4465c54` to `b9af80a`. This is because you pushed a new commit to this remote. Let's look at the SHA1 IDs of the last two commits on the math.carol repository.

TRY IT NOW Let's list the last two SHA1 IDs of your repository to confirm that they match what was reported in the `git push` command. Type the following:

```
git log -2
```

The `-2` switch is shorthand for `-n 2`, which limits the `git log` output to just two commits. This command will list something like the following.

Listing 13.4 `git log -2` output

```
commit b9af80a0d435ef74f5c72197311544c37a23ea91
Author: Rick Umali <rumali@firstfuel.com>
Date:   Thu Aug 21 20:30:27 2014 -0400

    Updated master/renamed_file

commit 4465c540dc79718076bcf66951d27fb65152a895
Author: Rick Umali <rickumali@gmail.com>
Date:   Wed Aug 6 08:54:56 2014 -0500

    A small update to readme.
```

You can see that the last two commits do correspond to the `git push` output `4465c54..b9af80a master -> master`. The `git push` announces the old SHA1 ID of the branch (`4465c54`), and then the new SHA1 ID (`b9af80a`) after the push.

13.1.3 Verifying a successful git push

Figure 13.4 attempts to visualize the push. When you make a commit (the short arrow labeled A), you add a change to the master branch of your local repository (math.carol, on the right). When you push this change to your math.git repository, the commit is sent to the remote (the arrow labeled B), along with the corresponding changes.

How can you tell that your new commit made it to the remote repository? In chapter 12, you learned about `git ls-remote`, a command that lets you query the branches on the report repository. You can use this to confirm that the remote has the latest SHA1 ID.

TRY IT NOW In the math.carol directory, type the following:

```
git ls-remote origin
```

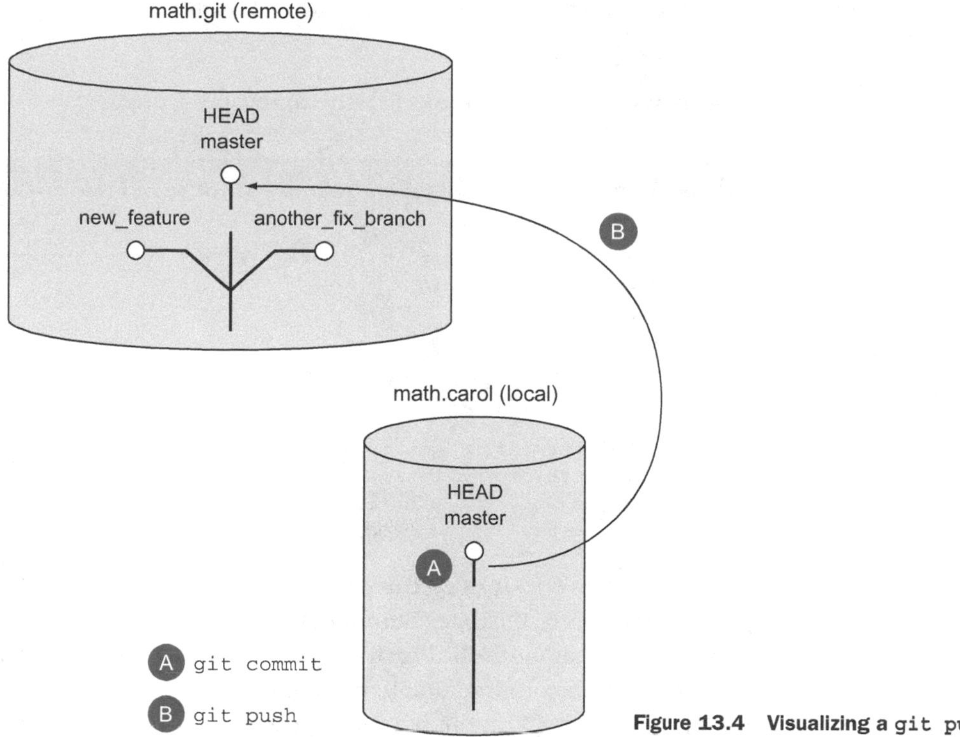

Figure 13.4 Visualizing a `git push`

Look at the listing returned from this command. Pay attention to the `refs/heads/master` reference. They should have the same SHA1 IDs as your latest commit.

When you run this command, you should see output like the following listing.

Listing 13.5 `git ls-remote` output

```
From /Users/rumali/math.git
b9af80a0d435ef74f5c72197311544c37a23ea91  HEAD
23d30770e5b8b0e42bc5927a0a348a6912963aff  refs/heads/another_fix_branch
b9af80a0d435ef74f5c72197311544c37a23ea91  refs/heads/master
fc85daffa32ce38362b28ed846cdd12fff5429c5  refs/heads/new_feature
f4b5a261dfdcdc5d9081b2ecc252a62f198b01c3  refs/tags/four_files_galore
ef47d3fd293bc13321270e88af284f63d6f85f84  refs/tags/four_files_galore^{}
```

Notice that `refs/heads/master` is the same SHA1 ID (`b9af80a`) as the last commit of math.carol. (Make sure to confirm this in your own repository, because your SHA1 IDs will be different from this text.)

Making comparisons in this fashion is a little tedious. A better way is to use the `git remote` command's `show` subcommand.

TRY IT NOW In the math.carol directory, type the following:

```
git remote -v show origin
```

This should show you output that looks like the following listing.

Listing 13.6 `git remote -v show origin` (annotated)

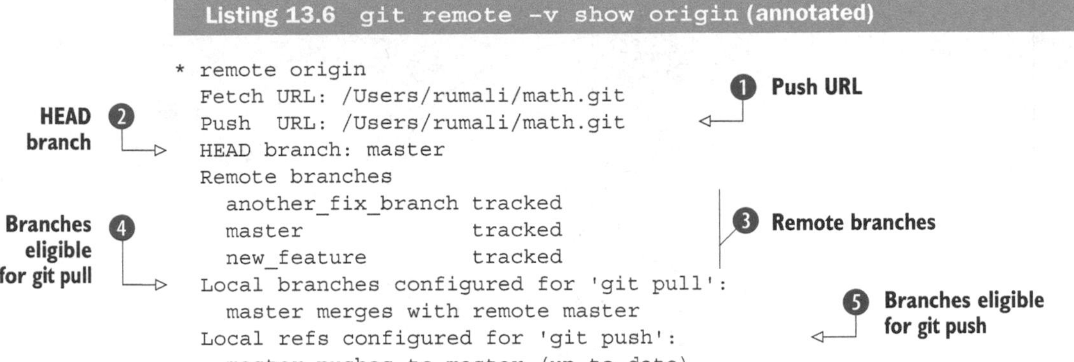

```
* remote origin
  Fetch URL: /Users/rumali/math.git
  Push  URL: /Users/rumali/math.git
  HEAD branch: master
  Remote branches
    another_fix_branch tracked
    master             tracked
    new_feature        tracked
  Local branches configured for 'git pull':
    master merges with remote master
  Local refs configured for 'git push':
    master pushes to master (up to date)
```


This output has a lot of detail, so let's go through it carefully. The push URL is ❶, the directory that you used in the `git clone` command that created this current repository.

If you try this in the math.github directory, you'll see that the push URL is the GitHub URL (https://github.com/rickumali/math.git).

The current HEAD branch ❷ on the remote repository is master. If you change the branch in the math.git directory, ❷ will change accordingly.

All the available branches on the remote ❸ are listed. Your remote has three branches: master, another_fix_branch, and new_feature. When you did the `git clone`, Git created remote-tracking branches for all the branches that were available at math.git.

The local branches, and what they merge to ❹, are discussed in the next chapter, but for now just recognize that this section has to do with `git pull`. Depending on what you've checked out, you may see more than one branch in this section.

The local branches, and what branch they'll push to on the remote ❺, will look like the list from ❹ because a `git clone` or a `git checkout` will associate the current branch ❷ with its corresponding remote branch.

The up-to-date label (in parentheses in ❺) shows that all your local changes are in sync with the origin. This should make sense, because you've made one change to math.carol, and you've pushed it up to math.git.

Let's compare this output with the math.bob directory, which also has its origin set to math.git. Remember that you haven't done anything in math.bob to retrieve the latest changes.

TRY IT NOW In the math.bob directory, type the following:

```
git remote -v show origin
```

You're typing the same command as the previous TRY IT NOW, but you're in a new directory, the math.bob repository. This should give output like the following listing.

Listing 13.7 `git remote -v show origin` (not in sync)

```
* remote origin
  Fetch URL: /Users/rumali/math.git
  Push  URL: /Users/rumali/math.git
  HEAD branch: master
  Remote branches:
    another_fix_branch tracked
    master             tracked
    new_feature        tracked
  Local branches configured for 'git pull':
    master      merges with remote master
Local refs configured for 'git push':
    master      pushes to master        (local out of date)
```

❶ Branches eligible for git push

❶ shows that the local master branch in math.bob is out of sync. Why? Because you haven't done any operation in math.bob to bring in the changes that are now present in the math.git repository thanks to the `git push` that you did on math.carol.

Figure 13.5 illustrates why math.bob's local master branch is out-of-date.

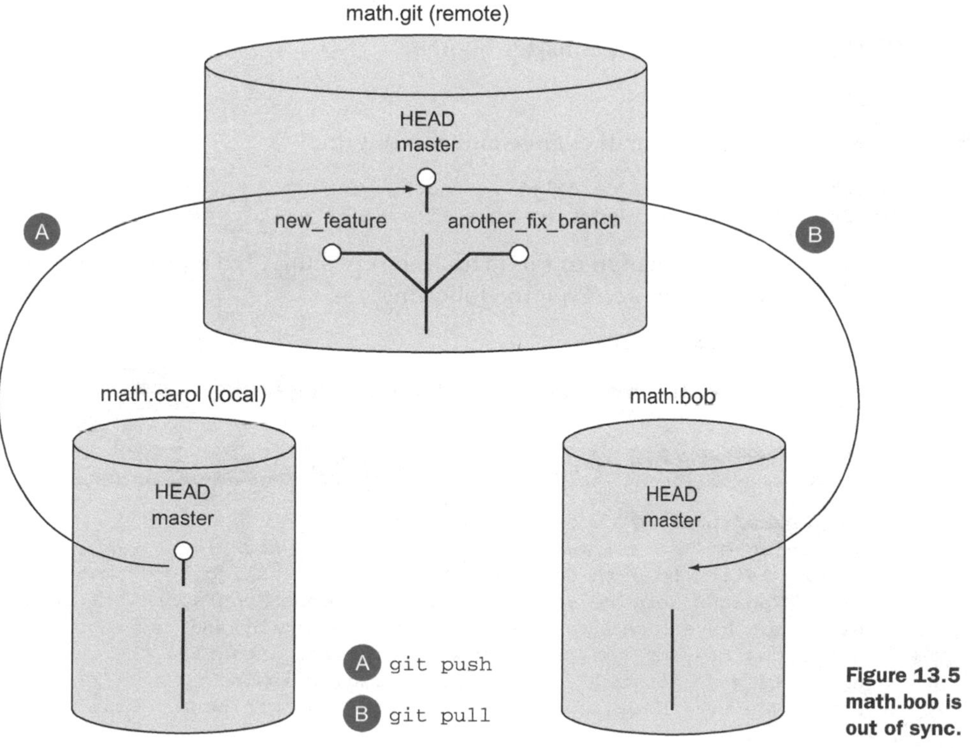

Figure 13.5 math.bob is out of sync.

In figure 13.5, the operation denoted by the arrow labeled A (the `git push` from math.carol to math.git) shows that a new commit is being added to math.git. But the `git pull` denoted by the arrow labeled B hasn't yet been performed to bring math.bob up-to-date with its remote. This is why the `git remote` command in math.bob shows that things are out of sync. You'll learn how to bring math.bob up-to-date in chapter 14.

Keeping your code base in sync (pushing your changes back up to the remote, or pulling down any new changes from the remote) is a circular effort. In this chapter, you focus on the push side, but keep in mind that's only half of the sync process. If you make contributions to a project, you'll be expected to push your changes. If the repository changes frequently, you'll have to pull in the new changes to your local working directory. But if you only track a project (for example, if you're the end user of a Git repository), `git pull` may be the only thing you need to do.

13.2 Understanding push conflicts

In your current situation, the changes to math.carol have been pushed to the math.git repo. But math.bob hasn't yet brought down these changes from math.git. math.bob is out of sync. Let's see what happens if you try to make a change to math.bob.

TRY IT NOW Let's commit a change in the math.bob directory. Type the following:

```
cd $HOME/math.bob
```

Make sure you're in the master branch:

```
git checkout master
```

Now let's make a small change and commit it:

```
echo "Small change to file" >> another_rename
git commit -a -m "Updating this file."
```

This commits a change to the math.bob repository. Let's publish this change to math.bob's remote. Type the following:

```
git push origin master
```

This produces the error message in the following listing.

Listing 13.8 `git push` error

```
To /Users/rumali/gitbook/math.git
 ! [rejected]        master -> master (fetch first)
error: failed to push some refs to '/Users/rumali/gitbook/math.git'
hint: Updates were rejected because the remote contains work that you do
hint: not have locally. This is usually caused by another repository pushing
hint: to the same ref. You may want to first integrate the remote changes
hint: (e.g., 'git pull ...') before pushing again.
hint: See the 'Note about fast-forwards' in 'git push --help' for details.
```

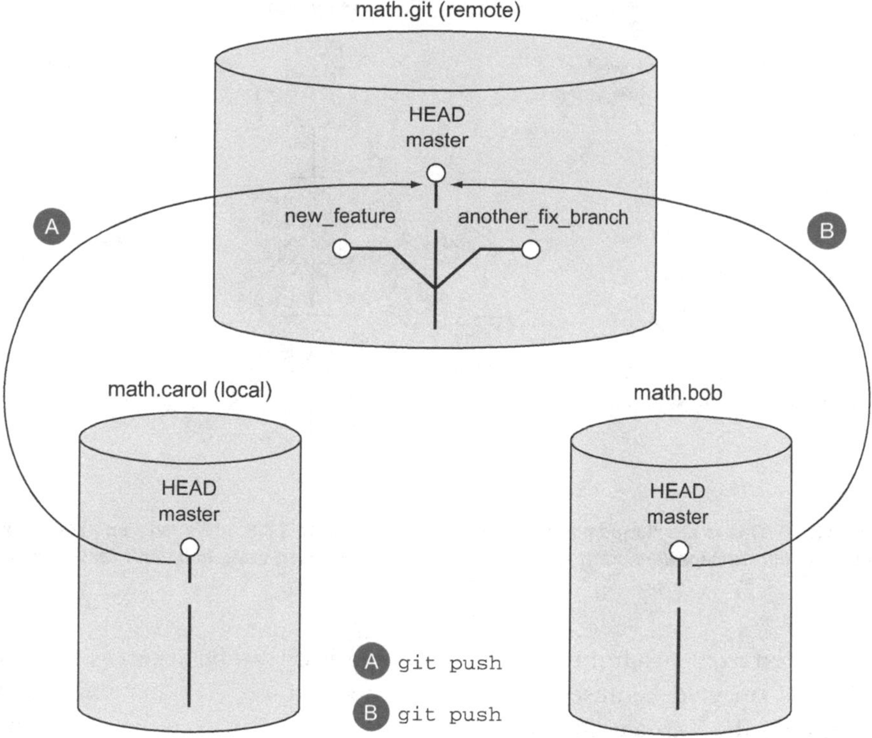

Figure 13.6 math.bob and math.carol are both pushing files to math.git.

Give this message a careful reading, especially these hints: `Updates were rejected because the remote contains work that you do not have locally. This is usually caused by another repository pushing to the same ref.` This describes the situation in figure 13.6.

You've pushed changes from math.carol to math.git first (the arrow labeled A). When math.bob does the same operation (the arrow labeled B), Git complains because math.bob isn't up-to-date with math.git. You can push changes to a remote only when those changes are a descendant of what is on the remote; `git push` is the equivalent of merging with the math.git repository.

If you need a refresher on merging, review chapter 10. Git allows you to push your branch only if the commit that you're pushing is a descendant of the remote's branch (a fast-forward merge).

The `git push` from the math.carol repository has this situation, as you can see in figure 13.7.

This push (the output is in listing 13.3) is allowed because the new commit C in math.carol is a descendant of commit B. Commit B is in the math.git repo.

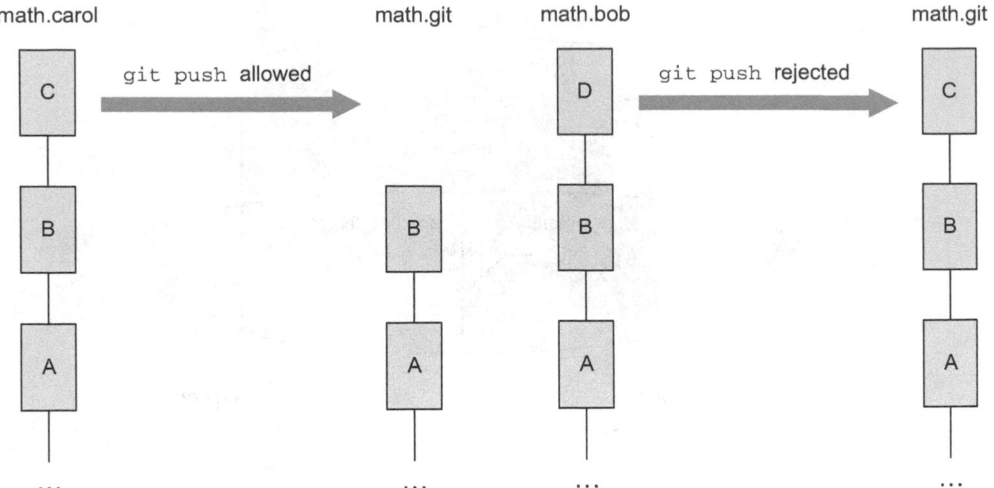

Figure 13.7 This is an allowed push, because math.carol can fast-forward math.git.

Figure 13.8 This isn't an allowed push, because math.bob can't fast-forward math.git.

The rejected `git push` from the math.bob repository can be drawn as in figure 13.8.

This is rejected because math.bob made a commit (D) that has a parent (B). math.git has the B commit, but its immediate child is C; math.bob needs to bring over math.git first. To fix this, you must use the `git pull` command, which you'll read about in detail in chapter 14.

In our lab, you'll be introduced to the `--force` switch of the `git push` command. This heavy hammer could also get you past this error, but it does blindly change the remote repository. When changes are pushed to the main repository, it's considered published, and using `--force` to overwrite those published changes will affect users when they pull down changes from the remote repository. After you push your changes, you shouldn't change them. And in the context of this error, if you encounter a conflict when pushing your changes to a repo, you must incorporate the changes from the repo first (you'll do that in the next chapter).

13.3 Pushing branches

For the past few sections, you've been running `git push` with two arguments: the source remote (typically origin) and the branch to push (typically master). In this section, you'll learn that you can omit these two arguments because `git checkout` automatically associates local branches with their corresponding remote-tracking branches, if their names match.

For example, your math.git repository has three branches: master, another_fix_branch, and new_feature. When you clone this repository into math.carol, `git clone` creates three remote-tracking branches: remotes/origin/master, remotes/origin/another_fix_branch, and remotes/origin/new_feature.

All the remote-tracking branches that are in your clone keep track of the upstream version. The word *upstream* means the original source. When you check out a local branch with the same name as one of the remote-tracking branches, Git assumes that this local branch will track the remote branch.

Let's check out a local branch that's one of our remote-tracking branches, so you can see this behavior for yourself.

> **TRY IT NOW** In the math.carol repository, check out a local branch of one of the remote-tracking branches. Type the following:
>
> ```
> cd $HOME/math.carol
> git checkout another_fix_branch
> ```
>
> This produces output that looks like the following listing.

Listing 13.9 `git checkout` output

```
Branch another_fix_branch set up to track remote branch another_fix_branch
➥ from origin.
Switched to a new branch 'another_fix_branch'
```

You analyzed this form of `git checkout` in chapter 11. Because another_fix_branch is the same name as an existing remote-tracking branch, this command is `git checkout -b another_fix_branch remotes/origin/another_fix_branch`. Now that you know about `git push`, you can better appreciate the message in listing 13.9. Because `git clone` sets up remote-tracking branches, and `git checkout` automatically associates these branches with local branches having the same name, you can call `git push` on this branch without any arguments.

> **TRY IT NOW** You'll now make a change to one of the files in another_fix_branch. Type the following:
>
> ```
> echo "Small change" >> another_rename
> git commit -a -m "Small change"
> ```
>
> This commits the change to your local branch. Observe the output from typing this:
>
> ```
> git status
> ```
>
> This command states that you can call `git push` to publish your local commits. Now type this:
>
> ```
> git push
> ```
>
> Observe that you didn't need to type the remote or local branch! (You may receive a warning message about push.default, as shown in the upcoming listing 13.14. The push will continue as you expect, but read section 13.6 to better understand the warning you might receive.)

This shortcut is set up by `git checkout`. When you call `git checkout` on a branch that has a corresponding remote-tracking branch, the ability to push to it using `git push` without any arguments is enabled.

But what if you create a new local branch? How do you go about publishing that new branch upstream? And how do you push to it automatically?

TRY IT NOW In the math.carol directory, create a new local branch from master:

```
git checkout -b new_branch master
```

This shortcut command was introduced in section 9.3.1. You made a new branch (specified by the `-b` switch and its argument `new_branch`), and based it off master. In figure 13.9, you see what happens to your two repositories when you type this command.

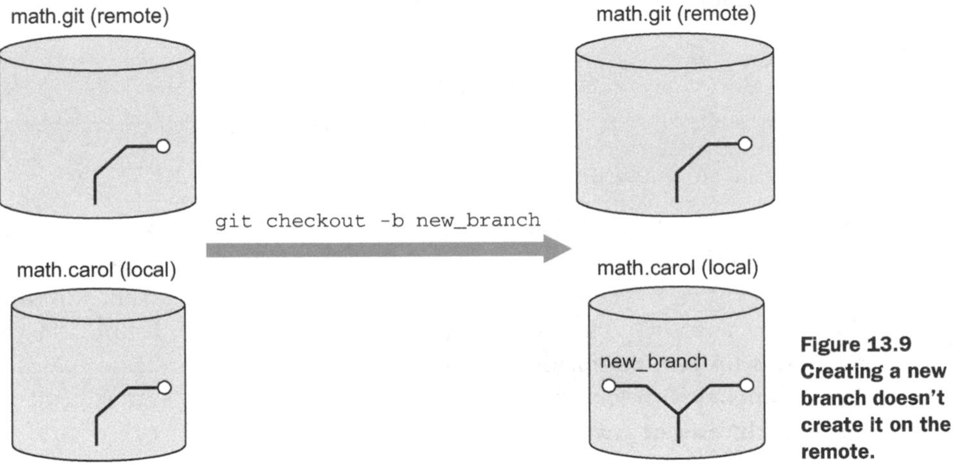

Figure 13.9 Creating a new branch doesn't create it on the remote.

The left side of figure 13.9 is your starting state. After the `git checkout` command, the right side shows that only the local repository is changed by the `git checkout` command. Now try to push this new branch upstream without any arguments. Type the following:

```
git push
```

You'll see a message like the following listing.

Listing 13.10 `git push` without an upstream configuration

```
fatal: The current branch new_branch has no upstream branch.
To push the current branch and set the remote as upstream, use

    git push --set-upstream origin new_branch
```

You can type the command exactly as suggested:

```
git push --set-upstream origin new_branch
```

The last command will not only push the branch, but also create the remote-tracking branch in your local repository. The result of this command is shown in figure 13.10.

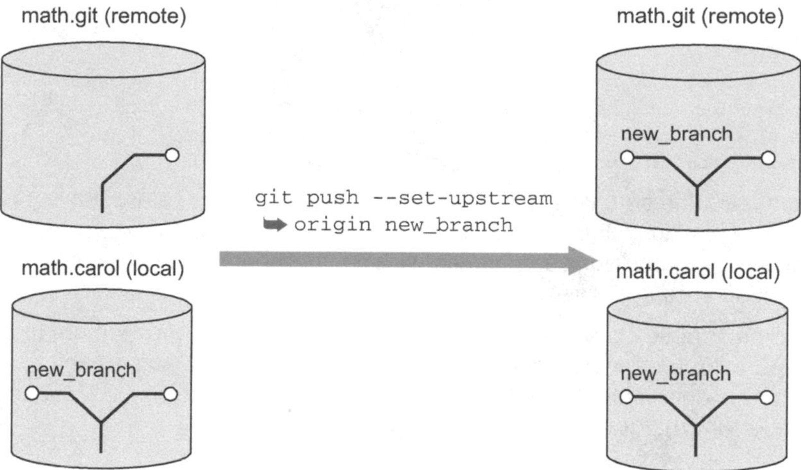

Figure 13.10 Creating the upstream branch via the `git push --set-upstream` command

The left side shows your two repositories before the `git push`. The right side shows that you've successfully pushed the new_branch to the remote repository.

After this, you'll be able to push to new_branch on the remote without any arguments. The output of the last command is shown in the following listing.

Listing 13.11 `git push` setting the upstream branch

```
Total 0 (delta 0), reused 0 (delta 0)
To c:/Users/Rick/Documents/gitbook/math.git
 * [new branch]      new_branch -> new_branch
Branch new_branch set up to track remote branch new_branch from origin.
```

You need to run `git push` with the `--set-upstream` switch only once. When you use this switch, Git writes this information into a configuration file, so you don't have to repeat it.

Above and Beyond

The `--set-upstream` switch sets the upstream branch and writes this information into a Git configuration file. The `git config` command accesses the Git configuration files, and you can use it to find out the values of any configuration setting.

In the math.carol directory, type the following:

```
git config --get-regexp branch
```

This command retrieves configuration settings that match the string branch. The meaning of `--get-regexp` is *get regular expression*, which means `git config` will match any configuration with the word *branch* in it. This should list settings like this:

```
branch.master.remote origin
branch.master.merge refs/heads/master
```

> ***(continued)***
>
> ```
> branch.another_fix_branch.remote origin
> branch.another_fix_branch.merge refs/heads/another_fix_branch
> branch.new_branch.remote origin
> branch.new_branch.merge refs/heads/new_branch
> ```
>
> Each configuration setting is followed by a space and then its value. Notice how each setting is of the form `branch.<name>.remote` and `branch.<name>.merge`. These allow separate values for each branch, which you specify in the `<name>` part of the configuration setting.
>
> The `branch.<name>.remote` configuration setting specifies which remote to push to. The `branch.<name>.merge` configuration setting specifies which branch to update. This setting is also controlled by the `push.default` setting. The `push.default` setting is so important that it's covered in section 13.6.

13.4 Deleting branches on the remote

In the preceding section, you created a new branch and pushed it back to your remote. This creates a new branch on the remote. What happens when you no longer need this branch?

In your local repository, you can delete a branch with the `git branch -d` command, as shown in figure 13.11. Recall from section 9.3.2 that you can delete a branch by using that command.

In figure 13.11, the left side are your local and remote repositories before you type the command. On the right side of figure 13.11, after typing the `git branch -d` command, the new_branch is removed from your local repository, but it's still on the remote. The command to delete a remote branch is `git push` with a colon (`:`) before the branch name. Let's try it.

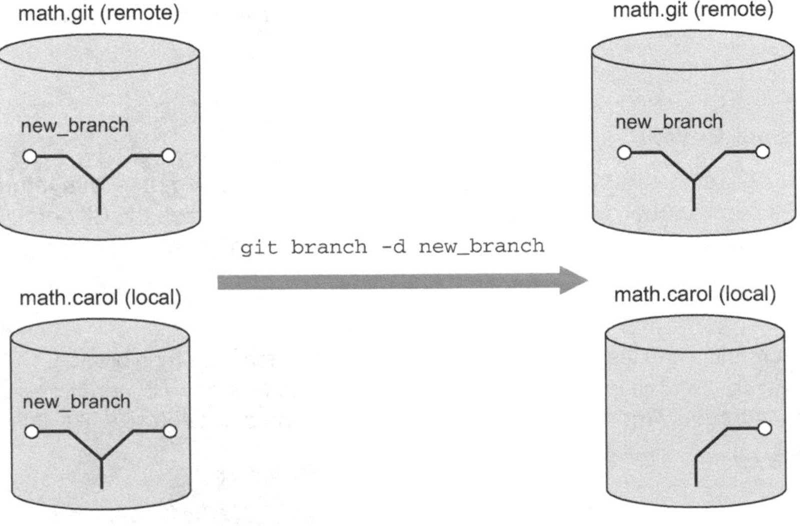

Figure 13.11 Deleting a local branch doesn't delete it on the remote.

TRY IT NOW In the math.carol repository, let's first delete the local branch. Type the following:

```
git checkout master
git branch
git branch -d new_branch
git branch
```

You first check out the master branch, which will enable you to delete the new_branch (you can't be in the branch that you're about to delete). Then you show the list of current branches (`git branch`), delete one of them (new_branch), and list the current branches again. This will help you confirm that new_branch is deleted locally. But you'll still be able to see it on the remote. Type the following:

```
git ls-remote origin
```

You'll see new_branch in the output of this command. It'll be listed as refs/heads/new_branch. To get rid of it on the remote, type this:

```
git push origin :new_branch
```

You should see the message shown in the following listing.

Listing 13.12 Output after deleting a branch

```
To /Users/rumali/gitbook/repo.git
 - [deleted]          new_branch
```

This gives you the picture in figure 13.12. The left side shows what your two repositories look like before you do the `git push`. After the `git push` command, notice that you've finally deleted new_branch from the remote repository.

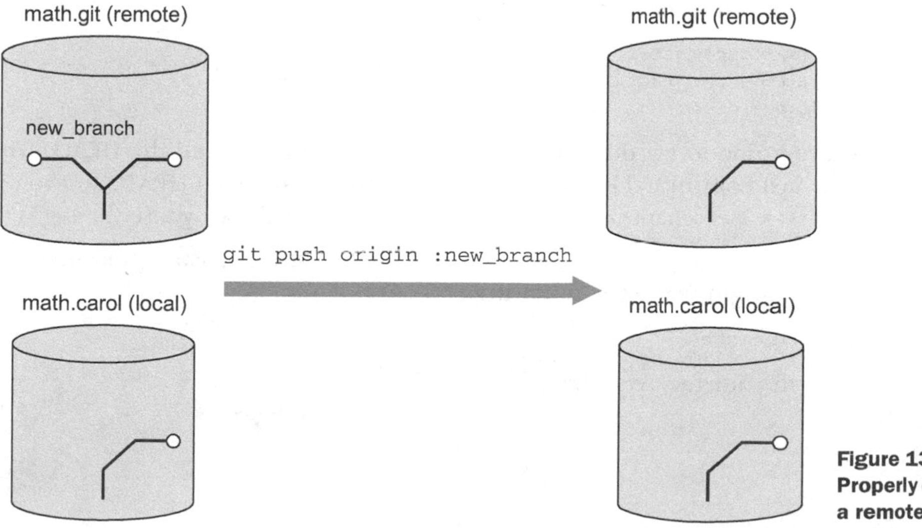

Figure 13.12 Properly deleting a remote branch

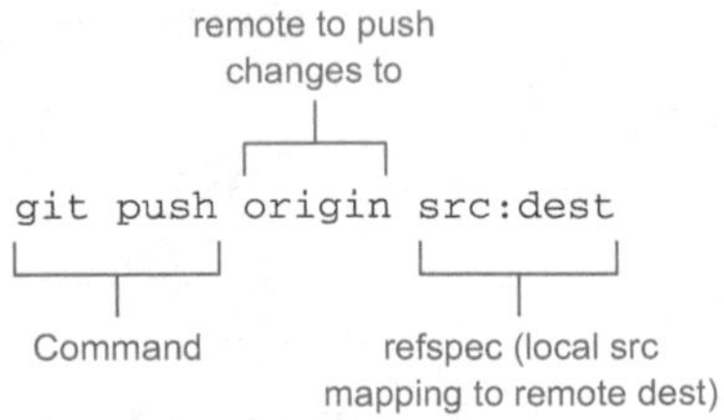

Figure 13.13 The more complete form of the `git push` command, showing the refspec

Using a colon causes the `git push` command to use the more complex meaning of the fourth argument. In figure 13.13, the fourth argument to `git push` is a *refspec*, a colon-separated string that describes the mapping between a source and destination branch. (Compare figure 13.13 with figure 13.3.)

Using `git push origin master` is equivalent to using `git push origin master:master`. Git pushes your local branch named master to the remote branch named master. When you omit `src` from the full form, leaving the string `:master`, you're telling Git to delete that remote branch.

Do be careful with this syntax and operation! After you've been granted permission to push to a remote, you can use this form of `git push` to delete any branch you want. Git doesn't have a safety net for this operation either.

13.5 Pushing and deleting tags

Tags behave just like branches, with regards to pushing to a remote or deleting from a remote. When you clone a repository, you receive all tags that were a part of the original repo. As you work on your local repo, you might make your own local tags. If you want to share your tags with your collaborators, you have to push them up to the remote separately.

TRY IT NOW In math.carol, let's look at our tags. Type the following:

```
git tag
```

You should see one tag: four_files_galore. Let's go ahead and create a tag in our repo. Type this:

```
git checkout master
git tag -a two_back -m "Two behind the HEAD" HEAD^^
git log --decorate --oneline
```

You're going to tag the commit that is two commits behind the HEAD (which is the last commit). The arcane syntax (`HEAD^^`) says from HEAD, go two commits back (see chapter 8). The `git log` command shows where the tag is.

Using `git ls-remote`, you can confirm that the tag isn't present on the remote origin. Type the following:

```
git ls-remote
```

To push your tag, type this:

```
git push origin two_back
```

You'll see output like the following listing.

Listing 13.13 Pushing tags

```
Counting objects: 1, done.
Writing objects: 100% (1/1), 166 bytes | 0 bytes/s, done.
Total 1 (delta 0), reused 0 (delta 0)
To /Users/rumali/gitbook/math.git
 * [new tag]         two_back -> two_back
```

Unlike the `git push` command with branches, you must explicitly state that you're pushing tags to the remote. There's no shortcut (though `git push` does have a `--tags` switch, but this sends over all the local tags that you've created, which might not always be what you want to do).

Deleting tags follows the same syntax as at the end of section 13.4.

TRY IT NOW In math.carol, confirm that you have the tag two_back in the remote:

```
git ls-remote
```

Delete this tag, and confirm that it's no longer on the remote:

```
git push origin :two_back
git ls-remote
```

Remember that this deletes only the tag from the remote. Your local tag still exists. To delete this, use `git tag` with the `-d` switch:

```
git tag -d two_back
```

13.6 Configuring simple pushes

Git 2.0, released in May 2014, changed a configuration that governs how `git push` works if you don't pass in any arguments. If you're on an older version of Git, and you use `git push` without arguments, you may see the lengthy warning in the following listing.

Listing 13.14 `git push` warning message

```
% git push
warning: push.default is unset; its implicit value is changing in
Git 2.0 from 'matching' to 'simple'. To squelch this message
and maintain the current behavior after the default changes, use:

  git config --global push.default matching

To squelch this message and adopt the new behavior now, use:

  git config --global push.default simple

See 'git help config' and search for 'push.default' for further information.
(the 'simple' mode was introduced in Git 1.7.11. Use the similar mode
'current' instead of 'simple' if you sometimes use older versions of Git)
```

```
Counting objects: 5, done.
Compressing objects: 100% (4/4), done.
Writing objects: 100% (4/4), 1.82 KiB | 0 bytes/s, done.
Total 4 (delta 0), reused 0 (delta 0)
To git@bitbucket:rickumali/gitbook.git
   d64950e..9751bdd  master -> master
```

To remove this warning, you must use the command specified in the warning: `git config --global push.default simple`. This command assigns the value `simple` to the configuration setting `push.default`. The `--global` switch indicates that this setting is global to any repository that you access.

You'll examine `git config` in great detail in chapter 20.

`git push` defines a `push.default` configuration setting that can take one of the values in table 13.1.

Table 13.1 Values for `push.default` configuration setting

push.default setting	Value
`nothing`	Don't push unless a source and a destination are specified.
`current`	Push current branch to update a branch with the same name.
`upstream`	Similar to `current`.
`simple`	Similar to `upstream`, but check that the branch name matches what is upstream.
`matching`	Push all branches that have corresponding branches on the remote.

Please read the `git config` documentation on `push.default` for more details. The value recommended for beginners is `simple`. To set this, type the following:

```
git config --global push.default simple
```

This setting prevents you from pushing a branch to the remote that you didn't intend to share. The developers of Git came to this newer setting in order to make things less confusing for beginners.

Git configuration is covered in more detail in chapter 20.

13.7 Lab

You use the `git push` command to publish commits from your repository. The following exercises help reinforce this idea, and introduce you to some interesting aspects of this command:

1. Read `git push` help, especially the Note About Fast-Forwards section.
2. Create another clone from math.git and, from that new clone, push up a few changes. Confirm from the other repositories that math.git is different.
3. The syntax to delete a branch or a tag uses a colon in front of a branch or tag name. This syntax is part of the refspec, a shorthand for specifying the source and destination branch or tag. Read about refspecs in the `git push` help.

4. You saw in section 13.2 that `git push` won't work if the local branch can't be merged with (isn't a fast-forward of) the remote branch. But there's a `--force` option. Experiment with it, but be advised that this should be used with caution. As you'll see in chapter 14, `git pull` is the recommended first step for getting past a failing `git push`.
5. Create multiple tags and then use `git push --tags` to push them all across to the remote.
6. Try to push from carol to bob. What does the error message mean?
7. Read `git config` help, especially the section Files and the configuration setting push.default.

13.8 Further exploration

You probably noticed that I didn't use much of Git GUI or gitk in this chapter. `git push` is somewhat supported by the Git GUI program. At the time of this writing, there isn't any push support in gitk.

In Git GUI, choose Remote > Push. From the dialog box that appears (shown in figure 13.14), push a branch back up to the math.git repository.

The Git GUI support for `git push` is adequate, but you need to have a good picture of the branches in your head. One thing Git GUI gives you is a listing of the available source branches.

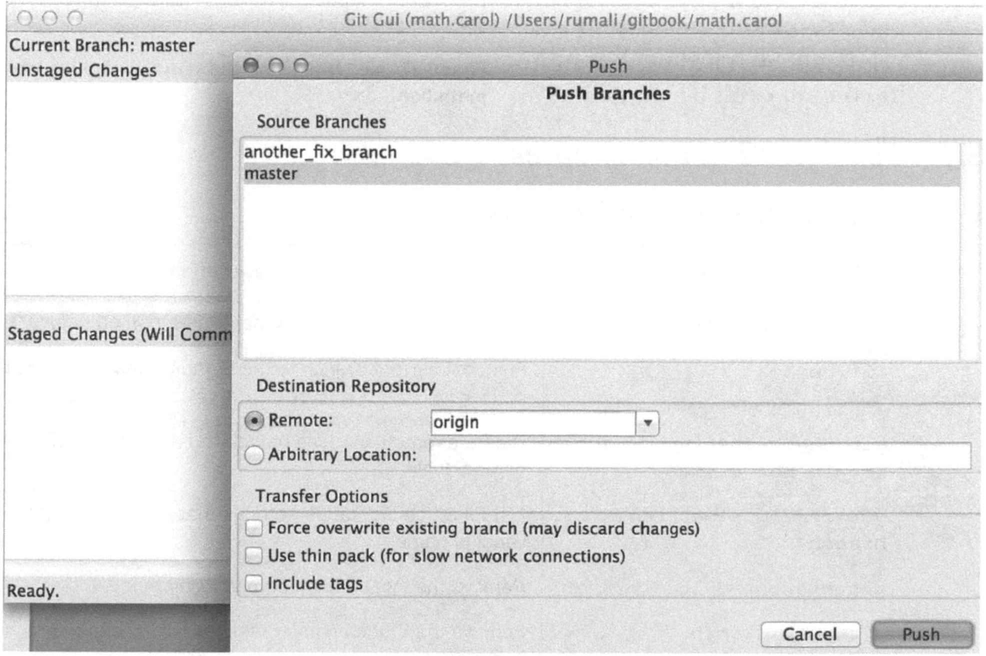

Figure 13.14 Git GUI support for `git push`

Gitk: edit view View 1 -- criteria for selecting revisions
View Name View 1 Remember this view
References (space separated list):
Branches & tags:
All refs All (local) branches All tags All remote-tracking branches

Figure 13.15 Enable the remote-tracking branches view.

Using gitk, and enabling the remote-tracking branches as figure 13.15 shows, allows you to see the remote-tracking branches.

When you enable this view and look at the math.bob repository, you'll see a view that looks similar to figure 13.16.

If you look at the first two lines (remotes/origin/master and master), you should be able to see that the master's last commit (`Updating this file`) isn't on the remote. If you pushed master to the remote, you'd get a rejection because its commit can't be merged.

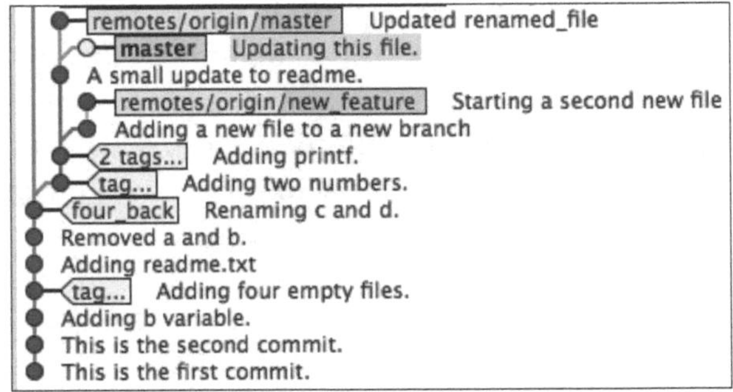

Figure 13.16 Observing the remote-tracking branches in math.bob

13.9 Commands in this chapter

Table 13.2 Commands used in this chapter

Command	Description
`git push origin master`	Push the master branch to the remote named *origin*.
`git push`	Push the current branch to the default remote-tracking branch set up by `git checkout` or `git push --set-upstream`.
`git push --set-upstream origin new_branch`	Create a remote-tracking branch to new_branch on the remote named origin.
`git config --get-regexp branch`	List all Git configuration settings that have the word *branch* in the variable name.
`git branch -d localbranch`	Remove the local branch named *localbranch*.
`git push origin :remotebranch`	Remove the branch named *remotebranch* from the remote named *origin*.

Table 13.2 Commands used in this chapter

Command	Description
`git tag -a TAG_NAME -m TAG_MESSAGE SHA1ID`	Create a tag to the SHA1ID with the name TAG_NAME and the message TAG_MESSAGE.
`git push origin TAGNAME`	Push the tag named TAGNAME to the remote named *origin*.
`git push --tags`	Push all tags to the default remote.
`git push origin :TAGNAME`	Delete the tag named TAGNAME on the remote named *origin*.
`git tag -d TAGNAME`	Remove the tag named TAGNAME from your local repository.
`git config --global push.default simple`	Set the `push.default` configuration variable to `simple` for all repositories that you have access to (globally).

14 Keeping in sync

Collaborating with a software project that uses Git requires that you keep your repository in sync with this remote repository. The focus of this chapter is the command that helps you keep in sync: `git pull`.

`git pull` is the opposite operation of `git push`. You'll learn how to use `git pull` to bring in changes that were made in the remote repository you're tracking. This is the step that keeps you in sync! You'll also learn that `git pull` consists of two commands: `git fetch` and `git merge`. You learned about `git merge` in chapter 10, but knowing how it's used in conjunction with `git fetch` will help you understand issues you might have with `git pull`.

14.1 Completing the cycle of collaboration

You've learned about clones (chapter 11) and how the `git remote` command (chapter 12) enables you to push to a remote (chapter 13). In figure 14.1, your clone (math.carol) can fetch (pull) from the remote (math.git), in addition to push.

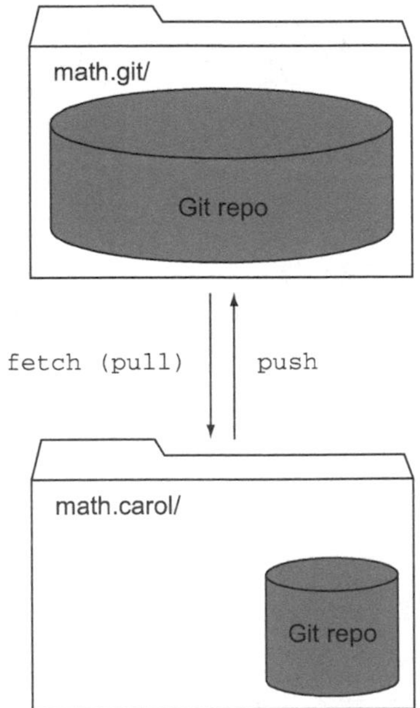

Figure 14.1 A cycle of collaboration between math.carol and math.git

When you make a clone of a repository, Git creates a remote that enables your clone to both push and pull from the repository that you just cloned. The remote sets up a cycle of collaboration. In section 14.2, you'll learn that `git pull` comprises two commands: `git fetch` and `git merge`. That's why figure 14.1 labels the arrow from math.git to math.carol as fetch (pull).

If you've been following along on your local computer, you've created a math.git directory, which you've designated as the official version, or official repository. It's the source repository that other collaborators can consider the official source code of your math project. Git doesn't have any special software for this designation: it's decided among collaborators on a project. An official repository of this sort is necessary in distributed version control systems because there's no centralized server.

As you make changes, you'll want to publish those commits to the official repository (using `git push`). But if you're working with many collaborators, you won't be the only person who'll be publishing changes. You saw in the preceding chapter that you can't push your changes to the remote unless your repository is in sync with the remote. This is where `git pull` comes in.

> **TRY IT NOW** You'll create a situation where two repositories are out of sync with the official repository. You'll then see how `git pull` helps bring these back in sync.
>
> The best way to set up your environment for this chapter is to run the script make_math_repo.sh to create another math directory. If you use your current math directory, some of the listings in this chapter won't match, which may be confusing.
>
> The script is in a zip file on the book's website. Delete your math directory (or rename it, if you want to preserve something you've done). Type the following after you have the script:
>
> ```
> cd $HOME
> bash make_math_repo.sh
> ```
>
> The make_math_repo.sh script leaves the math directory in a state described in section 9.4.2. To get out of that state, type this sequence:
>
> ```
> cd math
> git checkout -f master
> ```
>
> Now you can make the clones:
>
> ```
> cd ..
> git clone --bare math math.git
> git clone math.git math.carol
> git clone math.git math.bob
> ```
>
> In the $HOME directory, type this:
>
> ```
> git clone math.git math.bill
> ```

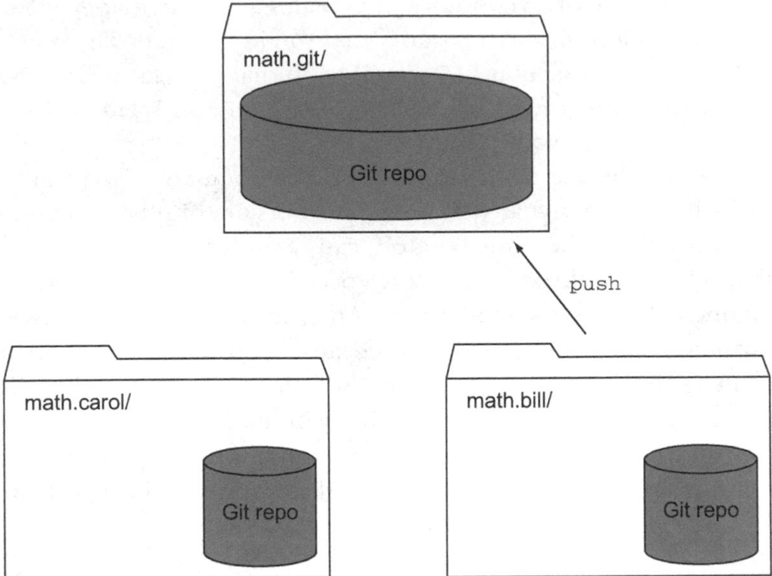

Figure 14.2 One half of the cycle (a push from Bill)

This creates a repository for Bill. Inside math.bill, you'll make a commit:

```
cd math.bill
echo "Small change" >> another_rename
git commit -a -m "Small change"
```

Now you'll push this change to math.git:

```
git push
```

You've created the situation in figure 14.2.

Figure 14.3 shows the commit, labeled small_change, in the math.bill repository. Once you do the `git push`, math.git is updated with the small_change commit. Note that math.carol hasn't been updated yet (the small_change commit isn't present in the math.carol repository).

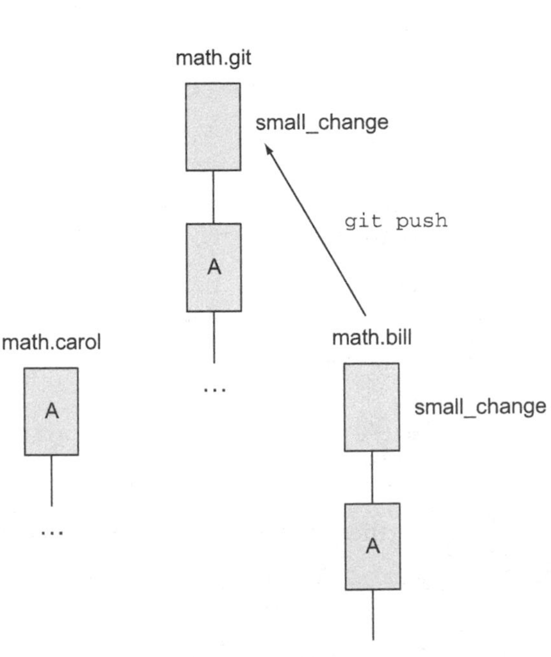

Figure 14.3 A look at the history after the math.bill push

Let's enter math.carol's point of view and confirm that math.git has changed. Type this:

```
cd $HOME/math.carol
git remote -v show origin
```

This should result in output like the following listing.

Listing 14.1 `git remote -v show origin` output

```
* remote origin
  Fetch URL: c:/Users/Rick/Documents/gitbook/math.git
  Push  URL: c:/Users/Rick/Documents/gitbook/math.git
  HEAD branch: master
  Remote branches:
    another_fix_branch tracked
    master             tracked
    new_feature        tracked
  Local branch configured for 'git pull':
    master merges with remote master
  Local ref configured for 'git push':
    master pushes to master (local out of date)
```

For the `Local refs configured for 'git push'` section, notice that master is marked as `local out of date`. Trying to do a `git push` will cause an error. Type the following:

```
git push
```

Confirm that you get the error shown in the following listing.

Listing 14.2 Attempting a `git push`, from math.carol

```
To /Users/rumali/gitbook/math.git
 ! [rejected]        master -> master (fetch first)
error: failed to push some refs to '/Users/rumali/gitbook/math.git'
hint: Updates were rejected because the remote contains work that you do
hint: not have locally. This is usually caused by another repository pushing
hint: to the same ref. You may want to first integrate the remote changes
hint: (e.g., 'git pull ...') before pushing again.
hint: See the 'Note about fast-forwards' in 'git push --help' for details.
```

math.carol hasn't made any changes to math.git, so a `git push` from math.carol shouldn't have done anything. But because Git checks that the repositories are in sync whenever you do a push, Git sees that math.carol isn't in sync with the changes on math.git, and therefore a push to math.git isn't allowed.

If the push succeeded (by using `git push --force`), math.git would lose the commit that you made from math.bill.

To sync the math.carol repository with math.git, type this:

```
git pull
```

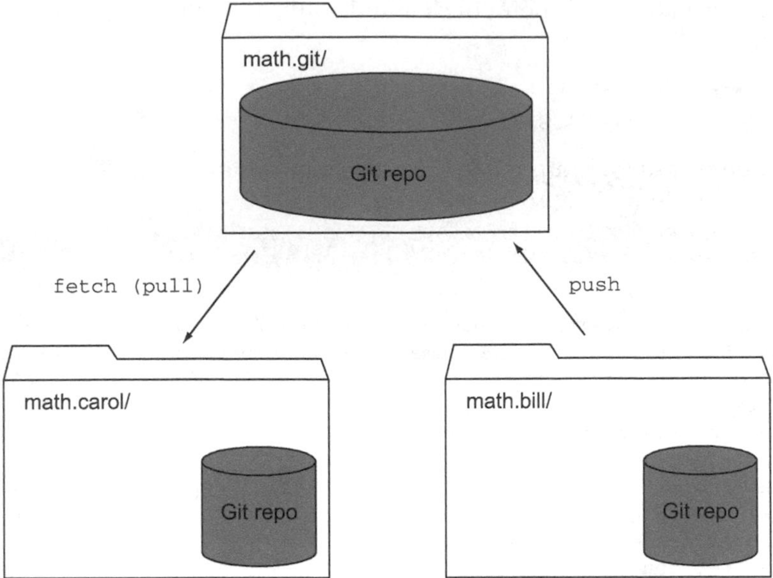

Figure 14.4 Completing the last half of the cycle

This produces the following output. (Note that the `git pull` command omits the .git suffix from the remote URL.)

Listing 14.3 A successful `git pull`

```
remote: Counting objects: 8, done.
remote: Compressing objects: 100% (2/2), done.
remote: Total 3 (delta 1), reused 0 (delta 0)
Unpacking objects: 100% (3/3), done.
From /Users/rumali/gitbook/math
   ab05274..e5c34ed  master     -> origin/master
Updating ab05274..e5c34ed
Fast-forward
 another_rename | 1 +
 1 file changed, 1 insertion(+)
```

Doing this last part allows you to complete the cycle, which you see in figure 14.4.

Figure 14.5 shows a picture of the completed cycle, using the commits.

14.2 Using git pull: a two-part operation

`git pull` is a two-part operation. In the first part, your local repository retrieves (fetches) the contents of the remote repository. In the second part, `git pull` attempts to make your repository look like the remote repository (which is now on your local machine) by performing a merge of the local branch and the remote-tracking branch.

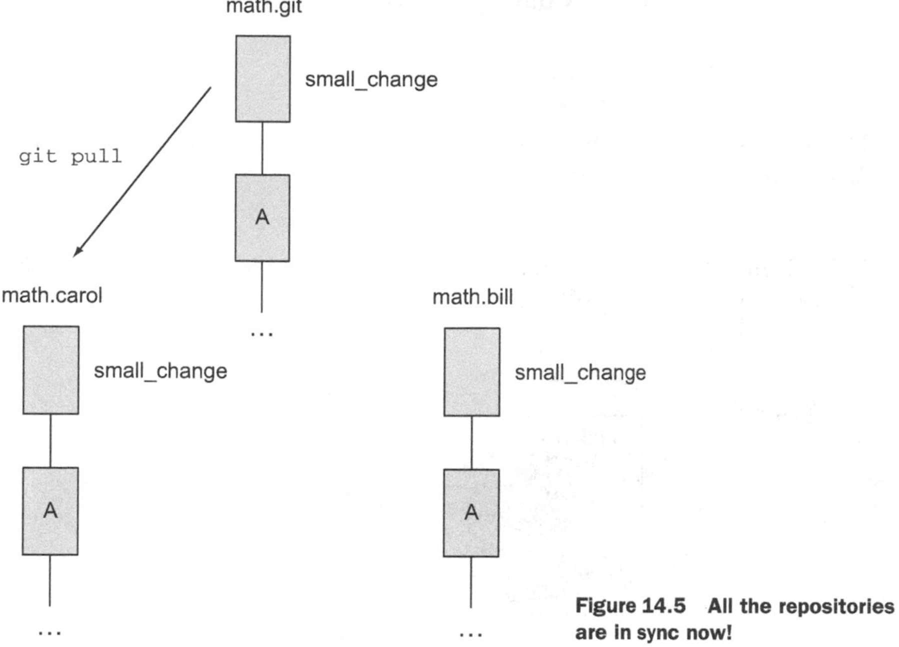

Figure 14.5 All the repositories are in sync now!

Performing the merge is what makes `git pull` much more complicated than `git push`. `git pull` is more than just the mirror operation of `git push`. `git push` pushes only changes that can be cleanly merged (fast-forwards). Anything more complicated than that requires intervention: you have to make the merge in a local repository and then push the result.

Reading the `git pull` help page, you'll see this sentence in the first paragraph: "In its default mode, `git pull` is shorthand for `git fetch` followed by `git merge FETCH_HEAD`." Let's examine these two steps carefully, because knowing these details will help you figure out why `git pull` sometimes produces conflicts.

14.2.1 Fetching files from a remote repository (git fetch)

`git fetch` is the first step of the `git pull` command. It's a completely separate command that retrieves files from one repository and incorporates those files into your repository. Specifically, `git fetch` retrieves references, which for you means branches or tags. When they arrive at your local repository, they are laid down on top of your repository, along with any files that they point to. These new references are tracked by using remote-tracking branches, which you've seen before in chapter 11.

> **TRY IT NOW** In this section, you'll create a change in one repository and then fetch those changes from another repository. You'll make your change in math.bill and then push it up to math.git. The math.git repository is the source repository (origin) that both math.bill and math.carol use.

In math.bill, as in the previous TRY IT NOW, you'll make one change:

```
cd $HOME/math.bill
echo "Tiny change" >> another_rename
git commit -a -m "Another tiny change"
```

This makes a commit that exists only in the math.bill repository. To publish this change, type this:

```
git push
```

This pushes the change to math.git. So far, you've done the first half (in figure 14.6) of the push/pull cycle.

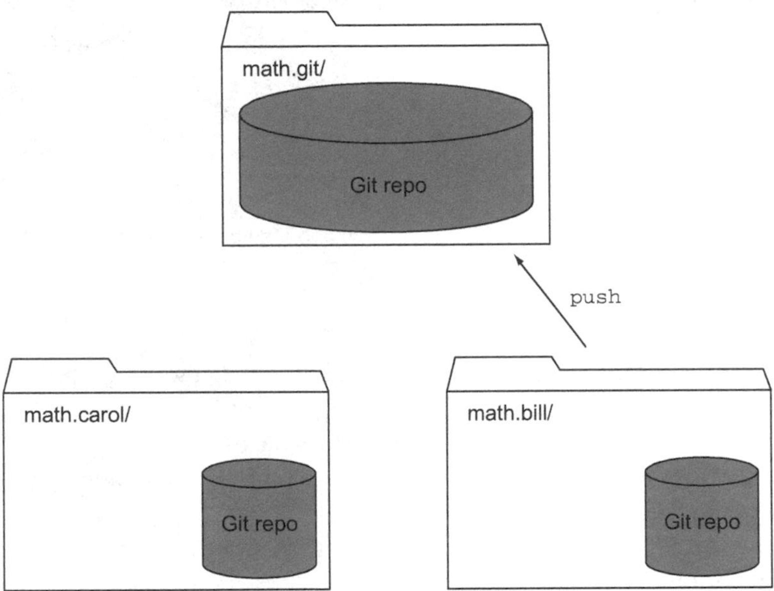

Figure 14.6 Doing a single push to math.git

From our previous TRY IT NOW, you know that to sync math.carol with math.git, you'll need to do a `git pull`. But this time, you'll do a `git fetch` so you can see exactly what `git pull` is doing. Before you do that, however, let's take a look at the current state of the math.carol repository.

TRY IT NOW In the math.carol repository, you're going to examine the log of commits. Type the following:

```
cd $HOME/math.carol
git log --decorate --oneline --all
```

Your output will look similar to the following listing (the difference being the SHA1 IDs).

Listing 14.4 `git log --decorate --oneline --all` **output**

```
195f2a1 (HEAD, origin/master, origin/HEAD, master) Small change
9517faf A small update to readme.
3682ea9 (origin/new_feature) Starting a second new file
eff9bb7 Adding a new file to a new branch
ebe9470 Adding printf.
133c8e4 Adding two numbers.
9a3e7f4 (origin/another_fix_branch) Renaming c and d.
a405b46 Removed a and b.
...
```

The `--decorate` switch adds key information to the `git log` output, and it's important to understand this output. From listing 14.4, let's focus on the following three lines:

```
195f2a1 (HEAD, origin/master, origin/HEAD, master) Small change
3682ea9 (origin/new_feature) Starting a second new file
9a3e7f4 (origin/another_fix_branch) Renaming c and d.
```

Each entry shows a SHA1 ID, a list of references in parentheses, and an excerpt of the commit log message. The references prepended with `origin/` are the remote-tracking branches.

The first line shows that the current commit in the math.carol repository is SHA1 ID `195f2a1`. Your SHA1 ID will be different. This line is what your working directory is pointing to, as indicated by the label `HEAD` (remember our analogy from chapter 8: HEAD is your playback machine, and it always points to your current branch).

The labels in parentheses (the decorations) from the first line further show that this is the master branch. Finally, this line shows that the remote repository (math.git) has at this time the same SHA1 ID for both origin/master and origin/HEAD remote-tracking branches. The key fact to note is that these remote-tracking branches aren't updated until you do a `git fetch` or `git pull`.

The second and third lines show the remote-tracking branches origin/new_feature and origin/another_fix_branch at their respective commits (`3682ea9`, `9a3e7f4`). In the listing, there's no local branch. Your listing may be different. The important point to note is that on the remote (math.git), new_feature and another_fix_branch are at these commits.

You can see this in the gitk program.

TRY IT NOW In the math.carol repository, type the following:

```
gitk
```

Figure 14.7 gitk configuration to look at all branches and remote-tracking branches

In the view configuration screen (figure 14.7), make sure to show all local and remote-tracking branches. If you omit the All Remote-Tracking branches toggle, you won't see the labels indicating the remote-tracking branches.

With this configuration, gitk should look like figure 14.8.

If your gitk view is different, check whether your view configuration window (figure 14.7) has the Simple History toggle button clicked on. You looked at this in section 11.1.2, and it greatly reduces the display.

Make sure to exit gitk.

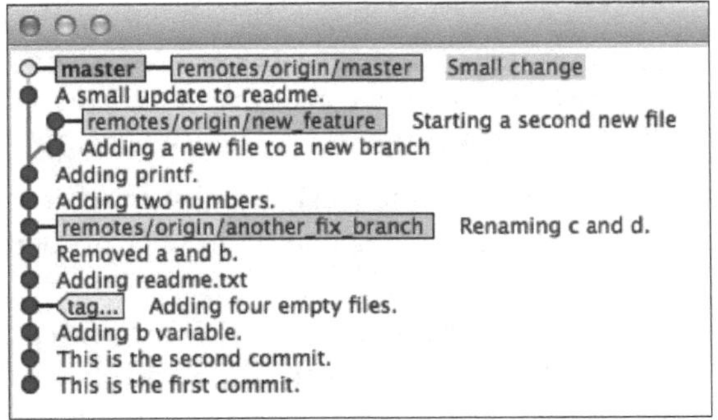

Figure 14.8 gitk showing the history, with labels for the remote-tracking branches overlaid on top

Now that you've examined the log of math.carol, you're going to perform a `git fetch`. This will update the math.carol repository, overlaying the new objects on top of your old one.

TRY IT NOW Type the following:

```
cd $HOME/math.carol
git fetch
```

This should give you the following output.

Listing 14.5 `git fetch` output

```
remote: Counting objects: 9, done.
remote: Compressing objects: 100% (2/2), done.
remote: Total 3 (delta 1), reused 0 (delta 0)
Unpacking objects: 100% (3/3), done.
From /Users/rumali/gitbook/math
   195f2a1..7746e35  master     -> origin/master
```

Now you'll repeat the steps from the previous TRY IT NOW to confirm that the remote branches have been updated. Type this:

```
git log --decorate --all --oneline
```

This should result in the following listing.

Listing 14.6 `git log` output after a `git fetch`

```
7746e35 (origin/master, origin/HEAD) Another tiny change
195f2a1 (HEAD, master) Small change
9517faf A small update to readme.
3682ea9 (origin/new_feature) Starting a second new file
...
```

You should be able to confirm that this listing is different from listing 14.4. Let's look at the first two lines:

```
7746e35 (origin/master, origin/HEAD) Another tiny change
195f2a1 (HEAD, master) Small change
```

From here, you can see that your current HEAD (master) is still at `195f2a1` on the second line, but the remote-tracking branch for master (origin/master) is `7746e35` on the first line. Let's take a look at this in gitk. In the math.carol directory, type the following:

```
gitk
```

You'll see a view like figure 14.9.

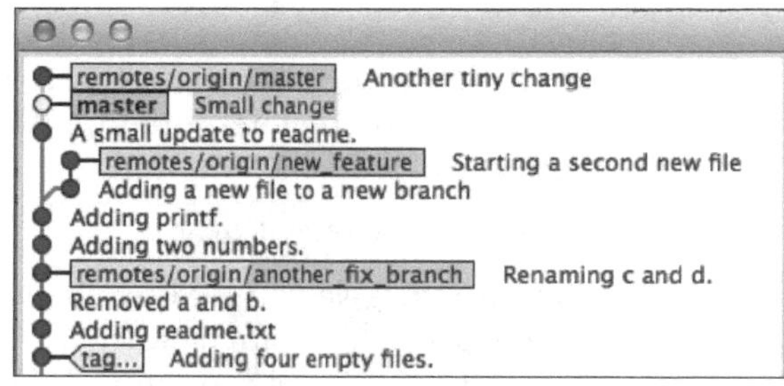

Figure 14.9 Notice that the remote origin/ master is ahead of your local master branch.

Hopefully it's clear by now that `git fetch` has brought in a new object (a new commit to master) and laid it right on top of your local repository. Another way to picture this is to use the branch diagrams from chapters 9 and 10, as shown in figure 14.10.

To make your local repository match the remote, you need to do a merge.

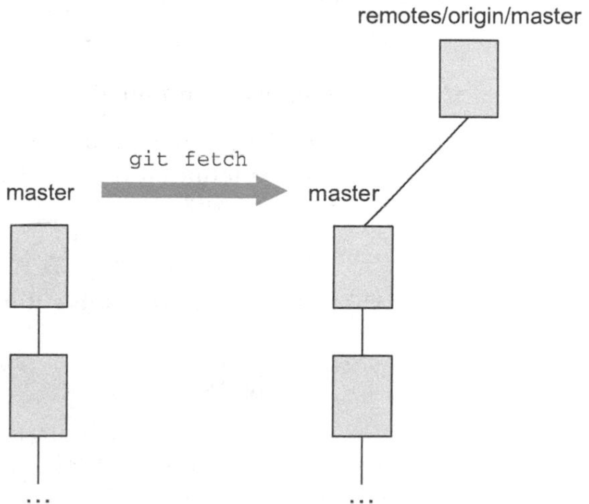

Figure 14.10 After performing the fetch, a new commit exists in your local repo.

14.2.2 Merging two branches (git merge)

The second step of `git pull` is to run `git merge FETCH_HEAD`. What is FETCH_HEAD? It's a reference to the remote branch that you just fetched in the previous section. Look at figure 14.11.

Compare figure 14.11 with figure 14.10. Every time you run `git fetch`, FETCH_HEAD will contain the SHA1 ID of the remote's HEAD, and `git merge` can use this to merge the change into your branch. Just as HEAD points to the current branch, FETCH_HEAD points to the most recent remote-tracking branch that was fetched.

Notice too that HEAD and FETCH_HEAD are capitalized. Git is case-sensitive regarding these particular names.

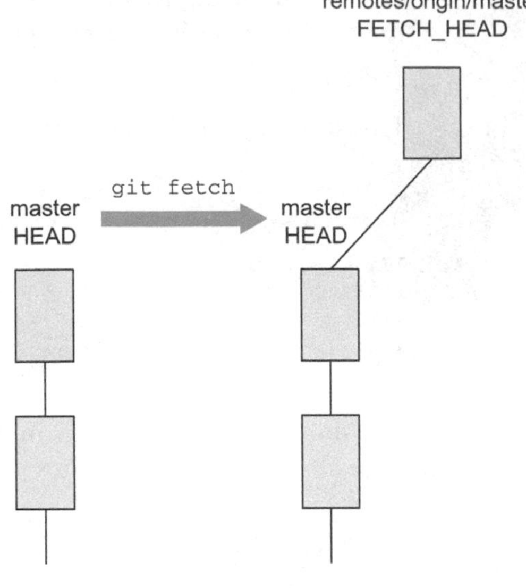

Figure 14.11 HEAD and FETCH_HEAD references labeled (compare with figure 14.10)

TRY IT NOW Let's examine FETCH_HEAD more closely and confirm that it's the same as remotes/origin/master (the remote-tracking branch). In math.carol, type the following:

```
git rev-parse FETCH_HEAD
```

This should give you a SHA1 ID that points to the latest commit from the remote's master branch. Keep in mind that this remote master branch is already in your local repository. You can access it by its special name, origin/master. Type the following:

```
git rev-parse origin/master
```

This last command should give you the same SHA1 ID as FETCH_HEAD.

Finally, because you have two commit IDs, you can ask Git to indicate what is different between these two branches, as you learned in chapter 10. Type the following:

```
git diff HEAD..FETCH_HEAD
```

This should give you output like the following listing.

Listing 14.7 Output from `git diff`

```
diff --git a/another_rename b/another_rename
index 86d347f..fb7f0ae 100644
--- a/another_rename
```

```
+++ b/another_rename
@@ -1 +1,2 @@
 Small change
+Tiny change
```

Finally, and most important, `FETCH_HEAD` is used as the argument to `git merge`.

TRY IT NOW Let's go ahead and perform the merge. Type the following:

```
git merge FETCH_HEAD
```

You should get the following output.

Listing 14.8 Output from `git merge FETCH_HEAD`

```
$ git merge FETCH_HEAD
Updating 195f2a1..7746e35
Fast-forward
 another_rename | 1 +
 1 file changed, 1 insertion(+)
```

Hopefully, that output was somewhat expected. math.carol hadn't made any changes, so you can just fast-forward this repo to the latest commit that math.bill performed (and pushed to math.git).

In figure 14.12, I've taken the branch that FETCH_HEAD points to (namely, remotes/origin/master), and I've merged it into the local master branch. Knowing that `git pull` is doing both a `git fetch` and `git merge` will help you understand the situations you'll run into with more-complicated merges and `git pull`.

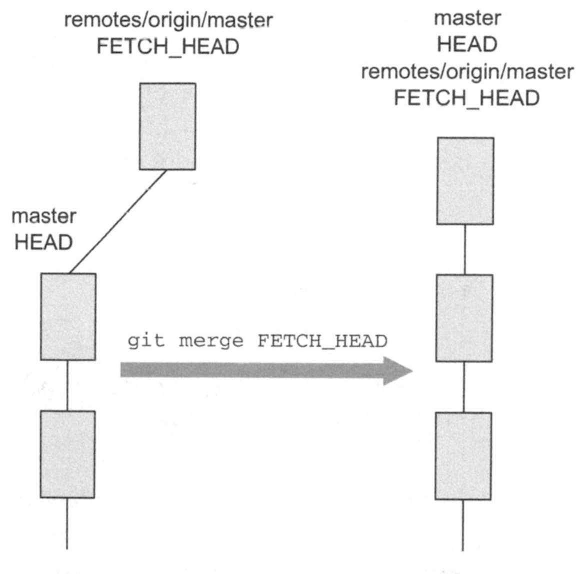

Figure 14.12 A `git merge` that results in a fast-forward

14.3 Merging a pull

The `git pull` cycle described in the previous section is the easiest kind of pull to perform. Why? Because when you made your change in math.bill and pushed it up to math.git, math.carol had no changes. When `git pull` is done in math.carol, the change from math.bill could be incorporated as a fast-forward merge.

Just like the merges you learned about in chapter 10, a `git pull`'s merge step might resolve cleanly or might result in a conflict.

14.3.1 Clean merge

A *clean merge* is one that Git can automatically resolve. For example, if the two repositories make the same change to the same file in the same line, Git's merge can resolve the difference by itself. But because `git pull` does an automatic `git merge`, you could be surprised at Git's behavior.

TRY IT NOW Make a commit in the math.carol directory:

```
cd $HOME/math.carol
echo "Small change 2" >> another_rename
git commit -a -m "Small change 2 from carol"
```

For the math.bill directory, do the same thing:

```
cd $HOME/math.bill
echo "Small change 2" >> another_rename
git commit -a -m "Small change 2 from bill"
```

Now let's push this change to math.git. Type this in the math.bill directory:

```
git push
```

It's important to recognize that both repositories have different commits (different SHA1 IDs) at the tips of their branches. Git will need to merge these two branches. You can see this for yourself by running gitk in both the math.carol and the math.bill directories. Their master branches will have the same change, but their SHA1 IDs will be different, as shown in figure 14.13.

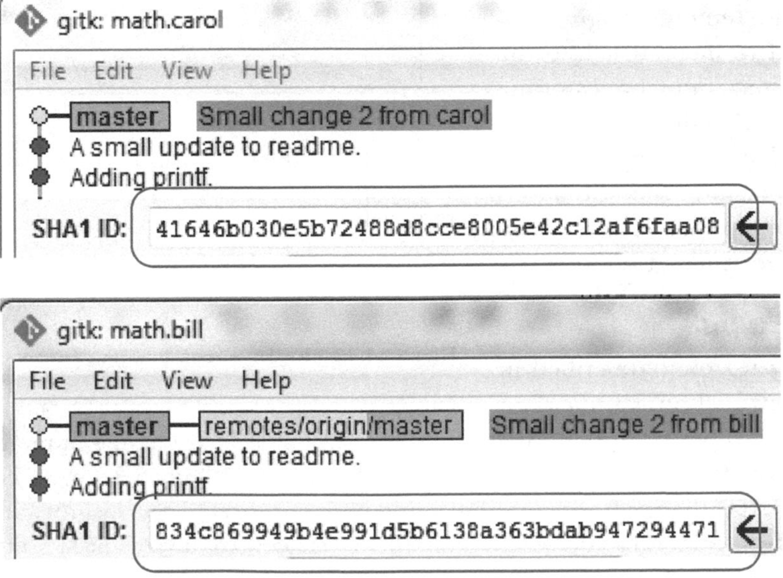

Figure 14.13 The same change still produces different commit IDs.

In figure 14.13, math.carol and math.bill have different SHA1 IDs for their last commits (indicated by the dashed arrow). This requires Git to merge these two branches. But because both files have the same change, Git can merge the files automatically. There's no conflict between the files.

> **TRY IT NOW** In math.carol, type the following:
>
> ```
> git pull
> ```
>
> You'll see one of two things: an output message as in listing 14.9, or a text file open in the Git's default editor, as shown in figure 14.14. (You'll learn how to change the default editor in chapter 20.) The first item is known as *a clean merge with an automatic commit message*. The second item is a *clean merge with a nonautomatic commit message*.
>
> The following listing is your clean merge with an automatic commit message.

Listing 14.9 `git pull` of a clean merge

```
remote: Counting objects: 9, done.
remote: Compressing objects: 100% (2/2), done.
remote: Total 3 (delta 1), reused 0 (delta 0)
Unpacking objects: 100% (3/3), done.
From c:/Users/Rick/Documents/gitbook/math
   db106c7..fe04975  master     -> origin/master
Merge made by the 'recursive' strategy.
```

Figure 14.14 is your clean merge with a nonautomatic commit.

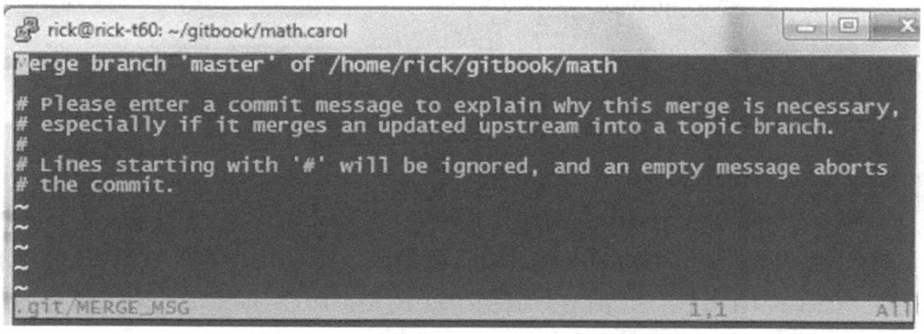

Figure 14.14 This clean `git pull` causes you to jump inside Git's default editor.

This second case can be confusing. What exactly happened? Let's break this second case down in the next section, 14.3.2, and the simpler, first case, in section 14.3.3.

14.3.2 Clean merge with nonautomatic commit

When you're in Git's default editor, it turns out that `git pull` has already performed the `git merge` cleanly, and is now asking you to write a commit message. This is why the editor window has appeared—for you to write a commit message. In figure 14.14,

Git does provide a default message. `git pull` will run `git commit` for you after you exit the editor.

To see this for yourself, let's cancel the merge by creating an empty commit message.

> **TRY IT NOW** The steps in this section can be followed only if your `git pull` resulted in a clean merge with a nonautomatic commit message. This means that your `git pull` put you inside the editor window. If this isn't the case, your `git pull` was a clean merge with an automatic commit message, and you can skip to section 14.3.3.
>
> If you're inside Git's default editor (vi), type the following exactly:
>
> ```
> :%d
> :wq
> ```
>
> The first line deletes all the lines from the current editor message. The second line saves the message. You've produced an empty commit message. Because the message is empty, the `git commit` that `git pull` tried to run will fail. Your screen should now look like the following listing.

Listing 14.10 A clean merge, after cancelling the automatic commit message

```
remote: Counting objects: 8, done.
remote: Compressing objects: 100% (2/2), done.
remote: Total 3 (delta 1), reused 0 (delta 0)
Unpacking objects: 100% (3/3), done.
From c:/Users/Rick/Documents/gitbook/math
   e150c19..834c869  master     -> origin/master
error: Empty commit message.
Not committing merge; use 'git commit' to complete the merge.
```

In the output, the key line is this:

```
e150c19..834c869  master     -> origin/master
```

This shows that the commit `e150c19` (the old commit) was updated to `834c869` (the new commit). The last line, which begins `Not committing merge`, is from the merge command. This means you're still in the middle of a merge. You can confirm all of this with `gitk` and with `git status`. Resolving merges is something that happens in other version-control systems.

> **TRY IT NOW** In math.carol, now that you've cancelled the automatic commit message, type this:
>
> ```
> git status
> ```
>
> You should see output as in the following listing.

Listing 14.11 Output from `git status` after cancelling the merge

```
On branch master
Your branch and 'origin/master' have diverged,
and have 1 and 1 different commit each, respectively.
```

```
  (use "git pull" to merge the remote branch into yours)

All conflicts fixed but you are still merging.
  (use "git commit" to conclude merge)

nothing to commit, working directory clean
```

This message is hard to understand at first glance. The first sentence of the status message (`Your branch and 'origin/master' have diverged`) means that these two branches aren't on the same line of development anymore. The next sentence (`1 and 1 different commit each, respectively`) says that your branch (you're in master) has one newer commit, and origin/master has one newer commit. To understand this line, look at the following line again:

```
e150c19..834c869  master     -> origin/master
```

This says that commit `834c869` was added to `e150c19` (or origin/master was added on top of master). This looks like figure 14.15.

master
remotes/origin/master
41646b03
834c869
e150c19
...

Figure 14.15 master is diverged with this pull. Note that the base is `e150c19`.

This can be visualized by running gitk in the math.carol directory, as shown in figure 14.16.

Finally, `git log` can draw the same type of graph, using the `--graph` switch.

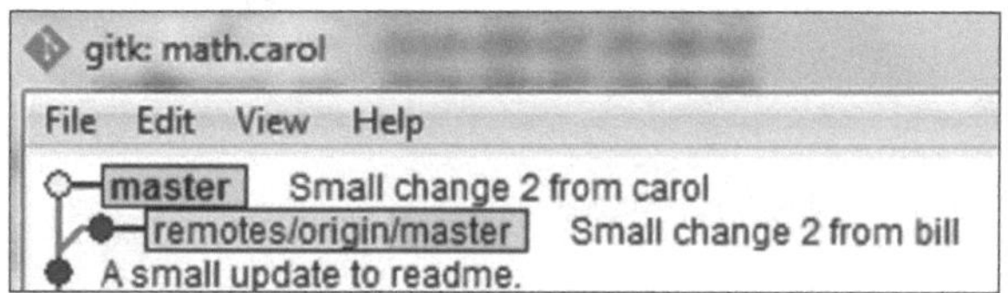

Figure 14.16 Two paths diverging from `A small update`.

TRY IT NOW In the math.carol directory, type the following:

```
git log --decorate --graph --oneline --all
```

The top part of the listing looks similar to the following.

Listing 14.12 A graph view from `git log`

```
* 41646b0 (HEAD, master) Small change 2 from carol
| * 834c869 (origin/master, origin/HEAD) Small change 2 from bill
|/
* e150c19 A small update to readme.
```

The first two lines can appear in any order. This is why you're in the editor in the first place. Git is allowing you to write a new commit message to replace the default message. This may be helpful to your fellow collaborators, who may be expecting a commit message explaining why a merge has taken place.

TRY IT NOW Let's complete the merge, which you interrupted with the empty commit message. Type the following:

```
git commit
```

You'll be thrown into Git's default editor (notice that you didn't pass a message via the `-m` switch). An autogenerated commit message is already entered, and to make things easy, let's just accept it. Type this:

```
ZZ
```

You'll be returned to the prompt, and you'll see a short message like the following listing.

Listing 14.13 Typical merge message

```
[master 655cbee] Merge branch 'master' of /home/rick/gitbook/math
```

14.3.3 Clean merge with automatic commit

In a clean merge with an automatic commit message, Git writes the commit message for you and commits it. You don't have to run `git commit` at all, but you also don't get the chance to edit the commit message. If your merge had an automatic commit, type `git log -1`, and you'll see the generated commit message, shown in the following listing.

Listing 14.14 A merge commit

```
commit 655cbee109eadd98d894639ad57f35ff7ce5bf59
Merge: 41646b0 834c869                              ❶
Author: Rick Umali <rickumali@gmail.com>
Date:   Sat Sep 13 14:39:11 2014 -0400

    Merge branch 'master' of c:/Users/Rick/Documents/gitbook/math
```

❶ The parents of this commit

This commit joins the two parents (`41646b0` and `834c869`, in ❶) into a new commit (`655cbee`). If you were to open gitk, you'd see this merge (see figure 14.17), as you saw in chapter 10.

The merge will be the same, regardless of whether you had an automated commit message or not.

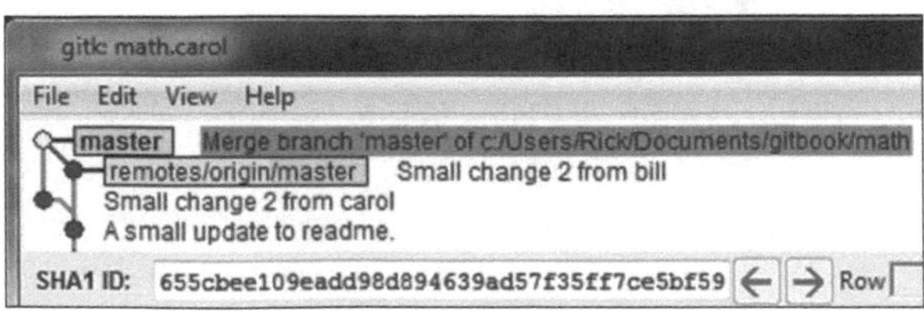

Figure 14.17 Merging the clean merge

14.3.4 Conflicted merges

There's one more level of complexity past what we've already discussed: conflicted merges. Recall from chapter 10 that Git's algorithms can't resolve all differences. If two branches have modified the same file in the same place but in different ways, Git has to ask how to resolve the conflict. This same situation can arise when you use `git pull` (remember, `git pull` is just `git fetch` and then `git merge`).

TRY IT NOW To set up for this TRY IT NOW, you'll push the changes you made in the previous section from math.carol back to the repo:

```
cd $HOME/math.carol
git push
```

Then, in the math.bill directory, you pull these changes:

```
cd $HOME/math.bill
git pull
```

Now both carol and bill are back in sync with one another and the main repository. As in the previous sections, you'll introduce a commit in the math.bill repository, push it to math.git, and then introduce a conflicting commit in the math.carol repository. You'll first make a commit in the math.bill directory, and push this to the central math.git repo:

```
cd $HOME/math.bill
echo "JKL MNO PQR" >> another_rename
git commit -a -m "JKL part of alphabet"
git push
```

These steps conclude with `git push`, shown in figure 14.18.

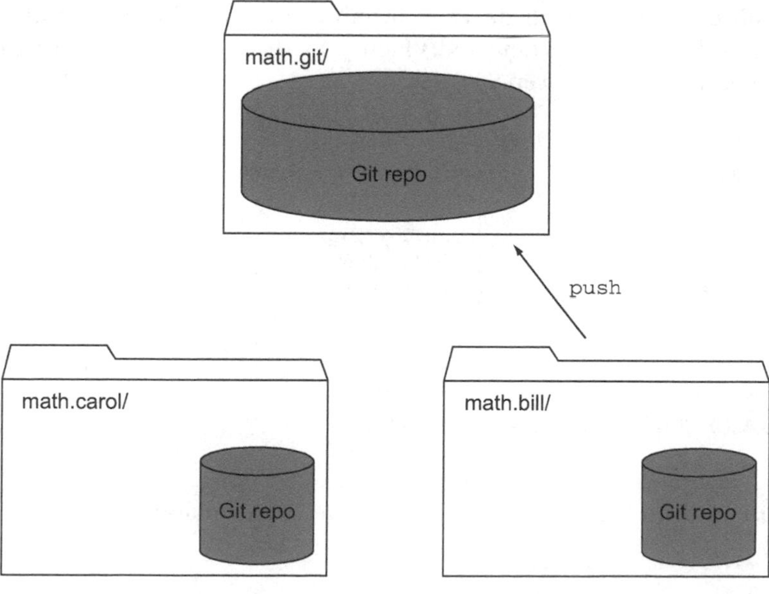

Figure 14.18 Pushing a change to math.git

Then you'll make a commit in the math.carol directory:

```
cd ~/math.carol
echo "ABC DEF GHI" >> another_rename
git commit -a -m  "ABC part of alphabet"
```

So far, you have the same picture as before (you didn't push your change from math.carol to math.git), but this time the files have conflicting edits on the same line.

You're now going to pull down the changes from math.git and encounter the merge conflict.

TRY IT NOW Type the following:

```
git pull
```

You should see the `CONFLICT` error shown in the following listing.

Listing 14.15 A `git pull` conflict

```
remote: Counting objects: 9, done.
remote: Compressing objects: 100% (2/2), done.
remote: Total 3 (delta 1), reused 0 (delta 0)
Unpacking objects: 100% (3/3), done.
From c:/Users/Rick/Documents/gitbook/math
   834c869..5b84cb4  master     -> origin/master
Auto-merging another_rename
CONFLICT (content): Merge conflict in another_rename
Automatic merge failed; fix conflicts and then commit the result.
```

You'll be returned immediately to the prompt. Git needs help resolving the conflicting changes in the file another_rename. (The next-to-last line shows the conflicted file, though do keep in mind that Git will enumerate all the conflicted files that it has detected.) Figure 14.19 shows the conflicting line in this file between the two repositories.

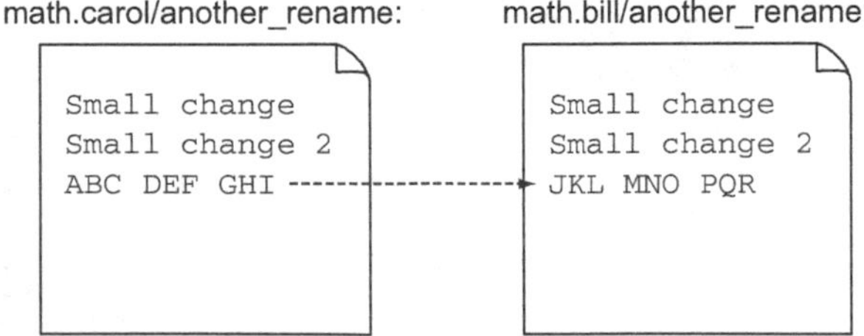

Figure 14.19 The last line is the conflicting line.

You can confirm the status with the `git status` command:

```
git status
```

You should see output like the following listing. Section 14.3.3 explains the `1 and 1 different commit each` message, but you may have skipped this section if your first `git pull` produced a clean merge with an automatic message.

Listing 14.16 `git status` for a conflicted merge

```
On branch master
Your branch and 'origin/master' have diverged,
and have 1 and 1 different commit each, respectively.
(use "git pull" to merge the remote branch into yours)

You have unmerged paths.
(fix conflicts and run "git commit")

Unmerged paths:
  (use "git add <file>..." to mark resolution)

     both modified:      another_rename

no changes added to commit (use "git add" and/or "git commit -a")
```

At this point, the file must be manually fixed. With this one `git pull`, Git has brought down the new changes (`git fetch`) and tried to merge them (`git merge`). As in chapter 10, the file containing the conflict is modified to show you the lines that Git had problems with. Use your favorite editor to open the file. Confirm that you see your file at the top, but that conflicted lines appear at the bottom of the file, as in the following listing. (You last looked at this in detail in section 10.3.2.)

Listing 14.17 The conflicted lines

```
<<<<<<< HEAD
ABC DEF GHI
=======
JKL MNO PQR
>>>>>>> 5b84cb4be250b2a748515d66da76bbad4314f455
```

TRY IT NOW You covered handling merge conflicts in chapter 10, so flip back to that chapter to review the details. The key with merge conflicts is to pick the right lines for the merged file. Typically, one file has the correct line, but there's nothing to prevent you from replacing the lines with anything that makes sense. Using your favorite editor, open the file named another_rename, remove the lines (shown in the box in figure 14.20), and replace them with `ABC DEF GHI JKL MNO PQR`.

After you've made the fix, type the following to commit the change in the math.carol repository:

```
git add another_rename
git citool
```

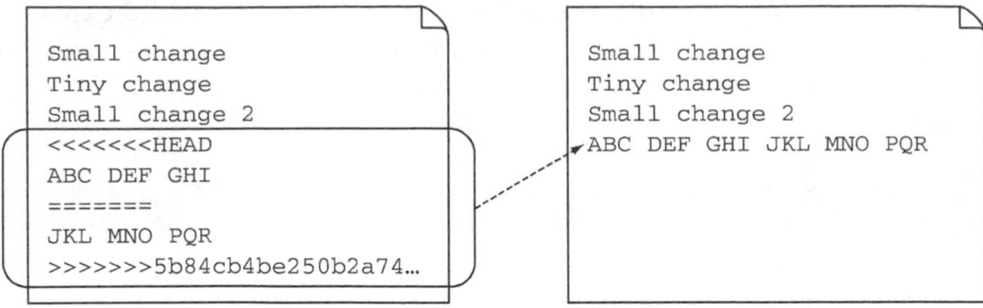

Figure 14.20 Fixing the merge by combining the changes. The line on the right is the new merged file.

The last command displays the Git GUI tool, shown in figure 14.21. There you can see that a commit message is automatically generated. Click the Commit button to finish.

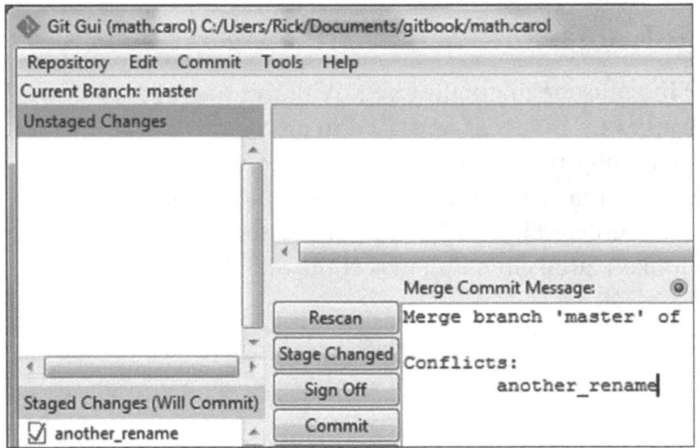

Figure 14.21 Committing the change with `git citool` (a.k.a. Git GUI)

In the previous TRY IT NOW, you could have used the `git commit` command directly with the `-m` switch to specify your own message. This would override the message that Git would have generated for you.

At this point, you must push your changes from math.carol back to math.git, and then have Bill pull these changes down from math.git.

TRY IT NOW This will sync up both the math.carol and math.bill repos with one another. Type the following:

```
cd $HOME/math.carol
git push
cd $HOME/math.bill
git pull
```

14.4 Restricting pulls to fast-forwards only

Clearly, fast-forward merges are the easiest to deal with. You can run `git pull` to incorporate only commits that are descendants of your current branch, using the `--ff-only` switch. This prevents Git from doing any automated work, unless it's a fast-forward merge.

> **TRY IT NOW** Repeat the setup steps between math.bill and math.carol to introduce a merge. Here are those steps again, listed concisely:
>
> ```
> cd $HOME/math.bill
> echo "ABC" >> another_rename
> git commit -a -m "Alphabet (on bill)"
> git push
> cd $HOME/math.carol
> echo "ABC" >> another_rename
> git commit -a -m "Alphabet (on carol)"
> ```
>
> These steps create the situation where two branches have to be merged. Now type this (from the math.carol directory):
>
> ```
> git pull --ff-only
> ```
>
> You'll see the following output.

Listing 14.18 git pull --ff-only

```
remote: Counting objects: 9, done.
remote: Compressing objects: 100% (2/2), done.
remote: Total 3 (delta 1), reused 0 (delta 0)
Unpacking objects: 100% (3/3), done.
From c:/Users/Rick/Documents/gitbook/math
   5b84cb4..1ac8efa  master     -> origin/master
fatal: Not possible to fast-forward, aborting.
```

If you're running Git 2.0 or greater, it's also possible to configure Git so that this is the default behavior. To do this, type the following:

```
git config pull.ff only
```

Setting this behavior means that you'll always need to manually perform the merge, even though Git might be able to automatically perform the merge.

At this point, you'll do the merge yourself, push the changes from math.carol back to math.git, and then have Bill pull these changes down from math.git.

> **TRY IT NOW** The first `git pull` is a repeat of the steps you saw in section 14.3:
>
> ```
> cd $HOME/math.carol
> git pull
> ```
>
> The next steps sync up the math.carol and math.bill repos with one another:
>
> ```
> git push
> cd $HOME/math.bill
> git pull
> ```

14.5 Using git fetch and merge instead of pull

`git pull` is one of those commands that Git developers consider troublesome for the beginner. The Git mailing list had a lively thread declaring `git pull` as evil. One of its shortcomings is that it doesn't allow you to see what will change when you perform the merge.

Instead of using `git pull` all the time, especially in an actively updated repository, consider performing the `git fetch` and the `git merge FETCH_HEAD` (section 14.2) commands, so you see exactly what files will be merged and how. This is helpful in an active repository with frequent commits.

TRY IT NOW Let's repeat the steps to set up a conflicted merge:

```
cd $HOME/math.bill
echo "ABC" >> another_rename
git commit -a -m "ABC (on bill)"
git push
cd $HOME/math.carol
echo "DEF" >> another_rename
git commit -a -m "DEF (on carol)"
```

If you were to perform a `git pull` on math.carol, you'd immediately get thrown into a conflicted state. But type this:

```
git fetch
git diff HEAD FETCH_HEAD
```

This displays the differences between the two commits, as shown in the following listing.

Listing 14.19 `git diff` between HEAD and FETCH_HEAD

```
diff --git a/another_rename b/another_rename
index cec2e13..c39bef6 100644
--- a/another_rename
+++ b/another_rename
@@ -3,4 +3,4 @@ Tiny change
 Small change 2
 ABC DEF GHI JKL MNO PQR
 ABC
-DEF
+ABC
```

`git diff` is showing you how to change the HEAD so it looks like FETCH_HEAD. Because the only difference is the last line (in math.bill, you had `ABC`, and in math.carol you had `DEF`), the `git diff` output says you'd remove the `DEF` and add `ABC` to make math.carol's another_rename file contents match the contents from math.bill's another_rename file.

Using `git pull` on a common branch can result in surprises. This is why `git pull` is considered evil: surprise merges! If you perform the two steps of `git pull` individually, you'll have a better chance of anticipating errors in the merge step by doing a `git diff` between your branch and the updated branch.

14.6 Lab

You've learned a lot in this chapter, and yet it feels like more could be covered. This lab enables you to explore some of these other areas:

1. Merge the last fetch in the previous section.
2. In the math.bill repo, make a change and commit it in every branch (master, another_fix_branch, and new_feature). Push these up to math.git by using `git push --all`, and then do a `git fetch` from math.carol. Can you confirm that the `git log --decorate --all --oneline` output shows that all these branches are advanced on the remote-tracking branches?
3. Confirm that the FETCH_HEAD file changes every time you do a `git fetch`. Use `git checkout` on every branch, and check FETCH_HEAD (with `git rev-parse`, or by directly looking at the file).
4. In all of the exercises of this chapter, you pushed from math.bill to math.git, simulating a centralized Git server. Try the following:

   ```
   cd $HOME/math.carol
   git remote add bill ../math.bill
   git branch --set-upstream-to=bill/master
   ```

   This sets up math.bill to be a remote to carol. You can now do all the exercises without a push to math.git. What are the implications of this approach? Note: Reset the math.carol repository to point back to origin by typing the following:

   ```
   git branch --set-upstream-to=origin/master
   ```

5. The repository contains two remote-tracking branches: another_fix_branch and new_feature. Merge these branches by typing this:

   ```
   cd $HOME/math.carol
   git checkout master
   git merge new_feature
   ```

   Did Git complain? Why? At this point, you could merge new_feature directly from the remote-tracking branch, by typing this:

   ```
   git merge origin/new_feature
   ```

   Or you could create a new local branch based on origin/new_feature. Try this!
6. Examine the contents of the FETCH_HEAD file. Compare it with the HEAD file (found in your working directory's .git directory). For example, type the following:

   ```
   cd $HOME/math.carol
   cat .git/FETCH_HEAD
   ```

14.7 Commands in this chapter

Table 14.1 Commands used in this chapter

Command	Description
`git pull`	Sync your repository with the repository that you cloned from (a.k.a. the upstream repository). This command comprises two commands: `git fetch` and `git merge`.
`git fetch`	The first part of `git pull`. This brings in new commits from the remote repository and updates the remote-tracking branch.
`git merge FETCH_HEAD`	Merge the new commits from FETCH_HEAD into the current branch.
`git pull --ff-only`	The `--ff-only` switch will allow a merge only if FETCH_HEAD is a descendant of the current branch (a fast-forward merge).

15 Software archaeology

When you start interacting with a new Git repository, you might want to understand its history. Git has a variety of commands that let you dig into these details. This chapter describes how to survey a repository's history by using `git log`, `git shortlog`, and `git name-rev`. The chapter also focuses on commands that help you understand the files in the repository: `git grep`, `git show`, and `git blame`.

If you're going to start work on an existing repository, especially one with lots of contributors or lots of history, you'll want to perform a kind of archaeological survey of the code base. You'll want to examine the code base at a high level and then narrow down to a particular set of files for a closer examination. These Git tools enable you to perform this survey of the code base. Let's get digging!

15.1 Understanding git log

`git log` is the command that displays your repository's timeline history. You've been using this throughout the book already, but if you look at its documentation, you'll see that it has a rich set of capabilities. You'll check out some of that functionality here.

15.1.1 Reviewing the basics of git log

You learned in chapter 2 that `git log` shows your timeline of commits, from the most recent commit to the root. `git log`'s basic output format is to show the entire commit message, so you learned about the `--oneline` switch to produce a more succinct form. In chapter 4, you learned about the `--stat` switch, which makes `git log` display the files that were updated in a commit.

TRY IT NOW In the math.carol directory, review your commit history by typing the following `git log` commands. The listing may be long, so Git may use the pager. Remember to press Q to exit the pager.

```
git log
git log --oneline
git log --stat
```

You should see familiar-looking output, as these were the commits you made over the past day or so.

A full commit consists of the parts in figure 15.1.

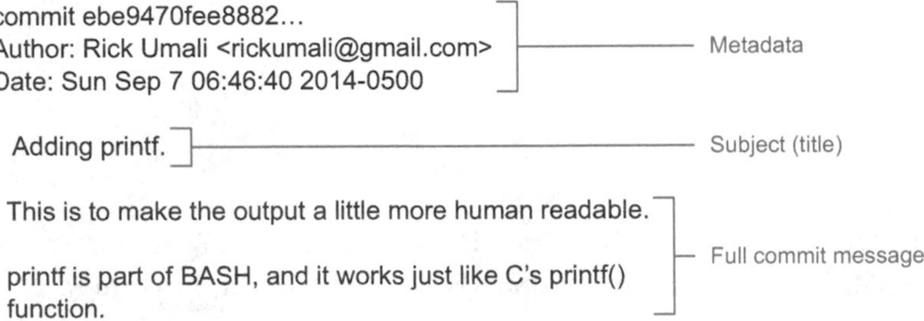

Figure 15.1 A breakdown of a commit message

When you write a commit message, the first line is considered the *title,* or subject, of the commit. This title is displayed when you use the `--oneline` switch. This title should be a sentence that is 50 characters or fewer, so Git commands that display this text next to other data won't be too crowded.

In chapter 8, you learned about the `--parents` switch. Now that you have merges, you can practice the `--merges` switch, which finds all the merges in your repository.

TRY IT NOW Let's identify from the `git log` output those commits that represent merges of two branches. In the math.carol directory, type the following:

```
git log --parents
git log --parents --oneline
```

Adding the `--oneline` switch makes the output much more concise. You should see something similar to the following listing. (Don't worry if the order is different, or if your repository has different commits. The key is to find a merge commit.)

Listing 15.1 An annotated listing of `git log --parents --oneline`

```
fb9fed6 c4ccf59 20a708a Merge branch 'master' of ../math.bill
c4ccf59 e150c19 JLK Part of Alphabet (on carol)
20a708a e150c19 ABC Part of Alphabet (on bill)
e150c19 f48c719 A small update to readme.
```

Three SHAI IDs ❶

```
f48c719 58ee0fc Adding printf.
58ee0fc d3ae3ea Adding two numbers.
d3ae3ea dd87c91 Renaming c and d.
dd87c91 11a90b4 Removed a and b.
11a90b4 12a7b37 Adding readme.txt
12a7b37 907b870 Adding four empty files.
907b870 56d7919 Adding b variable.
56d7919 c57cd5c This is the second commit.
c57cd5c This is the first commit.
```

You can eyeball the output and see that ❶ contains three SHA1 IDs. Look at this closely by typing the following:

```
git --no-pager log fb9fed6 -n 1
```

Replace `fb9fed6` with the SHA1 ID of the merge commit. The `-n 1` limits the `git log` to display only one commit entry. The `--no-pager` switch tells Git not to paginate the output. Your SHA1 ID will be different. Find an entry that has three SHA1 IDs.

This command results in the following output.

Listing 15.2 A log listing of a merge (with `--parents` switch)

```
commit fb9fed602170a079db5f5eeb6ee6477eb4fa3fca
Merge: c4ccf59 20a708a
Author: Rick Umali <rickumali@gmail.com>
Date:   Sun Sep 14 21:22:58 2014 -0400

    Merge branch 'master' of ../math.bill
```

This shows that commit `fb9fed6` is a merge of commits `c4ccf59` and `20a708a`.

Rather than finding these merges by hand, you can use the `--merges` switch:

```
git log --merges
```

You should see only the commits that have merges in them.

15.1.2 Limiting the display of commits

The switch `--merges` provides a way to limit (or reduce) the number of commits displayed. By default, `git log` shows all commits, from the most recent commit all the way back to the first commit. As you scan and survey your new repository, this full listing is helpful, but as you narrow down a particular time range or a particular set of files, you'll want to see specific parts of the history. The `git log` documentation has a section named Commit Limiting, which discusses how to reduce the number of commits based on criteria. These criteria can be combined, as you saw in the previous section, by combining the `--merges` switch and `-n 1`.

The following sections show how to specify different kinds of criteria to `git log`. This criteria reduces the number of commits displayed by `git log`.

LIMITING BY FILE

Let's learn how to limit your history to a specific file or set of files.

TRY IT NOW In the math.carol directory, you'll limit the `git log` output to the commits that affected one or more files:

```
git log --oneline readme.txt
```

You can do this for more than one file. Type the following:

```
git log --oneline readme.txt renamed_file
```

To prove that you're getting only the commits that pertain to these two files, type this:

```
git log --stat --oneline readme.txt renamed_file
```

This produces output similar to the following listing.

Listing 15.3 `git log --stat --oneline` output

```
8e07daf Adding a line to renamed_file
 renamed_file | 1 +
 1 file changed, 1 insertion(+)
d09f697 Merge branch 'new_feature' of /home/rick/gitbook/math into
  new_feature
7f16f04 Adding new line
 readme.txt | 1 +
 1 file changed, 1 insertion(+)
48c8718 A small update to readme.
 readme.txt | 1 +
 1 file changed, 1 insertion(+)
2624567 Renaming c and d.
 renamed_file | 0
 1 file changed, 0 insertions(+), 0 deletions(-)
50d36bc Adding readme.txt
 readme.txt | 1 +
 1 file changed, 1 insertion(+)
```

❶ Stat line for renamed_file

❷ Stat line for readme.txt

The files displayed by the `--stat` switch (❶ and ❷) correspond to the files that were entered on the command line. By using the `--stat` switch, `git log` indicates the files that have changed in the commit, showing only those files specified on the command line. (Leave out readme.txt and renamed_file to see the difference.)

FINDING SPECIFIC COMMITS

You can tell `git log` to show only those commit messages that match a particular string by using the `--grep` switch to search commit messages for that string. For example, if your project is in the habit of entering bug numbers in commit messages, you can use `git log --grep` for all commits that pertain to a particular bug number.

TRY IT NOW In the math.carol directory, type this:

```
git log --grep=change
```

This should list any commits that have the word *change* in the commit message.

LIMITING BY TIME RANGE

You can limit the commits to those that took place in a certain time range, using the --since and --until switches. For example, you might find documentation with a particular date and might want to see what code changes took place immediately after that date. This kind of exploratory digging is possible with these switches.

> **TRY IT NOW** In the math.carol directory, examine the dates of the commits. Now try to show only those commits that happened in a two-day period. You'll have to come up with dates yourself, but your command will look similar to this one:

```
git log --since 10/10/2014 --until 10/24/2014
```

LIMITING BY AUTHOR

You can also limit commits to a particular author, with the --author switch. Of course, you'll need a list of authors, so let's tackle that first. An easy way to get the list of authors from a repository is to use `git shortlog`.

> **TRY IT NOW** In the math.carol directory, type this:

```
git shortlog
```

This should display something like the following listing.

Listing 15.4 `git shortlog` output

```
Rick Umali (12):
      This is the first commit.
      This is the second commit.
      Adding b variable.
      Adding four empty files.
      Adding readme.txt
      Removed a and b.
      Renaming c and d.
      Adding two numbers.
      Adding printf.
      A small update to readme.
      Small change
      Another tiny change
```

To get the author list with email addresses, pass in the -e switch:

```
git shortlog -e
```

After you have an author's name or email address, you can restrict commits to that one author by passing it into the `git log --author` command.

> **TRY IT NOW** In chapter 12, you cloned the math repository from GitHub into the directory math.github. Let's go into that directory to practice the `git shortlog` and `git author` commands:

```
cd $HOME/math.github
git shortlog -e
git log --author="Rick Umali"
```

You can also search by a partial name or partial email address. (In Git, the author of a commit is a concatenation of a name and email address.) Type the following:

```
git log --author="Rick"
git log --author="gmail.com"
```

As you're digging for information about changes in the code base, and your clues point to a particular author, you can use this command to find other places where they made changes as well.

15.1.3 Seeing differences with git log

Each commit corresponds to a distinct version of the entire repository. A commit can consist of one change to one file, or multiple changes to multiple files. One question you'll often ask is what is the difference between two commits? Perhaps while debugging a problem, you'll learn that the code base was functioning properly in one commit but began failing after another commit. `git log` can be used to see the differences between those two commits.

TRY IT NOW Use this command to display the number of files that have changed between the most current commit (`HEAD`) and its immediate predecessor (`HEAD^`):

```
git log --stat HEAD^..HEAD
```

The `--stat` switch shows a list of files that have changed.

Two interesting pieces of syntax are at play. The first is `HEAD^` (HEAD appended with the caret symbol), which signifies the parent of HEAD. HEAD without the caret symbol represents the current commit. You can substitute any reference (for example, a branch or tag) or even a SHA1 ID in place of HEAD. The second interesting syntax is the double-period. It specifies a revision range. Ranges specify a set of commits. In this case, the two commits are next to each other (`HEAD^` is immediately before HEAD).

If you wanted to see what changed in the files since the last commit, try this:

```
git log --patch HEAD^..HEAD
```

The `--patch` switch shows the contents that have changed. You'll see output like the following listing.

Listing 15.5 Output of `git log --patch`

```
commit 7746e35930e562304e347ac69929aa276ed345dc
Author: Rick Umali <rumali@firstfuel.com>
Date:   Sun Sep 7 21:51:15 2014 -0400

    Another tiny change

diff --git a/another_rename b/another_rename
index 86d347f..fb7f0ae 100644
```

```
--- a/another_rename
+++ b/another_rename
@@ -1 +1,2 @@
 Small change
+Tiny change
```

It's possible to combine `patch` and `stat` with `--patch-with-stat`, though you probably want to do `--stat` first, to see how many files were affected, and then run `git diff` on those files individually. I discussed `git diff` extensively in chapters 6 and 7.

Keep in mind that unless you limit the files, `git log --patch` shows all the files that are different between the two versions. If multiple files are changed, multiple files will be displayed. Files by themselves don't have versioning information. Git tracks changes across the entire set of files in the repository.

TRY IT NOW Let's go to the math.github directory and look at a difference that spans multiple files. Type the following:

```
cd $HOME/math.github
git log --patch ef47d3f^..ef47d3f
```

Notice the use of the caret (^) after a SHA1 ID. This indicates the version immediately before `ef47d3f`. This `git log` command produces output that looks like this listing.

Listing 15.6 `git log --patch` revealing multiple files

```
commit ef47d3fd293bc13321270e88af284f63d6f85f84
Author: Rick Umali <rickumali@gmail.com>
Date:   Sat Aug 2 18:54:56 2014 -0500

    Adding four empty files.

diff --git a/a b/a
new file mode 100644
index 0000000..e69de29
diff --git a/b b/b
new file mode 100644
index 0000000..e69de29
diff --git a/c b/c
new file mode 100644
index 0000000..e69de29
diff --git a/d b/d
new file mode 100644
index 0000000..e69de29
```

The listing shows that four files have been added. Each one looks like this entry:

```
diff --git a/a b/a
new file mode 100644
index 0000000..e69de29
```

Git uses two indicators to show that this is a new file:

- `new file mode`, which indicates that this is a new file with permissions mode `100644`
- `index 0000000`, which indicates that a previous version for this file didn't exist

You can limit the output by typing the following:

```
git log --patch ef47d3f^..ef47d3f -- a
git log --patch ef47d3f^..ef47d3f -- a b
```

In these commands, you have to separate the files with a double-dash. When you type the commands, you'll see that the `git log` output is limited to the files you specified.

15.1.4 Using git name-rev to name commits

If your repository contains lots of branches, you'll want to use limiting arguments to make sure you're not overwhelmed. In this section, you'll try to find some interesting artifacts among a large number of branches. You'll learn how to use `git name-rev` to give a human-readable name to any commit, which can help specify revisions.

TRY IT NOW To create your testing area, download the zip file containing the code for this book from the book's website (www.manning.com/umali). In that zip file is a script named make_lots_of_branches.sh. This script creates a repository full of branches. Unzip the contents into your $HOME directory. Then type the following:

```
cd $HOME
bash make_lots_of_branches.sh
```

You may have run this earlier, in the chapter 9 exercises, but if you haven't, be advised that this script may take a minute to run. It's creating many branches. After you've done that, type the following:

```
cd $HOME/lots_of_branches
git branch
```

This output shows a long listing of all the branches in the repository. You can use `git branch`'s `--column` switch to make the listing more manageable. It displays the branch list in columns, as shown in the following listing. Type this:

```
git branch --column
```

Listing 15.7 `git branch --column` output

```
branch_01   branch_08   branch_15   branch_22   branch_29   branch_36
branch_02   branch_09   branch_16   branch_23   branch_30   branch_37
branch_03   branch_10   branch_17   branch_24   branch_31   branch_38
branch_04   branch_11   branch_18   branch_25   branch_32   branch_39
branch_05   branch_12   branch_19   branch_26   branch_33   branch_40
branch_06   branch_13   branch_20   branch_27   branch_34 * master
branch_07   branch_14   branch_21   branch_28   branch_35
```

If you were to run the git lol command that you created in the previous section, it would display all the commits for each one of these branches. This is because you created the git lol command with the --all switch. You can limit the output by passing one or more branches as arguments to the git log command.

TRY IT NOW Type the following:

```
git log --graph --decorate --oneline --all
```

This shows all the branches. Because there are so many, you'll need to use the pager to view all the contents. Try running git log with the --no-pager switch.

```
git log --graph --decorate --oneline branch_03
```

This command limits the git log output to one branch, which is easier to see. Now type this:

```
git log --graph --decorate --oneline branch_03 branch_10 master
```

This form passes in three branches to the git log command. You'll see a tree listing of the branches, but notice how the listing is easier to understand because you limited the commits to a few branches (instead of showing the full set of branches).

The following listing shows the output of the last command on my machine. Your output will be different, because the script randomizes the files in each branch. The master branch should have exactly the same commit log messages, however, with different SHA1 IDs.

Listing 15.8 Annotated output of `git log --graph` with branch limiting

```
* c524ab7 (branch_03) Commit for file_15360
* ae8c666 Commit for file_28769
| * 9ae4e58 (branch_10) Commit for file_23500
| * 7a1adec Commit for file_5795
|/
* ec2c398 (HEAD, tag: four_files_galore, master) Adding four empty files.
* d7bf074 Adding b variable.
* c4df58a This is the second commit.
* a9c7ba1 This is the first commit.
```

❶ Commit ae8c666 belongs to branch_03

❷ Commmit 7a1adec belongs to branch_10

Each commit's SHA1 ID is followed by a decoration, or label. This label shows the branch that the commit belongs to. Commit ae8c666 ❶ belongs to branch_03, and commit 7a1adec ❷ belongs to branch_10. Git has commands that show which branch a commit belongs to, in case you're ever asked to look at a specific SHA1 ID and need to know which branch to check out.

TRY IT NOW You'll use the math.github repository, so you can specifically reference a known SHA1 ID. Type this:

```
cd $HOME/math.github
git log --graph --decorate --oneline
git log 80f5738 -n 1
```

The last command should show you the commit `Removed a and b`. What branch does this commit belong to? You can find out by typing this:

```
git name-rev 80f5738
```

The `git name-rev` command produces a name for a commit, based on the nearest branch. This is one way to find out the branch that a commit belongs to.

Another way to determine what branch a commit belongs to is to use `git branch --contains`:

```
git branch -r --contains ce051a3
```

This command outputs the string `origin/new_feature`, meaning that this SHA1 ID is part of the remote-tracking branch new_feature. If you check out the new_feature branch and type `git branch --contains ce051a3`, you'll see the output `new_feature`.

The switch `--contains` is a useful switch to `git branch` when you're given a SHA1 ID and want to know what branch it belongs to.

15.2 Understanding gitk view configurations

Working on the command line to dig through a repositories history is noble, but ultimately, it can break down. The repository for Git software (known as git-core) contains over 30,000 commits and 500 tags. The `simplify-by-decoration` switch cuts this down to only 500 lines. Even utilizing the pager, it's hard to look at that many commits, tags, or references in the terminal window.

In this section, you'll look at the GUI tool gitk and see how its graphical design may make your code research a little easier.

15.2.1 Showing only specific branches in gitk

Like `git log`, the gitk GUI, by default, shows only the commits of the current branch. You can pass the `--all` switch to `gitk` to get all the branches, but if you want to isolate your view to specific branches, you can do so from two places: the view configuration and the command line.

TRY IT NOW Go into the lots_of_branches directory and type the following:

```
gitk --all
```

In gitk, click View > Edit View, as shown in figure 15.2.

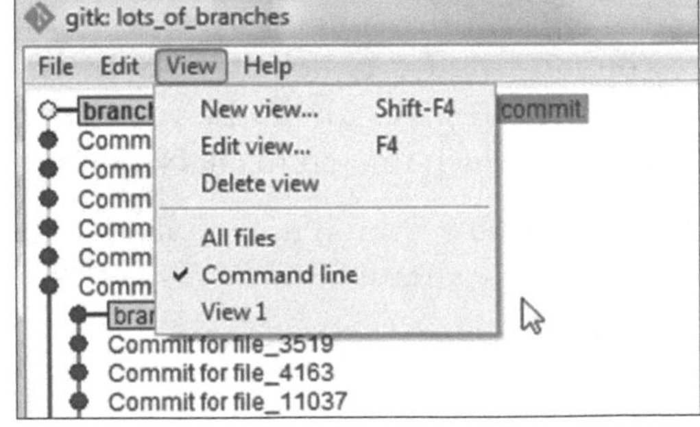

Figure 15.2 The view configuration from `gitk --all`

Gitk: edit view Command line -- criteria for selecting revisions
View Name Command line   Remember this view
References (space separated list):
Branches & tags: branch_03 branch_10 master
All refs   All (local) branches   All tags   All remote-
Commit Info (regular expressions):

Figure 15.3 Selecting the refs by using the Branches & Tags field

Notice that this lengthy configuration window lets you specify which branches you'd like to look at. First, deselect the All Refs check box, as highlighted in figure 15.3. Then type `branch_03 branch_10 master` in the Branches & Tags text field. (Each branch is separated by a space.)

Clicking the OK button changes the view, limiting the branches to what was specified. (The OK button is at the bottom of the tall configuration window.) You can see how this new view compares with the previous view in figure 15.4. On your computer, you'll be able to confirm that the view containing three branches is faster to scroll through because there are only three branches, instead of all branches.

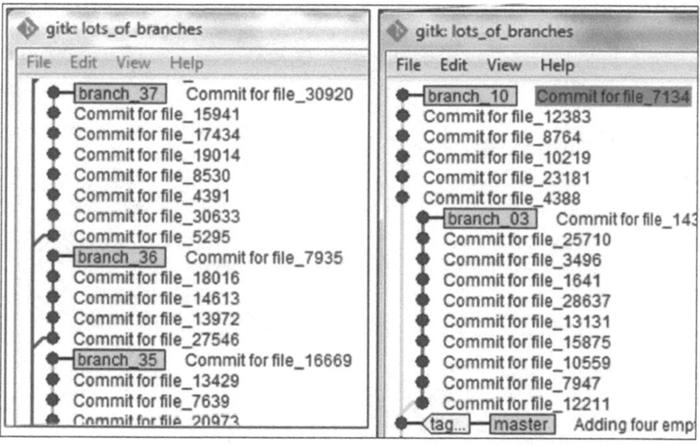

Figure 15.4 Comparing two gitk views (the left includes all the branches, and the right shows just three branches)

The GUI in figure 15.4 was started by typing `gitk --all` at the command line. You then limited the view by accessing the view configuration. If you know ahead of time that you want to view a specific set of branches, you can pass those into gitk via the command line.

Pay attention to the branches shown in the gitk tool. Controlling the selection of branches can be done in either the view configuration or on the command line.

> **TRY IT NOW** Exit the gitk window from the previous TRY IT NOW section. On the command line, from the lots_of_branches directory, type this:
>
> ```
> gitk branch_03 branch_10 master
> ```

When gitk appears, you should observe that it contains only the three branches that you specified on the command line, as shown in figure 15.5.

Figure 15.5 A simplified view of your branches

15.2.2 Working with simplified views

Let's construct a simplified view of all the branches. This view is going to be the same as what you saw in section 11.1.2 with the `--simplify-by-decoration` switch. Namely, it will show only commits that are at the tip of the branch.

> **TRY IT NOW** Stop and restart the gitk program. From the View menu, choose New View, as shown in figure 15.6.
>
> When the view configuration screen appears, click the following:
>
> - All Refs (toggle on)
> - Simple History (toggle on)
>
> The Simple History option is under the Miscellaneous section of the view configuration, as shown in figure 15.7.

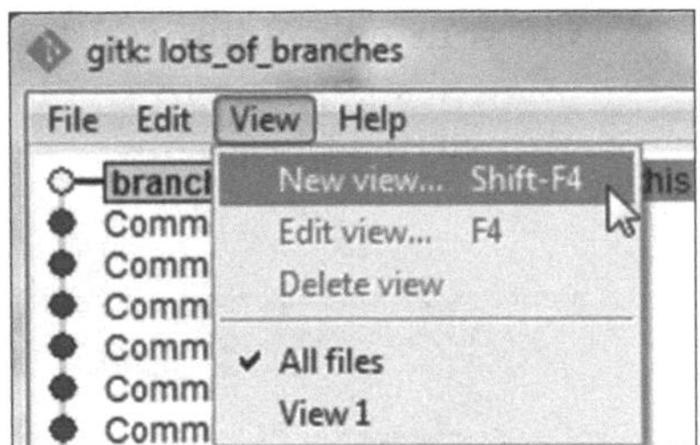

Figure 15.6 Creating a new view

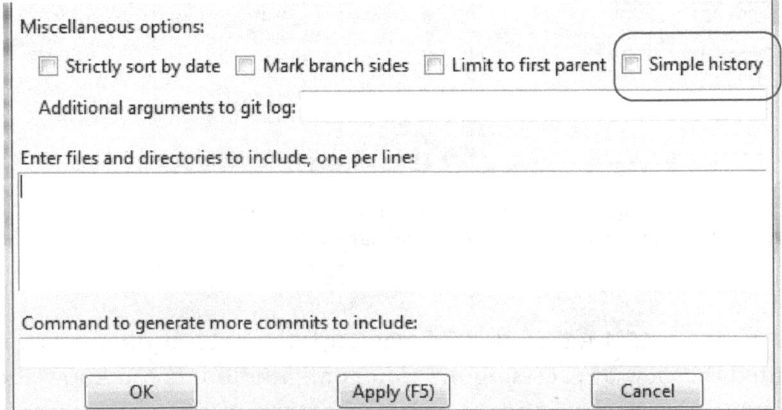

Figure 15.7 The Simple History configuration in the view configuration window

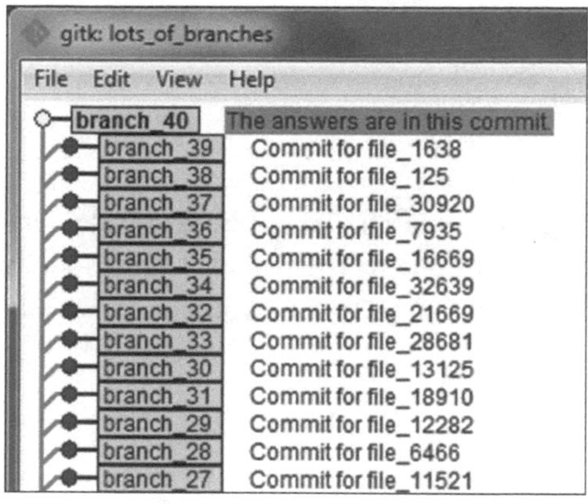

Figure 15.8 gitk's simplified listing of all branches

When you click Apply, the gitk view changes, as in figure 15.8.

The view that you configured can be saved for future use. At the top of the view configuration window (which doesn't go away even after you click Apply), in the View Name text box, type `Simplified All`. Then click the Remember This View check box (see figure 15.9).

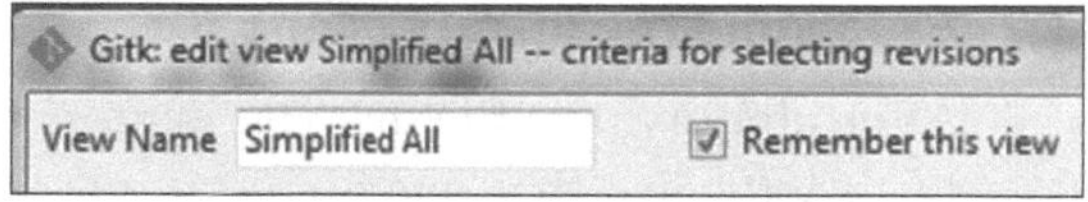

Figure 15.9 Creating a new view called Simplified All

When you click OK to close the view configuration window, this view becomes available to gitk as a menu item. Confirm this by clicking the View menu (figure 15.10).

You can exit gitk at this time.

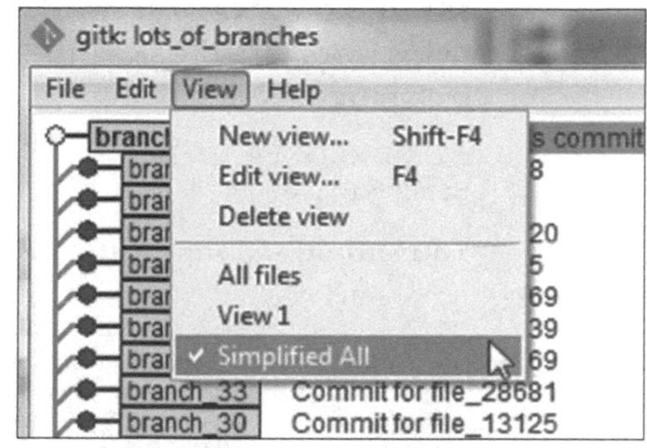

Figure 15.10 You saved your view to the menu.

15.3 Studying files

Looking through a repository's history shows timelines and files together. This remains the most important view of a project's history. But after you get the lay of the land, you'll need to focus on a specific file or set of files with which to work. To continue the archaeology analogy, after you've surveyed the land, you pick one or two places to start digging. For developers, this means studying individual files.

15.3.1 Finding files of interest (git grep)

In section 15.1.2, you learned how to limit your history to those commits that contained a specific string. You did this by using the `git log --grep` command. But this searches only the text in the commit messages. If you're looking for files that contain a particular string, you'll need to use the `git grep` command.

TRY IT NOW In the math.carol repository, suppose you wanted to find all the files that contained the word *change*. Type the following:

```
cd $HOME/math.carol
git grep change
```

This should produce output like the following listing.

Listing 15.9 `git grep` output

```
another_rename:small change
```

The output shows the filename containing the match (in this example, it's another_rename), the matching line, and the matching text. You don't have to specify the filename to search; `git grep` searches all the files in your repository for a matching string.

15.3.2 Examining the history of one file

After identifying a file that you want to study further, you'll often want to examine the history for just that one file. The gitk window lets you examine how a particular file has changed between commits. I covered the basics of this in chapter 8, but now let's dive into this feature more closely.

TRY IT NOW Navigate to the math.github directory. Using this repository helps you share the SHA1 IDs. You can isolate your view to just the math.sh file by typing this:

```
cd $HOME/math.github
gitk math.sh
```

You should see something like figure 15.11.

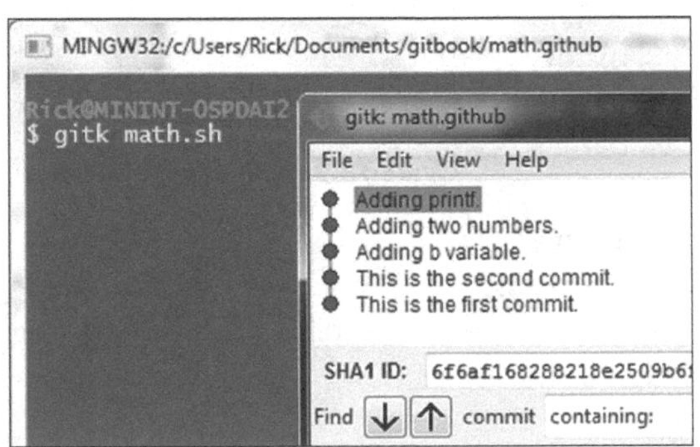

Figure 15.11 Isolating gitk to study one file

When you start gitk like this, it shows only the commits that touch the math.sh file. If you open the view configuration, you'll notice math.sh listed in the text box labeled Enter Files and Directories to Include, One per Line, as in figure 15.12.

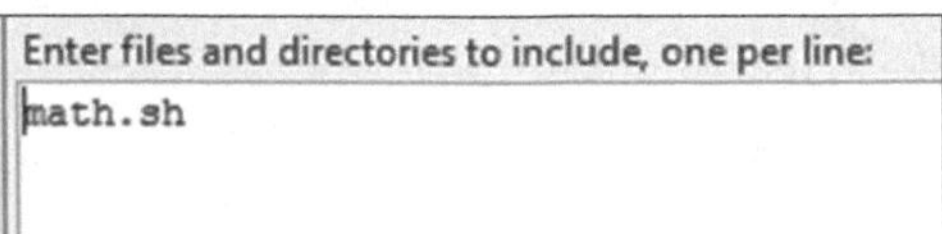

Figure 15.12 The view configuration for viewing one file in gitk

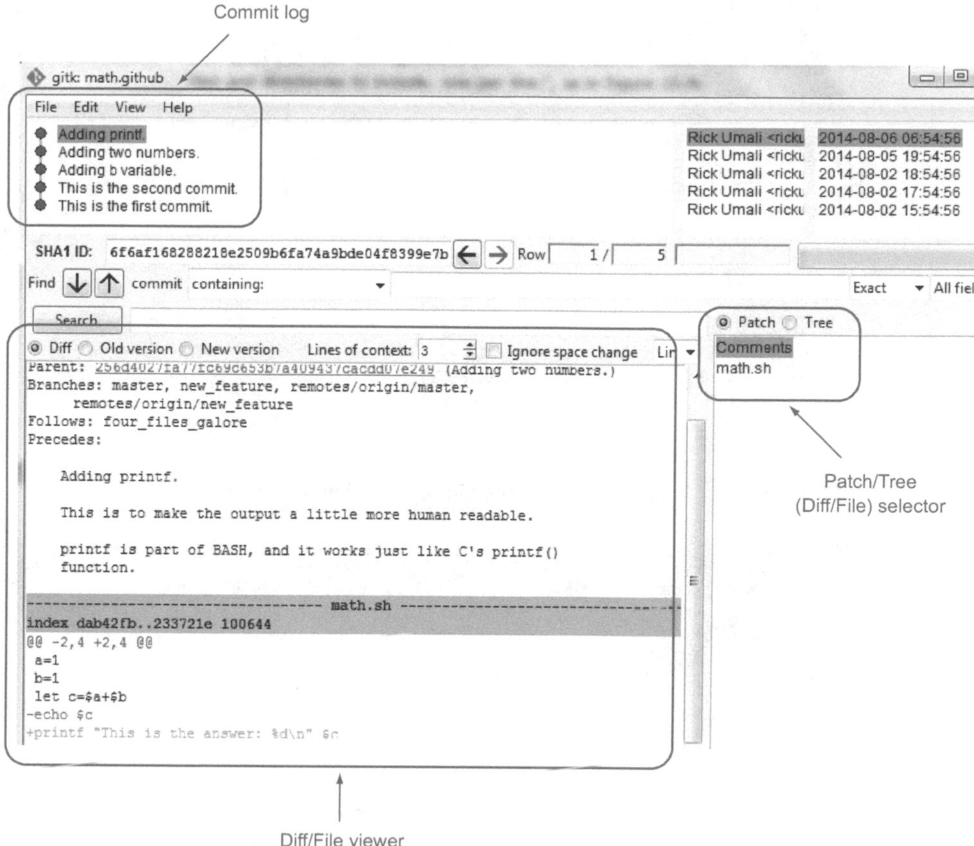

Figure 15.13 The diff window and commit log of gitk

TRY IT NOW Let's step through each commit and observe the Diff/File viewer window. (See figure 15.13 for the different windows on this screen.)

In the Commit Log window, go to the commit that reads `This is the first commit`. Make sure that Patch is selected in the Patch/Tree selector. This turns the Diff/File viewer into a Diff (or patch) viewer, and you can see the diff indicating that this is a file creation.

In the Diff/File viewer, click through the three toggles labeled Diff, Old Version, and New Version. The viewer displays these views, as shown in figure 15.14. (The Diff/File viewer can only show one view at a time. Figure 15.14 is a composite of all three views.)

Doing this changes what is shown in the Diff/File viewer. In Diff view, you'll see a diff. In this case, in the first commit, the diff is between an empty file and the first commit of this file. In the Old Version view, the file view is empty. In the New Version view, it shows the current file.

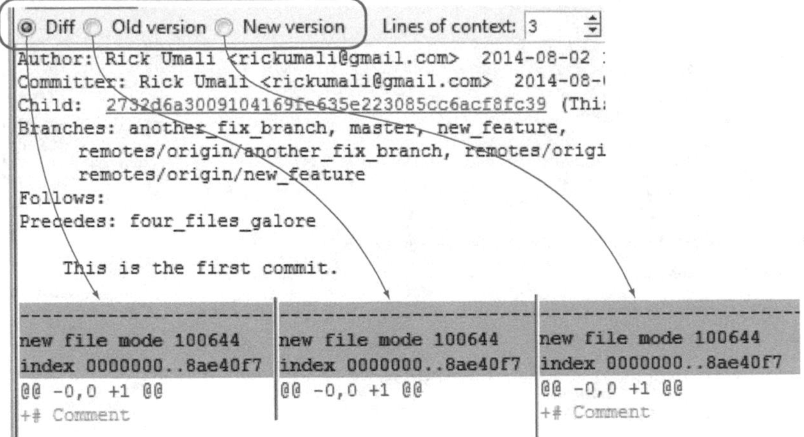

Figure 15.14 Toggling between the Diff, Old Version, and New Version views

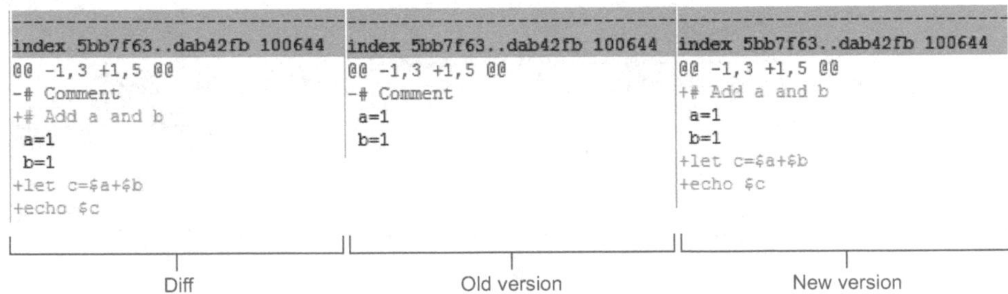

Figure 15.15 Different views from the commit `Adding two numbers`

Now click the commit log `Adding two numbers`. The viewer displays these views in figure 15.15.

Take a look at these three views so you can convince yourself that each view is appropriate. You can see how the file math.sh changed between the commit `256d402 (Adding two numbers)` and its parent.

Each file change is in the context of a commit, which itself considers the entire repository. The Patch/Tree selector window shows how other files might have changed for a particular commit (though in our math.github repository, only one file has been changed).

Above and Beyond

In figure 15.15, each view of the patch (Diff, Old Version, New Version) shows a line that starts with `index 5bb7f63..dab42fb`. Each SHA1 ID is a file being compared (a particular version of math.sh).

> You can view the contents of any SHA1 ID by using the `git show` command. This is what you'll see:
>
> ```
> % git show 5bb7f63
> # Comment
> a=1
> b=1
>
> % git show dab42fb
> # Add a and b
> a=1
> b=1
> let c=$a+$b
> echo $c
> ```
>
> Git contains every version of every file that's committed into the repository, and `git show` lets you call up those versions.

15.4 Finding which revision updated a specific line of code

The gitk Diff/File viewer is one way to see how a file has evolved. As you click through each commit, you can see how the file has changed between each call to `git commit`. Another way to understand how a file has evolved is to look at the file's `git blame` output. The `git blame` command examines a file and announces which commit contributed to each line. For source files, this lets you find out what Git commit contributed to a specific line of code.

15.4.1 Running git blame as a GUI

The `git gui` command supports `git blame` output. You'll see this first, but if difficulties arise with this on your computer, see the next section for an alternative way to obtain this listing on the command line.

> **TRY IT NOW** Using the math.github directory, let's use the same gitk program and limit it to math.sh. Use the steps from the preceding TRY IT NOW to show the math.sh in the Diff/File viewer for the commit `Adding two numbers`.
>
> In the Diff/File viewer, bring up the context menu (by right-clicking, or using a two-finger mouse click on a Mac). You'll see a menu like figure 15.16.

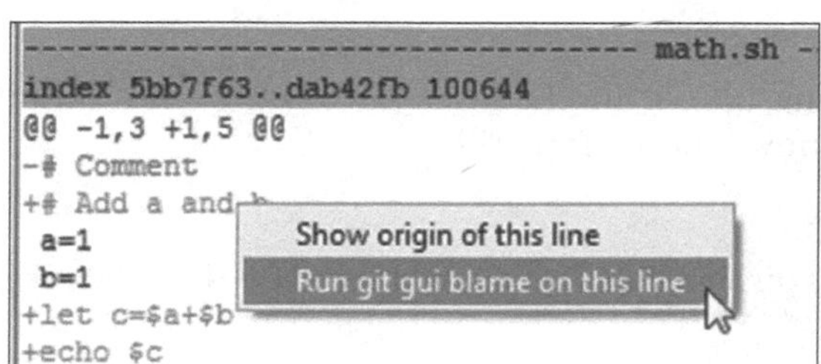

Figure 15.16 Context menu to start `git gui blame`

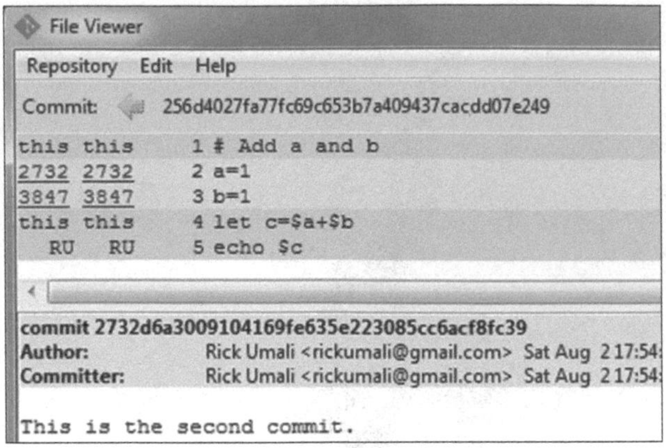

Figure 15.17 `git gui blame` output

Click the Run Git Gui Blame on This Line item. This brings up a separate window, as in figure 15.17.

This new window shows all the lines of the math.sh file for this particular commit. (For you, it's commit `256d402`, in the yellow section of figure 15.17. This is the log entry `Adding two numbers`.) Each line of this file is now annotated with the SHA1 ID of the commit that added the particular line.

This view is a powerful way to figure out the context for a particular set of lines in computer source code. How many times have you wondered why a particular line was added, or who might have inserted a bug into a piece of code? With `git blame`, you can answer those questions.

This window is interactive. You can click through each commit to go backward in time. When you click any line outside the SHA1 ID columns, the lower half of the `git gui blame` window changes, showing the commit log message.

TRY IT NOW The `git gui blame` functionality is part of the Git GUI program. You can start it directly from the command line. Close gitk and Git GUI at this time. Now type this from the math.github directory:

```
git gui blame math.sh
```

This brings up the version math.sh for the current branch. To specify a version, type the following:

```
git gui blame 256d4027 math.sh
```

Alternatively, you can bring up the file browser. This lists all the files in the current branch that Git knows about. Type this:

```
git gui browser master
```

See figure 15.18 for what you should see. You can also type the following:

```
git gui browser HEAD
```

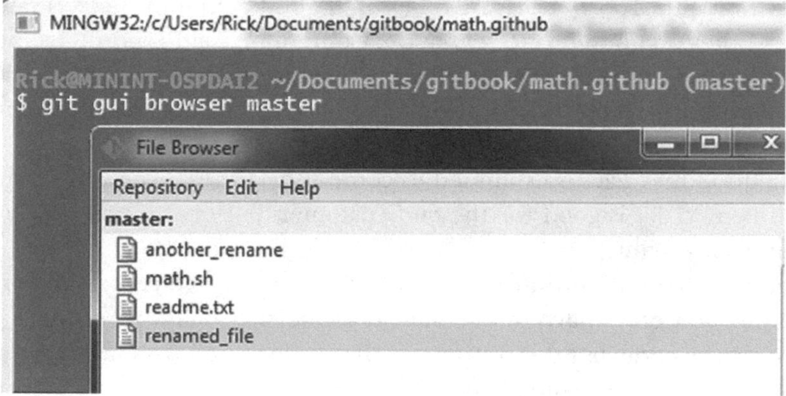

Figure 15.18 Running `git gui browser master` from the command line brings up a file browser.

15.4.2 *Using git blame on the command line*

You may run into problems getting the Git GUI program to run `git blame`. If so, you can use a command-line mechanism.

TRY IT NOW Finally, `git blame` can also be run from the command line, in a far less interactive program. On the command line, type this:

```
git blame math.sh
```

You'll see output like the following listing.

Listing 15.10 `git blame` output

```
256d4027 (Rick Umali 2014-08-05 18:54:56 -0500 1) # Add a and b
2732d6a3 (Rick Umali 2014-08-02 16:54:56 -0500 2) a=1
3847b0be (Rick Umali 2014-08-02 17:54:56 -0500 3) b=1
256d4027 (Rick Umali 2014-08-05 18:54:56 -0500 4) let c=$a+$b
6f6af168 (Rick Umali 2014-08-06 05:54:56 -0500 5) printf "This is the
   answer: %d
```

Because source files will be long, you might consider saving the output of `git blame` to a file by typing the following:

```
git --no-pager blame math.sh > math-annotate
```

Recall that the > symbol redirects the output to a file. Now you can view this file in your favorite editor.

15.5 *Leaving messages for those who follow*

The Git tools you've explored in this chapter show lots of ways that you can learn more about your code base. But these tools are only as helpful as the messages that developers leave behind.

You saw that `git log --oneline` relies on a good commit title. You saw that `git log --grep` relies on good text to search on. Commit messages should be messages to your future self, and to those who have to maintain and work on this code. When you make a commit, you're the expert of that commit. When you make a commit to the repository, you should share the pertinent details that led to this change.

Leaving good messages isn't limited to commit messages. Consider strong branch names and tags. v1.37 is good for the end customer, but maybe branch_fix_memory_leak is better internally.

I named the chapter "Software archaeology" because if you're working on an existing repository, you often have to decipher the clues left behind. You have to understand not only the code, but the context for that code as well. You might learn context from the comments in the source code, but a lot of context might come from the Git commit messages. Commit messages aren't tweets: you can provide elaborate explanations in commit messages. You should cite external sources such as requirements, specifications, documentation, and even bug reports.

15.6 Lab

`git log` has more than 130 command-line switches. Take an afternoon or an evening reading what's available. The switches are subdivided into several sections (commit limiting, formatting, diff options, and so forth). You'll need to explore this documentation to get through the following questions.

1. The `git log --merges` command shows all those commits that are the result of merges. What switch (or switches) is the `--merges` switch shorthand for?
2. In the math.carol repository, create a new branch named `another_rename`. Now type `git log another_rename`. You'll get an error message like the following:

   ```
   fatal: ambiguous argument 'another_rename': both revision and filename
   ```

   Follow the instructions to separate the path from the branch name.
3. When you made our own version of `git log --oneline`, you implemented everything except the color highlighting of the SHA1 ID. How would you add color to the SHA1 ID portion of the output?
4. Where is your Simplified All menu item saved? (Check the help for gitk.)
5. In section 15.2.1, you learned how to make gitk display information for a single file, by typing `gitk filename` on the command line. Can you do this after you've started gitk?
6. What `git blame` switch limits the output to a particular set of lines?

15.7 Further exploration

The `git notes` command is another helpful feature that lets you attach arbitrary notes to Git commits. In Git, after you make a commit, you freeze the commit log message along with the change into your repository's timeline. But let's say a bug was identified as originating from this particular commit. You could attach a short note to the commit that states this fact, without changing the commit itself. Think of `git notes` as yellow sticky notes for your commits.

TRY IT NOW Let's add a note to the last commit in the math.carol repository. Start by typing this:

```
cd $HOME/math.carol
git log -n 1
```

The preceding steps put you in the math.carol working directory and print out the most recent commit. Now type the following:

```
git notes add -m "This is an attached note"
```

This command adds a note to your commit. Type this:

```
git log -n 1
```

You'll see this output.

Listing 15.11 Output of `git log` after attaching a note

```
commit 7746e35930e562304e347ac69929aa276ed345dc
Author: Rick Umali <rumali@firstfuel.com>
Date:   Sun Sep 7 21:51:15 2014 -0400

    Another tiny change

Notes:
    This is an attached note
```

Like most Git commits you've seen, you can add a longer note with an editor by typing `git notes add`. The gitk tool recognizes commits with these extra notes by drawing a little yellow square next to the commit summary (as shown by the circled area in figure 15.19).

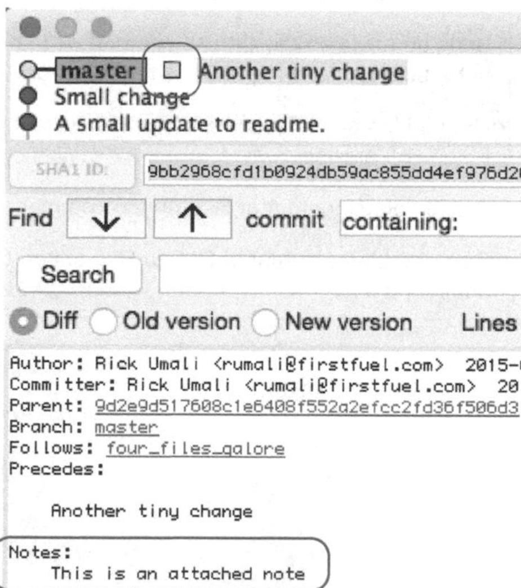

Figure 15.19 `git notes` shown in the gitk tool

15.8 Commands in this chapter

Table 15.1 Commands used in this chapter

Command	Description
`git log --merges`	List commits that are the result of merges.
`git log --oneline FILE`	List commits that affect FILE.
`git log --grep=STRING`	List commits that have STRING in the commit message.
`git log --since MM/DD/YYYY --until MM/DD/YYYY`	List commits between two dates.
`git shortlog`	Summarize commits by authors.
`git shortlog -e`	Summarize commits by authors (and show email address).
`git log --author=AUTHOR`	List commits by AUTHOR (name or email).
`git log --stat HEAD^..HEAD`	List commits (with files) between the current commit and its immediate parent.
`git log --patch HEAD^..HEAD`	List commits (with text changes) between the current commit and its immediate parent.
`git branch --column`	List all branches in columns
`git name-rev SHA1_ID`	Print a name for the specified SHA ID, based on the closest branch.
`git branch -r --contains SHA1_ID`	Similar to the preceding command, in that it will identify all the branches that contain this SHA1 ID (`-r` specifies remote-tracking branches; omit this to print local branches).
`git grep STRING`	Find all files that contain STRING.
`git gui blame FILE`	Bring up a FILE in the Git GUI showing `git blame` output (each line showing what commit it's from).
`git gui browser REV`	List all files at REV (use HEAD for the current directory) in the GUI browser.
`git blame FILE`	Display blame output of FILE on the command line.
`git --no-pager blame FILE > FILE-annotate`	Save the blame output of FILE to FILE-annotate on the command line.

16 Understanding git rebase

The `git rebase` command is one of the most powerful commands in Git. It has the ability to rewrite your repository's commit history, by rearranging, modifying, and even deleting commits. Trying to understand all its capabilities might take you the rest of the book. Instead, you'll focus on the two primary reasons for using `git rebase`: keeping up with the repository you've cloned, and cleaning up your branch before you merge it.

When you clone a repository, you have a copy of that repository. But your collaborators will often add changes to the original repository. You can use `git pull` to refresh your clone, but if you've created a local branch to isolate your development, you may need to resync your branch to incorporate the changes from the original repository. The `git rebase` command handles this, and if you're collaborating on an active repository, you'll find yourself using this technique to keep up with the upstream branch. (Remember, the upstream branch is the original source that you branched from.)

The `git rebase` command also allows you to clean up and edit your commits before you perform a `git merge` of your local branch. In this chapter, you'll use `git rebase --interactive` to exercise full control over how you clean and edit the commits on your local branch.

Finally, you'll study the `git reset` command, which lets you revert your repository to a previous known working state. Along with the `git reflog` command, to help you find the previous known working state, `git reset` is the command that you'll reach for if you make mistakes with the `git rebase` command.

16.1 Examining two git rebase use cases

The `git rebase` command is almost always used in the context of collaboration with other developers. If you're a solo developer, you'll probably never need to do a `git rebase`. But if you're working with even one other person, you may find yourself needing to reparent your branch (to keep up with changes made to the original repository), or cleaning up your branches before you merge them back. This section dives into these use cases at a theoretical level. In section 16.2, you'll perform the `git rebase` steps.

One thing that will help in this chapter is to pretend that the master branch is being actively developed and committed to by other people. You're working on the new_feature branch, but pretend that others are actively working on master, which is your upstream branch.

The `git rebase` command is often scorned because it rewrites history, but think of `git rebase` as a way to keep your local branches up-to-date. It's also a way to edit and polish your local commits before you merge them with the main repository. Keep these two use cases in mind as you dive into the mechanics of `git rebase`.

Notice too that `git rebase` is meant only for your local repository and, even more specifically, your local branch. You'll see that `git rebase` creates new SHA1 IDs whenever you use them, so if you've pushed your work to a public repository, you shouldn't be using `git rebase`.

16.1.1 Keeping up with the upstream by using git rebase

You learned in chapters 9 and 10 how to branch and merge your code. You create local branches to isolate your work from the main branch (which is typically master). This is a best practice. Your branch code is new development. When you branch your code, you implicitly create a starting point for your work. Examine figure 16.1.

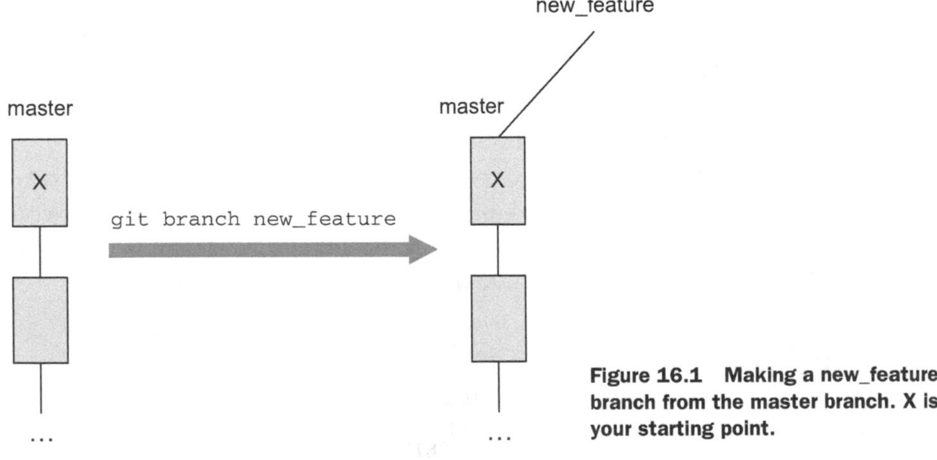

Figure 16.1 Making a new_feature branch from the master branch. X is your starting point.

In figure 16.1, the master branch is to the left of the arrow. When you type `git branch new_feature`, you create the branch new_feature from the commit labeled X. The starting point of your work is on the master branch. When you're in the new_feature branch (by using `git checkout new_feature`), the master branch is considered the upstream, or originating, branch.

You make commits to new_feature, and other people make commits to master. Soon you'll have a situation like figure 16.2: one commit has been added to the master branch (commit Y), and two commits have been made in the new_feature branch (commits A and B).

In chapter 10, you learned how to merge these two separate branches back to master, using `git merge`. A merge joins master and new_feature, bringing the two branches together. But what if new_feature isn't yet ready to merge, but you want the newest commit from master to be part of new_feature? If you want to bring the newest commit or commits from master into new_feature, you can use `git rebase`. Bringing in the newest commit is a way to keep up with the upstream repository. The `git rebase` command resets your branch's starting point from X to Y (in figure 16.2). This is the first use case of `git rebase`. In section 16.2, you'll go over this simple use case.

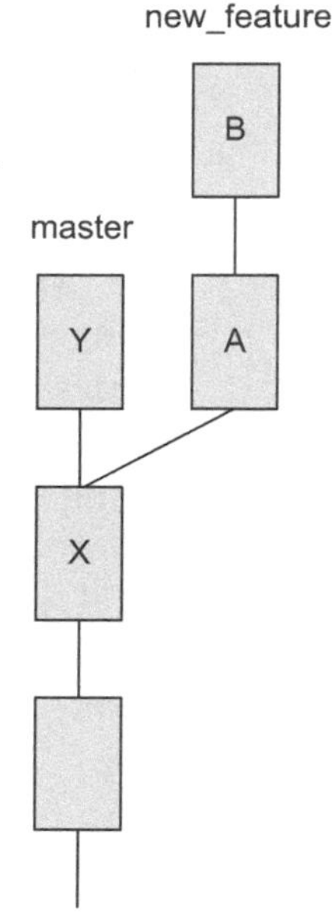

Figure 16.2 Work performed on both master and new_feature branches

16.1.2 Cleaning up history by using git rebase

When you work with others, you'll eventually establish some conventions for your shared repository. You'll read about this topic in the next chapter, but making your commits fit the conventions used by your fellow collaborators is one reason to use the capabilities of `git rebase`.

As with the previous section, let's consider a repository in which your work is in a local branch called new_feature. Let's assume that the seven commits in the `git log` output, shown in the following listing, represent your work on that new_feature branch.

Listing 16.1 `git log` output

```
b8f3239 Fixed last error. This compiled and passed tests!
9afd8a0 Added comment about global.
fb6f863 Incorporated new data structure.
8748134 Fixed spelling mistake.
92ea3ac Fixed script error.
ffcc1f8 Fixed failure on building machine.
bb6ac5e Committing final change for build. Last syntax error!
```

Some of the commit subjects look significant. The commit log `Incorporated new data structure` might be an important one to study, especially for a new developer coming into a project, for example. Some of the commit subjects suggest more day-to-day work, such as fixing a spelling mistake or adding a comment. Remember, each commit represents a complete working directory. When you look at the differences between SHA1 ID `92ea3ac` (fixing a script error) and `8748134` (fixing a spelling mistake), the only difference you'll see is the correction of a spelling mistake. This commit might be an unnecessary detail.

With `git rebase`, you can squash these detailed commits. When you squash commits with `git rebase`, you keep their changes but remove their commit log entries. You'll hear the word *squash* among Git users often. Squashing commits (combining them from one or more commits) is a form of editing your commit history, and you'll examine this technique closely in section 16.4. With `git rebase`, you can modify your commit history, as in figure 16.3.

In figure 16.3, you've reduced the number of commits from seven to four, presumably removing the unnecessary commits' log entries but leaving their changes. Note too that `git rebase` changes the base of the new_feature branch from A to Y.

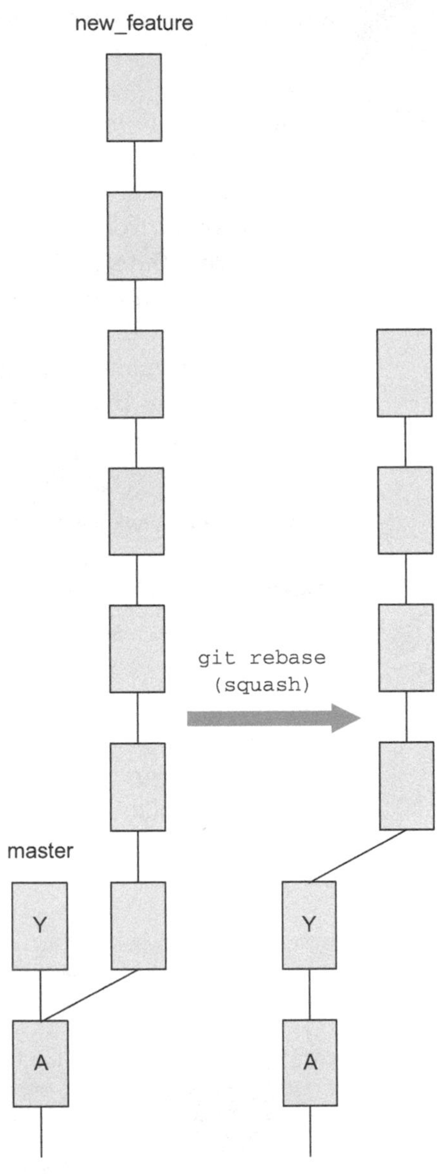

Figure 16.3 Using `git rebase` to squash some unhelpful commits

16.2 Examining use case 1: keeping up with the upstream

The word *rebase* means to give a new parent to a branch. This is what you're doing when you use `git rebase` to keep up with the upstream repository. The most important reason for using `git rebase` is to change the starting point of your local branches.

Remember, in your first use-case, you want to incorporate the new changes from the upstream's master branch into your local new_feature branch. Moreover, you

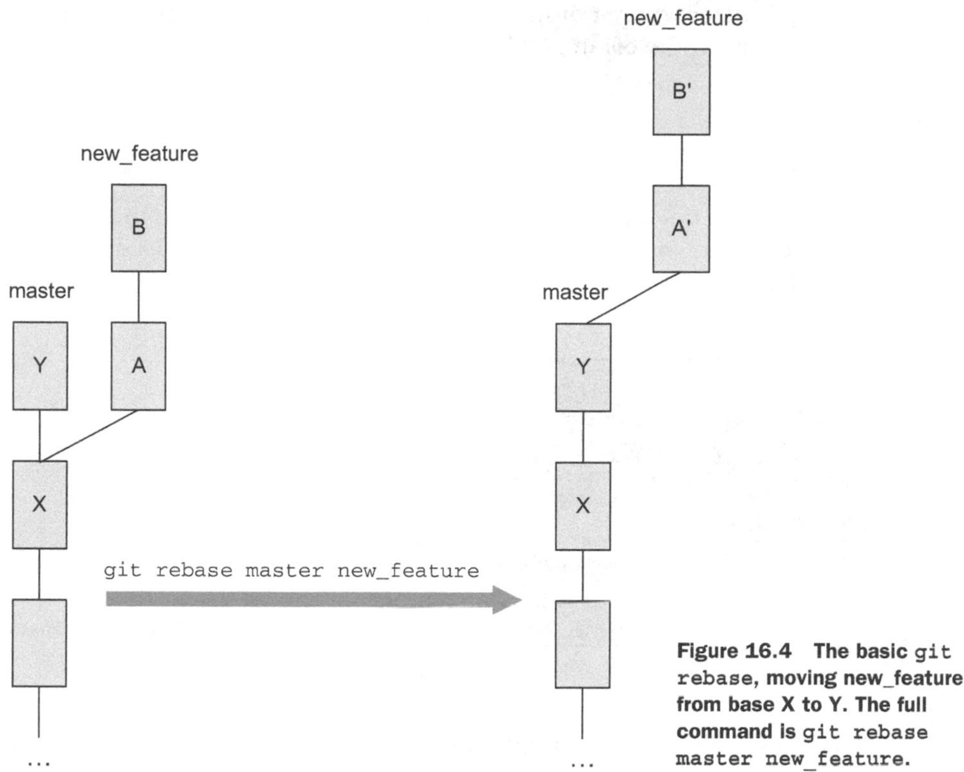

Figure 16.4 The basic `git rebase`, moving new_feature from base X to Y. The full command is `git rebase master new_feature`.

don't want to merge new_feature because it isn't ready to be merged yet. This last requirement is a key consideration: sometimes you may be working on your new_feature for a long time before it's ready to be merged. The `git rebase` command keeps your new_feature branch up-to-date.

In figure 16.4, you start with your repository at the left of the arrow. You're developing a new feature on a separate branch. You based that new feature on the repository at commit X. Commit X is what you branched from, and you could think of it as your base, or starting point. Observe that this base commit, X, has advanced by one commit, Y. If you wanted to incorporate the changes of Y into your new_feature branch without doing a merge, you could choose to rebase your new_feature commits on top of Y.

In our examples, the commit represented by Y is a trivial change to a file. But in the real world, the commit in Y might represent code that you'll need to use in order to continue your development of new_feature. In the real world, Y may consist of multiple commits.

TRY IT NOW To try the most basic form of the `git rebase` command, you'll re-create the math directory by using a script called make_rebase_repo.sh. This script is a variation of the make_math_repo.sh script introduced in chapter 9. After you've downloaded this script into your home directory, type the following at the command line:

```
cd $HOME
rm -rf math
bash make_rebase_repo.sh
cd math
```

If you were to open gitk at this time (with All Refs chosen from the gitk view configuration), you'd see something like figure 16.5. Notice its similarity with figure 16.4.

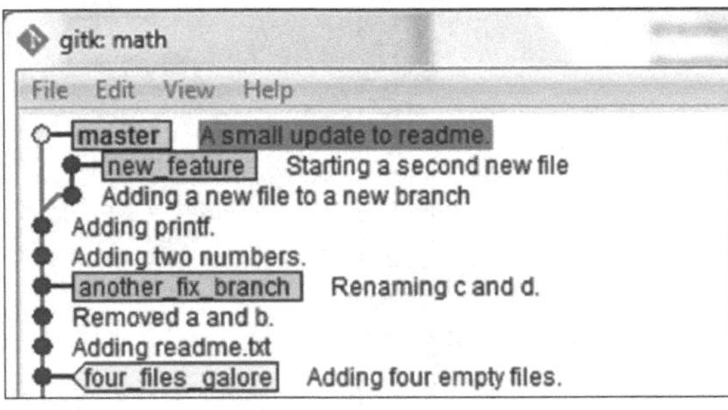

Figure 16.5 Your initial repository

What you'll now do is rebase the new_feature branch on the master branch. This will replay the commits from new_feature on the latest commit from master (Y, in figure 16.4). Confirm that you're on the new_feature branch in the math directory. Type the following:

```
git checkout new_feature
```

Take note of the SHA1 IDs of the two commits of the new_feature branch. Type this:

```
git log --oneline master..new_feature
```

This should list two commits. The `..` syntax shows the commits between master and new_feature. These are the two commits in new_feature that will be rebased. To perform the rebase, type the following:

```
git rebase master
```

Remember, you must be on the new_feature branch. The command indicates to rebase the current branch (new_feature) with the latest commit from the master branch. The following listing shows the output.

Listing 16.2 The output of the `git rebase` master command

```
First, rewinding head to replay your work on top of it...
Applying: Adding a new file to a new branch
Applying: Starting a second new file
```

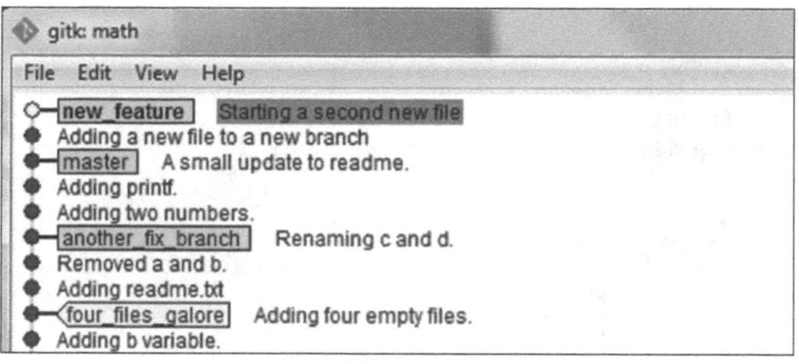

Figure 16.6 The original two commits were moved on top of master.

When you look at gitk, you'll notice that new_feature is on top of master (figure 16.6).

You should also notice that the SHA1 IDs of these two commits are different. Confirm this within gitk, or with `git log --oneline master..new_feature`, or with `git log --oneline -n 2`.

Figure 16.7, an excerpt of our earlier diagram, shows that the original A and B commits that were part of the new_feature branch have changed to A' and B' (read this as *A-prime* and *B-prime*). These are the same changes (with the same dates), but they now have new SHA1 IDs. More important, the new_feature branch has a new starting point, labeled Y in figure 16.7. Any changes in Y are now part of new_feature.

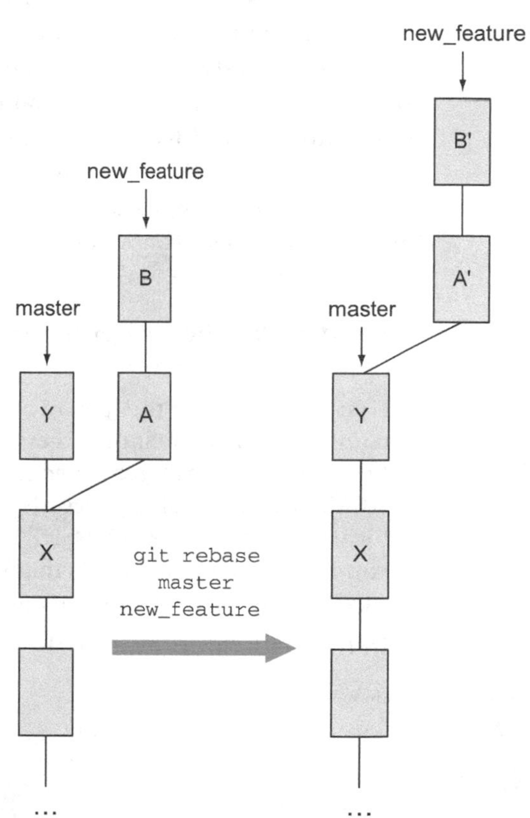

Figure 16.7 Commits A and B are different from A' and B'.

16.3 Using git reflog and git reset to revert your repo

You might inadvertently make a mistake with the `git rebase` command. Before discussing our second use case, which is cleaning up your commits before merging, let's look at how to recover from a mistake with `git rebase`.

Git has a way to reset your local repository back to its earlier state before the `git rebase` command, with the command `git reset`.

Remember from chapter 8 that HEAD always points to the branch (or commit) that Git (and you) are looking at. After the previous section, you can picture your

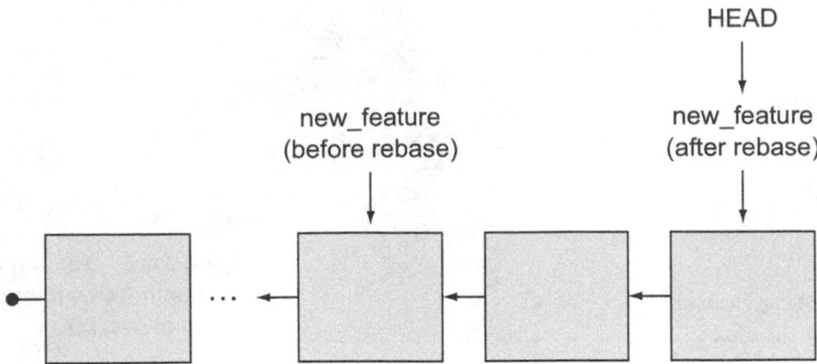

Figure 16.8 Our timeline diagram before and after running `git rebase`

timeline of the new_feature branch to look something like figure 16.8. (The finish of `git rebase` is shown two commits after the original commit, which matches figure 16.7, but remember that this can vary.)

If you wanted to go back to the earlier new_feature, before you ran `git rebase`, you could reset the HEAD pointer to the commit that points to the earlier new_feature. You'd have to identify that SHA1 ID, but you can't use `git log` for this, because the `git rebase` command removes the original SHA1 ID from the history! To find the earlier new_feature SHA1 ID, you'll need to use the `git reflog` command. Let's try out these commands to reset the repository back to the way things were before the rebase.

TRY IT NOW The goal of this session is to reset your local repository to the state before `git rebase`.

First, you must use the `git reflog` command to find the correct name for the commit. This command accesses an internal list of every change to HEAD. Every time you run `git checkout` or perform a `git rebase`, you change the HEAD, and `git reflog` records this by using the reflog. Think of the *reflog* as a local history of sorts. The `git reflog` command accesses this history, letting you reference old SHA1 IDs that may no longer exist. This is an important feature because `git rebase` changes the SHA1 IDs of the branch you're rebasing.

In the math directory, type the following:

```
git reflog
```

The first few lines should look roughly like the following listing.

Listing 16.3 First few lines of `git reflog`

```
9488bc2 HEAD@{0}: rebase finished: returning to refs/heads/new_feature
9488bc2 HEAD@{1}: rebase: Starting a second new file
1a7aa0d HEAD@{2}: rebase: Adding a new file to a new branch
094e6b3 HEAD@{3}: rebase: checkout master
f883bbd HEAD@{4}: checkout: moving from master to new_feature
```

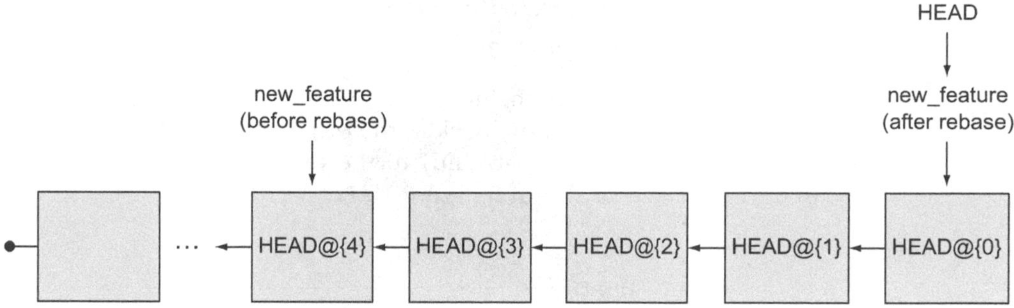

Figure 16.9 Our timeline diagram, overlaid with the `git reflog` names for each spot that HEAD visited

Each row of the `git reflog` output shows a SHA1 ID, an alias for this SHA1 ID, and an explanatory note. The alias is what you're looking for. This listing represents all the individual steps the HEAD has taken. Unlike the `git log` listing, this isn't a chain of commits, with each commit pointing back to its parent. Instead, it's a recording of every branch that HEAD has been set to. Look at figure 16.9.

This figure, along with listing 16.3, shows that HEAD traveled four spots (`HEAD@{3}`, `HEAD@{2}`, `HEAD@{1}`, and `HEAD@{0}`) from the version of new_feature before the `git rebase`. You can assume that this is the right version because the first part of our TRY IT NOW has you performing a `git checkout new_feature`, which you see in step `HEAD@{4}`. Also, `git rebase` seems to take up the first four rows of your reflog.

Finally, before you do the reset, confirm how your commit history looks with gitk. Type the following:

```
gitk
```

You should see a screen that looks like figure 16.10 (with All Refs chosen in the gitk view configuration).

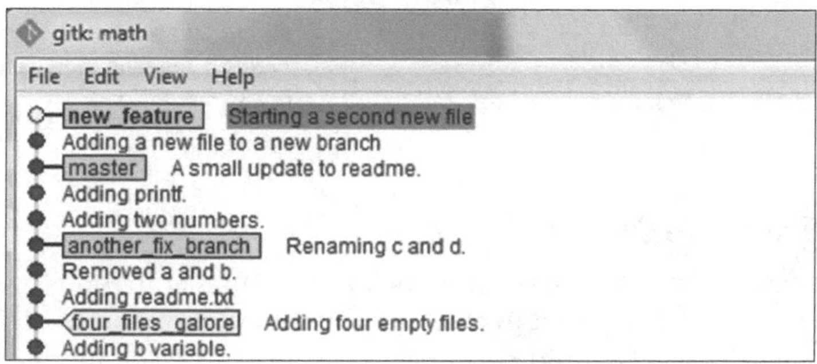

Figure 16.10 Your current repository, after `git rebase`

To reset your repository back to the earlier state, type this:

```
git reset --hard HEAD@{4}
```

The `--hard` switch resets both the staging area and the working directory. This properly sets the repository back to the original state. After you type this command, HEAD is moved (rewound) to the version of new_branch before you used `git rebase`, as shown in figure 16.11.

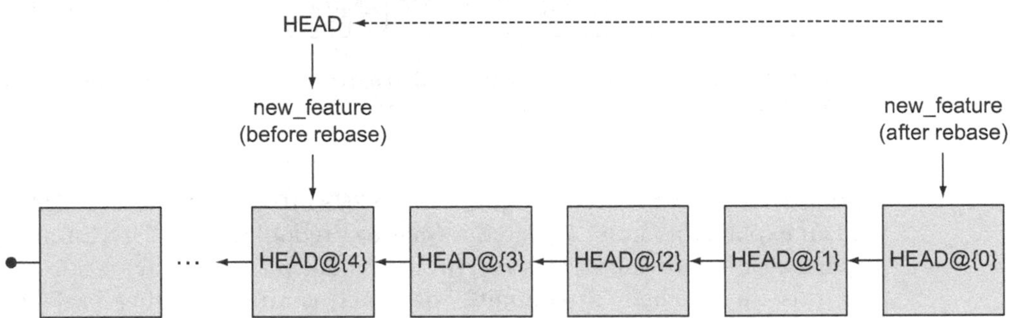

Figure 16.11 Moving HEAD backward in reflog history (to `HEAD@{4}`)

You can confirm this. Type the following:

```
gitk
```

Now confirm (after making sure your gitk view configuration has All Refs) that your repository looks like figure 16.12. You've gone backward in time by using `git reset`.

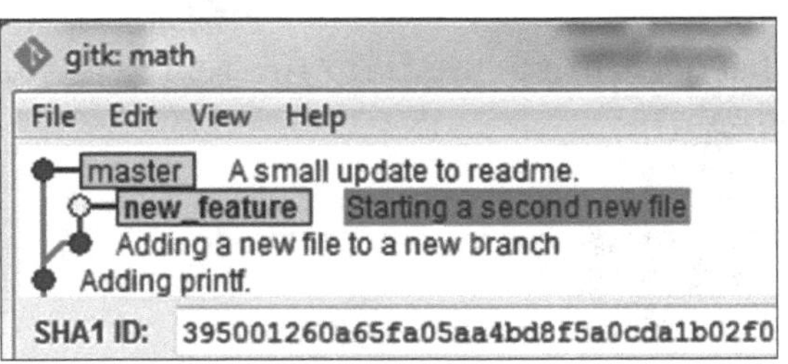

Figure 16.12 After using the `git reset` command, new_feature is back where it was.

16.4 Examining use case 2: cleaning up history

The second use-case for `git rebase` is to clean up commits. If you're collaborating with others, you might be required to clean up your commits before merging them to the master branch. Typically, you'd clean up your commits before merging your local

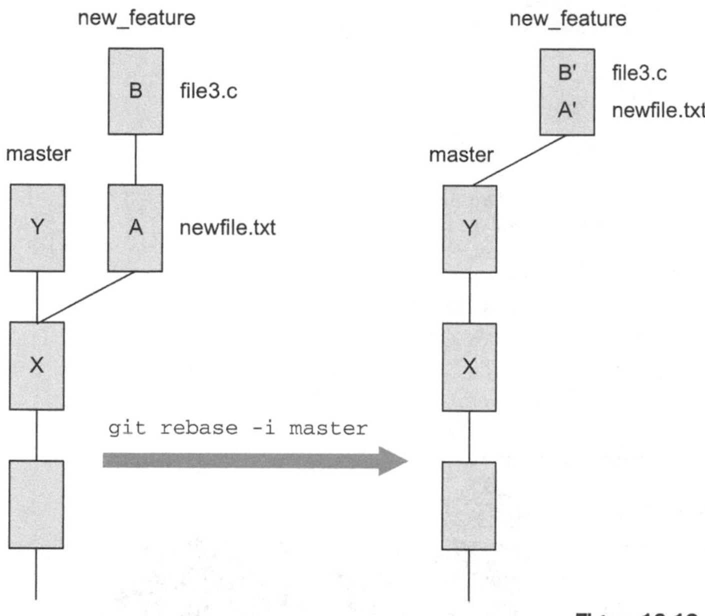

Figure 16.13 Your goal for this session

branch back to the main branch. If your repository doesn't follow any convention, you can merge your branch back to master, but cleaning up your local commits is a good practice. Consider figure 16.13.

Look at the two commits for new_feature. Each commit adds a new file to the repository. What if you wanted to make one commit to add both of these files?

You might want to squash multiple commits in this fashion because some repositories follow a convention of wanting to manage the number of commits that appear in the main branch. In this convention, you're encouraged to make as many commits as you want in your local branch, but when you're ready to share your work, you must squash your entire local branch before merging!

> **TRY IT NOW** Your repository should be in the correct state, thanks to the previous TRY IT NOW session. What you'll do now is confirm that you're in the right branch, and then you'll use `git rebase --interactive`. Type the following:
>
> ```
> git branch
> git log -n 2 --stat --oneline
> ```
>
> Confirm that you're in the new_feature branch, and that the previous two commits add both file3.c and newfile.txt. The two commands should look like the following listing.

Listing 16.4 Output from your `git branch` and `git log` commands

```
% git branch
  another_fix_branch
  master
* new_feature

% git log -n 2 --stat --oneline
3950012 Starting a second new file
 file3.c | 1 +
 1 file changed, 1 insertion(+)
b402970 Adding a new file to a new branch
 newfile.txt | 1 +
 1 file changed, 1 insertion(+)
```

Now type this:

```
git rebase --interactive master
```

Your command-line window changes to a text file inside Git's default editor, vi. (You'll learn how to change this default editor in chapter 20.) The window looks roughly like figure 16.14.

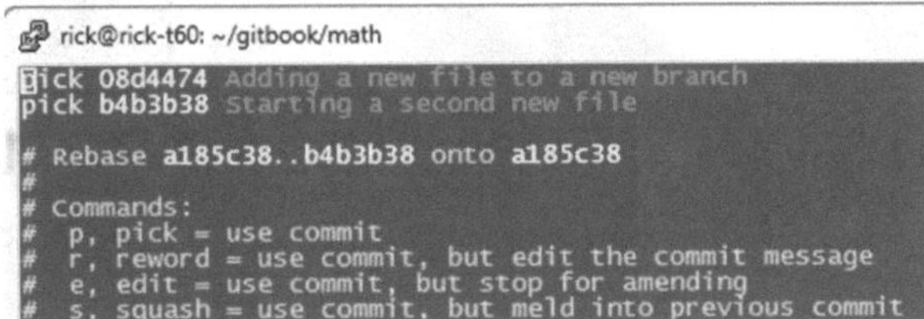

Figure 16.14 The `git rebase` interactive screen. (It's that vi editor window.)

The full text file inside the vi editor is shown in the following listing.

Listing 16.5 Annotated text of `git rebase --interactive`

```
pick 08d4474 Adding a new file to a new branch
pick b4b3b38 Starting a second new file

# Rebase a185c38.. b4b3b38 onto a185c38
#
# Commands:
#  p, pick = use commit
#  r, reword = use commit, but edit the commit message
#  e, edit = use commit, but stop for amending
#  s, squash = use commit, but meld into previous commit
#  f, fixup = like "squash", but discard this commit's log message
#  x, exec = run command (the rest of the line) using shell
#
# These lines can be re-ordered; they are executed from top to bottom.
#
# If you remove a line here THAT COMMIT WILL BE LOST.
#
# However, if you remove everything, the rebase will be aborted.
#
# Note that empty commits are commented out
```

1 The list of commits that you'll rebase

2 Instructions for commands to perform on each commit

The text file is divided into two parts: the list of commits available for the `git rebase` command ❶, and the instructions for how to interact with the list of commits ❷. Each line ❶ is preceded by an instruction (`pick`, `reword`, `edit`, `squash`, `fixup`, or `exec`).

Next you'll edit this file, specifying the instruction for each commit. This produces figure 16.15.

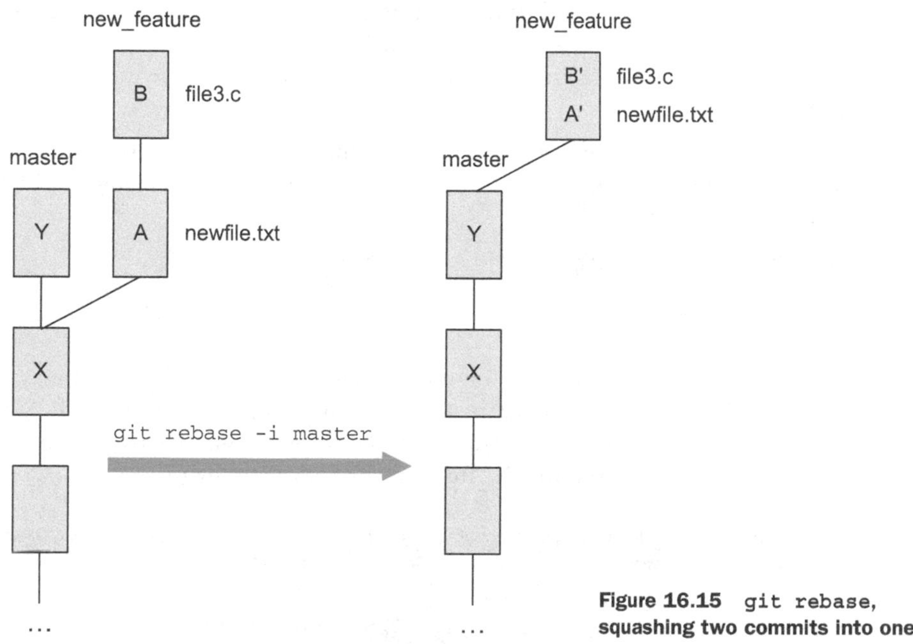

Figure 16.15 `git rebase`, **squashing two commits into one**

To do this, you have to edit the file. You'll change the second line's command from `pick` to `squash`. This change looks like figure 16.16.

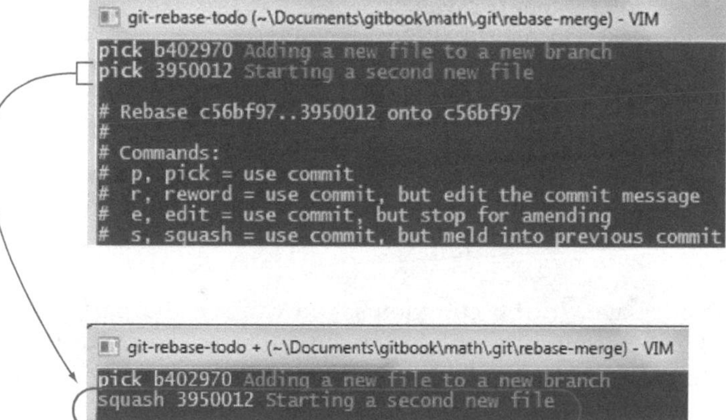

Figure 16.16 Your editing session to squash the two commits

After your change, the first two lines of your text file look like the following listing.

Listing 16.6 Your rebase file after you change the second line

```
pick 08d4474 Adding a new file to a new branch
squash b4b3b38 Starting a second new file
```

Each commit is processed one line at a time, from top to bottom. This order is reversed from the order presented with the `git log` command. When you enter the word `squash` in the second commit, Git melds it into the previous commit.

In the vi editor, arrow down to the first character of the second line (on the letter `p`). Then delete the word `pick` and insert the word `squash` in its place, as in figure 16.16.

Now type the following:

```
<ESC>
:wq
```

Pressing the Esc key ensures that you're not in vi's insert mode. The `:wq` saves the text file. The editor disappears, and the terminal window indicates that it's doing some work. After a moment, it presents some more text in another vi editor. The text contains the lines in the following listing.

Listing 16.7 Text for commit message after squashing two commits

```
# This is a combination of 2 commits.
# The first commit's message is:

Adding a new file to a new branch

# This is the 2nd commit message:

Starting a second new file


# Please enter the commit message for your changes. Lines starting
# with '#' will be ignored, and an empty message aborts the commit.
# rebase in progress; onto c56bf97
# You are currently editing a commit while rebasing branch
# 'new_feature' on 'c56bf97'.
#
# Changes to be committed:
#       new file:   file3.c
#       new file:   newfile.txt
#
```

This is a new commit message. In the comments, you can see the part that indicates the changes to be committed: file3.c and newfile.txt. To make the rebase easy to finish, let's use this commit message as is. Type the following:

```
:wq
```

This saves the current message and exits `git rebase`. The final output of this `git rebase` command should look like the following listing.

Listing 16.8 Output of `git rebase`

```
[detached HEAD 00ee93f] Adding a new file to a new branch
 2 files changed, 2 insertions(+)
 create mode 100644 file3.c
 create mode 100644 newfile.txt
Successfully rebased and updated refs/heads/new_feature.
```

To confirm that the new commit adds both file3.c and newfile.txt, type this:

```
git log -n 1 --stat
```

This command should show the output in the following listing. More important, both files will be present in the math directory.

Listing 16.9 Output of `git log` to confirm squash worked

```
commit 00ee93f1de6bd214ee25a1672d826adeef1b37da
Author: Rick Umali <rickumali@gmail.com>
Date:   Sat Oct 4 03:10:15 2014 -0500

    Adding a new file to a new branch

    Starting a second new file

 file3.c     | 1 +
 newfile.txt | 1 +
 2 files changed, 2 insertions(+)
```

From figure 16.15 and the previous `git log`, it should be apparent that squashing is a way of combining commits. This is one way to remove intermediate commits when their commit log entry isn't needed. You might consider using squash to remove the checkpoint commits where you fixed spelling mistakes, for example.

In the end, editing and refining your commits to fit the conventions of your repository are the marks of a good contributor.

16.5 Lab

It's probably worthwhile to review the TRY IT NOW exercises again. The `git rebase` command is complicated because it demands a good understanding of how the branches and commits work within a repository. Practicing the TRY IT NOW exercises will help you recognize when you should use `git rebase`.

1 Find the math.bob repository and then do a `git rebase` on the remote-tracking branch. Type the following:

```
git rebase master origin/new_feature
```

What branch are you on when this command finishes?

2. Reset the math repository as in the previous exercise, and then try to squash the first commit listed. You should get an error. Why?
3. Read the `git reflog` documentation. What command is it shorthand for? Try this other command, and confirm that it gives the same output as `git reflog`.
4. Try another `git rebase --interactive`, and this time explore using the `reword` command next to a commit (see listing 16.5). What does this let you do?
5. You used the syntax `master..new_feature` to see the commits reachable from new_feature and excluding master, as explained by the `git rev-parse` documentation. What is another way to express `master..new_feature`?

16.6 Further exploration

In this chapter, you covered how to manipulate your commit log with `git rebase`. You can explore two intriguing techniques on your own that are related to manipulating your history.

16.6.1 Cherry picking

Sometimes you might want to take a branch and copy it onto a new starting point. Your work on one branch is possibly applicable elsewhere in the repository, and you might want to copy this code to this new starting point. Git can do this with the `git cherry-pick` command. In figure 16.17, see that the commit labeled `b` is copied on top of master.

To explore this on your own, create a repository and a second branch. In the second branch, make a commit that adds one file. Check out the master branch, and

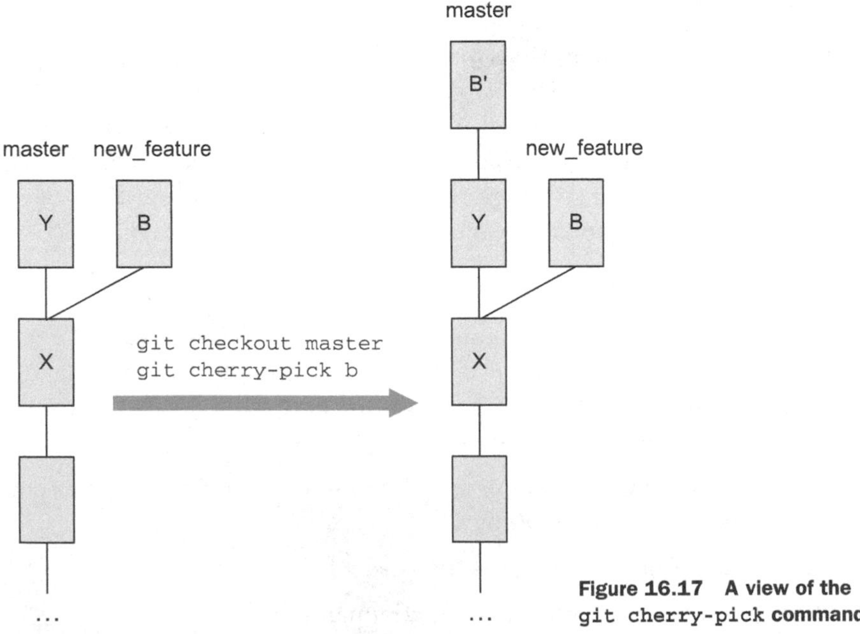

Figure 16.17 A view of the `git cherry-pick` command

then use git cherry-pick, specifying the SHA1 ID from the branch that added the file. A session that did this might look like the following listing.

Listing 16.10 git cherry-pick example

```
% git checkout master

% git log --graph --all --oneline --decorate
* ef05774 (HEAD, master) Updating the contents of bar
| * 01fdf3a (branch01) Updating baz
| * 8a1a317 Adding baz file
|/
* e9fb7a4 Adding foo file
* f171aa9 Adding bar file

% git cherry-pick 8a1a317
[master a798ed7] Adding baz file
 1 file changed, 0 insertions(+), 0 deletions(-)
 create mode 100644 baz

% git log --graph --all --oneline --decorate
* a798ed7 (HEAD, master) Adding baz file
* ef05774 Updating the contents of bar
| * 01fdf3a (branch01) Updating baz
| * 8a1a317 Adding baz file
|/
* e9fb7a4 Adding foo file
* f171aa9 Adding bar file
```

The git cherry-pick command makes a copy of the single SHA1 ID. Depending on circumstances, this technique may be helpful.

16.6.2 Commit deleting

If you've created a commit (or a set of commits) that need to be deleted, git rebase can do this for you, using the --onto switch. To explore this on your own, use the another_fix_branch commits from the math repository, as shown in the following listing.

Listing 16.11 The another_fix_branch commits

```
$ git log --oneline --graph --decorate
* 90e94aa (HEAD, another_fix_branch) Renaming c and d.
* 637439d Removed a and b.
* a374d84 Adding readme.txt
* 9252019 (tag: four_files_galore) Adding four empty files.
* c916ada Adding b variable.
* 5a5e48a This is the second commit.
* 5a8ca2f This is the first commit.
```

Let's say that you wanted to get rid of the commit that removes a and b (637439d in listing 16.11).

First, examine the output of this command:

```
git log --oneline another_fix_branch~1..another_fix_branch
```

This command shows one commit, `90e94aa (Renaming c and d)`. This is the commit at the tip of another_fix_branch. This is the commit that will be rebased.

Now, per the `git rebase` documentation, you can type this to delete `637439d`:

```
git rebase --onto another_fix_branch~2 \
  another_fix_branch~1 another_fix_branch
```

The commit that you'll rebase onto is `Adding readme.txt` (`a374d84` in listing 16.11). By rebasing on this commit, you effectively delete the commit that comes after it (`637439d` in listing 16.11). The commit that you're rebasing is the single commit specified by the range another_fix_branch~1 and another_fix_branch (which you saw with `git log`).

It's an interesting technique, but it's one that I hope you never have to use!

16.7 Commands in this chapter

Table 16.1 Commands used in this chapter

Command	Description
`git log --oneline master..new_feature`	Show the commits between the master branch and the new_feature branch.
`git rebase master`	Rebase your current branch with the latest commit from master.
`git reflog`	Display the reflog (the internal history of all the times that you changed HEAD).
`git reset --hard HEAD@{4}`	Reset HEAD to point to the SHA1 ID represented by `HEAD@{4}`. The `--hard` switch says to reset both the staging area and the working directory.
`git rebase --interactive master`	Interactively rebase your current branch with the latest commit from master. This opens an editor, allowing you to pick and choose which commits will be included in the rebase.
`git cherry-pick SHA1 ID`	Copy the commit to the current branch that you're on.

17 Workflows and branching conventions

Over the past 16 chapters, you've learned a great deal about Git. You now know how to create and update your local repository, and you also know the fundamentals of collaborating. The various Git mechanisms (branches, tags, commit messages) are just that: mechanisms. They don't impose any kind of policy or convention.

In this chapter, you'll discuss matters of policy and convention. It's important to at least get a feel for some of these, because when you collaborate with others, knowing the common conventions will prevent you from making mistakes with the code base. Your code base will be examined not just by other developers but also by automated testing systems, QA, support, and possibly even documentation. Conventions help all these audiences.

You'll first look at the common Git features that require good conventions. These open-ended commands don't impose any rules, so it's up to you to figure out how to use them sensibly and consistently. What's needed is a workflow, a sequence of steps you can follow for common source-code control tasks.

You'll then survey two popular workflows: git-flow and GitHub flow. These common workflows appear whenever you search the Internet about Git workflows. You'll use the appropriate Git commands to implement these flows, so you'll get a taste for which might suit your group, or how to adapt them for your specific needs.

17.1 The need for Git conventions

When multiple people are collaborating on a shared set of code, you need conventions. *Conventions* are like traffic signs and signals. They're there to enforce the rules of the road and to prevent accidents. Traffic signs and signals enforce an orderly flow of traffic, and serve as reminders about what particular driving technique to use (right on red, yield, merge, speed limits).

Software developer/manager Philip Chu wrote in his book *Technicat on Software*, "The fate of your company could depend on the ability to make a clean build." Conventions can be instrumental to an organization's ability to have clean builds.

In the following sections, you'll look at a few Git commands that require conventions. These are all commands that you've seen, but because of their open-ended nature, it would be helpful to set conventions on how they should be used in a particular repository. These conventions may change! Some of the conventions must be strict (when and where to push), but some can be fluid (commits).

Think of these conventions as rules of the road for a car. In this case, Git is your vehicle, but you need some rules of the repository so you can drive in a predictable and orderly fashion. In *Code Complete*, author Steve McConnell argues for conventions so programmers can handle the details of programming consistently. This is the same with Git. Many Git commands can be run in an arbitrary fashion. Deciding on how to use Git in these cases makes it easier for people joining the project.

17.1.1 Conventions for commits

Making a commit means updating your local timeline. When you make a commit, you're stating that the current directory of code represents a certain state. Your commit log message should indicate that current state, and how the change brings us there. For example, a good commit log subject is `Fixed bug 17414, the data shift issue`. This means that the changes introduced by this commit fix this particular bug.

Because Git is a distributed version-control system, each individual user can make commits at any time to their local repository. Some organizations might stipulate that a commit must not break the build, or that a commit must be accompanied by a unit test. You might even see some places try to enforce how the commit message should be written.

The only guideline to try to follow is to keep the Git commit subject (that first line in a commit message) under 50 characters. This way, the output of `git log --oneline` won't be truncated or wrapped.

Your local commits do become public when you push code, however.

17.1.2 Conventions for pushing code

Conventions for pushing code to a shared repository must be agreed upon. Because Git is a distributed version-control system, there might be more than one remote, but most places do use a shared server where the official code base resides. This shared server is often the location that automated testing, building, or deployment software will reference.

Organizations have a wide range of options when it comes to the release process. Some companies have departments that focus on release tasks. Some companies, adopting continuous integration principles, follow a more automated approach. Releasing code often includes preparing a release candidate from a branch that builds cleanly.

Because Git allows everyone to have the entire repository, it may make sense to limit who can push to where. Individual developers can commit as often as they'd like to their local repository, but the group or automation that deploys or builds the release may want to build up their own repository by pulling from individual repositories. Allowing everyone the ability to push to the official code base can get messy.

17.1.3 Conventions for branching

Some organizations may want to control the number of branches in the shared repository. Users working on their local repository should feel empowered to create as many local branches as they need to stay organized, but pushing such branches to a shared server might require conventions.

One such convention could standardize the names of branches. Because branches can be given any name, you might establish a convention requiring each branch to have a folder-like name such as feature/new_field, which means a feature branch for the new_field feature. Or the branch name could include the ID of the owner, which makes it easy to detect who is doing what on a repository.

Branching is the heart of the workflows that you'll be looking at later in this chapter. Because a branch represents a snapshot of your code, different parts of a large organization may choose to have a repository with two or more active lines of development. One branch would be used by the developers, one by the automated tests, another representing the deployed software, and so on.

17.1.4 Conventions for using rebase

Do you want your code to have every incremental commit? Or is a single commit representing the entire work for a feature or bug fix sufficient? If so, your organization will want to set guidelines around when and how to rebase. Conversely, if your organization wants to maintain a complete and unaltered history of development, warts and all, it may prohibit developers from using the `git rebase` command.

Rebase is helpful for individual developers to use on their own branches to prepare them for sharing. As discussed in chapter 16, cleaning up your work and making a presentable set of commits that represent it is welcome.

17.1.5 Conventions for tagging

The `git tag` command allows you to make bookmarks to any part of the commit history timeline. But like commits, tags are considered local until they are pushed to the main repository. As with branches, your company may want to impose a tag name convention. You may want to use it only to mark code that has been released.

17.2 Two Git workflows

This section introduces two popular Git workflows: git-flow and GitHub Flow. These are representative of the kinds of workflows that take advantage of Git's features. They're by no means the only two workflows, and Git is open enough that it can be made to fit different conventions. Table 17.1 presents key criteria you might think about for a workflow, and outlines how git-flow and GitHub Flow address those features.

Table 17.1 Workflow criteria addressed by git-flow and GitHub Flow

Feature	git-flow	GitHub Flow
Use of `git tag`	Yes	No
Long-lived branches	Two: develop and master (and possibly more)	One: master
`git merge --no-ff` required	Yes	No
Release numbering (for example, V1.0)	Important	Not a focus
Delete feature branches	Yes	No
Hotfix branch	Yes	No
Support scripts	Yes (see "Further exploration"), helpful but not required	Some, but not required

As a developer, joining any new development team requires you to learn the policies and conventions of the new group. Just as most teams have guidelines on when to take a vacation or conduct reviews, these same teams will have conventions that spell out how the repository should be used.

Computer source code has many audiences: developers, QA automation systems, release engineering, and even technical support. Workflows help people understand how the code is organized in relation to the developers and how to add code safely.

17.3 git-flow

git-flow is probably the most popular workflow you'll read about when it comes to Git. It was written by Vincent Driessen (a.k.a. nvie) and published in January 2010. It proposes that your repository be divided into two infinite branches: master and develop. The URL that describes git-flow is http://nvie.com/posts/a-successful-git-branching-model/.

The master branch contains released, production-level code. This is what the public can see, perhaps on a deployed website or in some released software that they've downloaded from you. The develop branch contains code that is about to be released. This is depicted in figure 17.1.

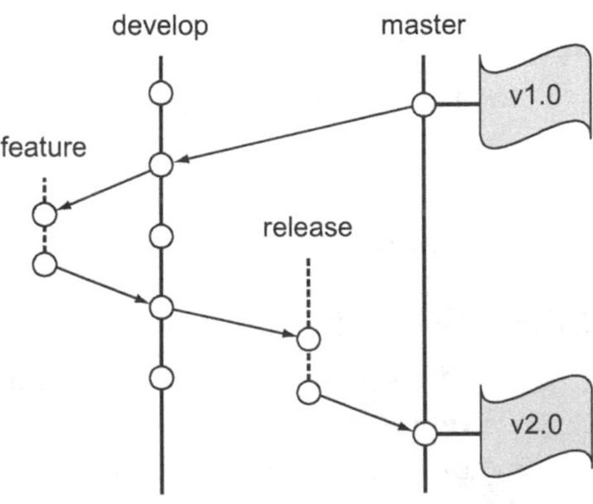

Figure 17.1 The two main branches in a git-flow repository. Other branches such as feature and release are created temporarily and then deleted when finished.

When the code in the develop branch becomes mature enough, and a release is required (or declared), then develop is merged into master, and master is tagged with a release number. Both develop and master are considered long-lived branches, or infinite branches, because they are never deleted in git-flow.

The workflow defines what Git commands to run when you want to transition your code from a new feature idea to mature code that should be merged to develop. Likewise, the workflow defines the Git commands to run when you want to release code. Both of these transitions use temporary branches. Let's explore these two aspects of the workflow: working with a feature branch and then releasing that code. To do this, you'll first create a new repository that is correctly set up for git-flow.

17.3.1 Making a feature branch

Imagine you're adding a feature to add two numbers. This will be a part of your released software. In git-flow, all feature development starts by branching off the develop branch. Let's explore this workflow. The book's website has a script that runs through the steps in this section, and the next, so don't worry if you get lost!

TRY IT NOW In your $HOME directory, let's create a new repository, and set it up so that you can use the git-flow workflow with it.

```
cd $HOME
mkdir nvie
cd nvie
git init
```

This gives you an empty repository. Remember that Git supplies a master branch already. For git-flow, you now need to make a branch named *develop*. Here's a fast way to do this:

```
git commit --allow-empty -m "Initial commit"
git branch develop
```

You can probably guess that the `--allow-empty` switch allows for a commit with no files, and it's a handy mechanism in this case. The master branch must contain an initial commit before you can create the develop branch from it.

After these two steps, you have two branches: master and develop. These are the long-lived branches of this workflow. Both branches initially are at the same starting point. Confirm this by typing the following:

```
git branch
```

You should see that you have two branches: master and develop. Git should indicate that you're still on the master branch. To see that both master and develop are starting at the same commit, type the following:

```
git log --decorate
```

The `--decorate` switch shows the branches that are part of this commit. See the following listing for this output.

Listing 17.1 Confirming that branch and master start on the same commit

```
commit 3bfe0c0e3991b598ffecce6626d2cbed8dfaf0e9 (HEAD, master, develop)
Author: Rick Umali <rickumali@gmail.com>
Date:   Mon Dec 22 21:47:39 2014 -0500

    Initial commit
```

You're now ready to make a feature. Your repository roughly looks figure 17.2. There's one initial empty commit and two branches.

In the git-flow workflow, you don't work on develop directly. Instead, you make a feature branch off it.

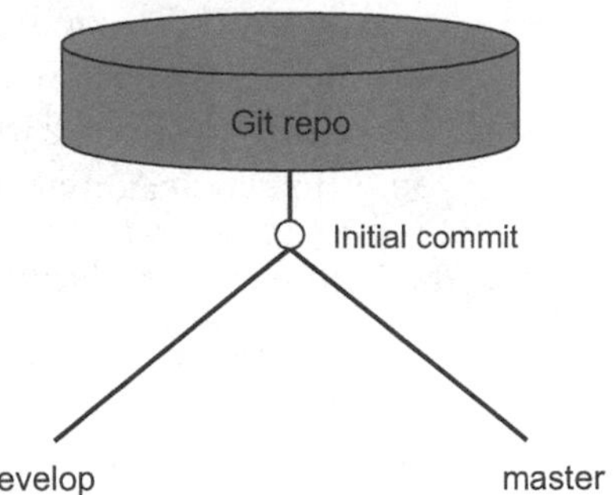

Figure 17.2 Your repository at the start of using git-flow

> **TRY IT NOW** In the nvie directory, type the following:
>
> ```
> git checkout -b feature/sum develop
> ```
>
> Notice that the name of your new branch is feature/sum. (Remember from chapter 9 that the -b switch to `git checkout` causes a new branch to be created.) The / character isn't important to Git, but some third-party tools will display this branch in a folder named *feature* because of the use of the / character. Remember, this is a convention. Git lets you name your branch practically anything.

At this point, you have a separate feature branch that you'll work on until your feature is ready. You'll now commit some code into it.

> **TRY IT NOW** In the nvie directory, confirm that you're on the feature/sum branch by typing the following:
>
> ```
> git branch
> ```
>
> This command should result in the output *feature/sum*. Let's add a simple program to this branch. Using your favorite editor, create a file called `sum.sh` and have it contain these lines:
>
> ```
> # Add a and b
> a=1
> b=1
> let c=$a+$b
> printf "This is the answer: %d\n" $c
> ```
>
> These lines may be familiar: they're the same as the math.sh file you used earlier. Remember that you can run this by typing the following:
>
> ```
> bash sum.sh
> ```

Now add your sum.sh file to the repository by typing this:

```
git add sum.sh
git commit -m "The sum program"
```

This commits your code to the feature/sum branch. Your repo should look roughly like figure 17.3.

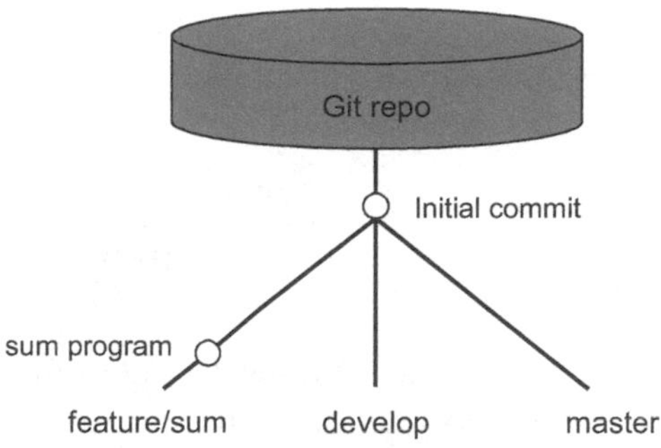

Figure 17.3 The repository after you make a commit to the feature/sum branch

In the conventions for git-flow, when your code is deemed ready, you have to merge it into develop before it can be released. The develop branch is the long-lived branch that represents the code that is coming up for release.

TRY IT NOW To merge a branch into develop, you must first be on the develop branch. Merges branch into us, remember. Type the following:

```
git checkout develop
```

Observe that no files are in the working directory! This is because no work has been made in the develop branch yet.

The git-flow convention says to use the `--no-ff` switch to `git merge`. This prevents a fast-forward merge (see chapter 10) by creating a merge commit. This preserves the history of the merge. Type the following:

```
git merge --no-ff feature/sum
```

If Git's default editor appears, accept the autogenerated merge message by typing this:

```
:wq
```

To confirm that the merge commit has been created and to view its message, type the following:

```
git log -1
```

This shows the last commit, which will be your merge commit, as in the following listing.

Listing 17.2 A merge commit with an autogenerated merge message

```
commit 22428bb3f6228c2cf3b54ebcce483722275de127
Merge: 3bfe0c0 d942d42
Author: Rick Umali <rickumali@gmail.com>
Date:   Mon Mar 16 21:29:03 2015 -0400

    Merge branch 'feature/sum' into develop
```

Now that the code is part of the develop branch, you no longer need the feature/sum branch. The workflow has you deleting these unneeded branches. Type the following to delete the feature/sum branch:

```
git branch -d feature/sum
```

What you've done is depicted in figure 17.4.

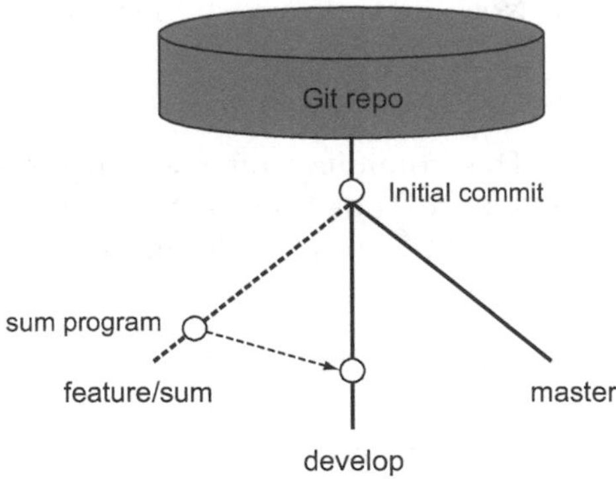

Figure 17.4 You've merged your program into develop.

Other developers will be merging code into the develop branch as well. Further, your repository will probably have set up a remote that you should push changes to. That's the workflow from the developer side. Let's see how to deliver this code into a version 1.0 release.

17.3.2 Making a release branch

After the develop branch is worthy to become a full release, you must merge develop into master, by way of a release branch. Remember that in this workflow, master always represents your released code. The release branch might be considered the candidate. Testing can be performed on this branch, as well as fixes, but it's expected to become a release. The convention also demands a tag, which will be created to mark this release.

TRY IT NOW Let's release the code that you have in develop. To start, you merge develop into a release branch. Type the following:

```
git checkout -b release-1.0 develop
```

Now you'll bump the version. This is also a convention in this workflow. A file will be updated to include your release number. Use your favorite editor to add this comment to the top of the file sum.sh:

```
# Version 1.0
```

Now commit this change:

```
git commit -a -m "Bumping to
  version 1.0"
```

Figure 17.5 shows what your code looks like now.

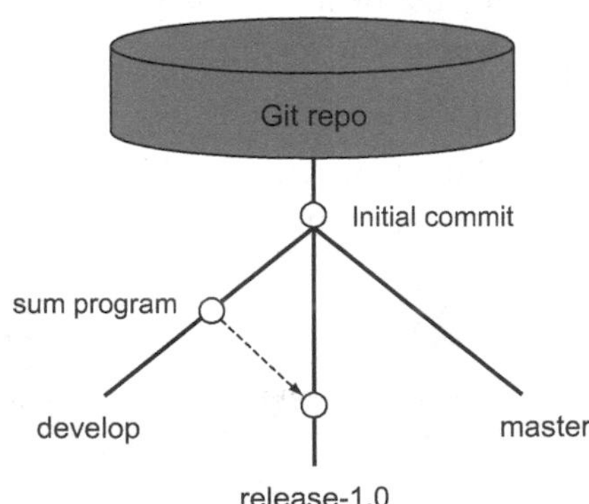

Figure 17.5 Now you have a release branch, but it's only temporary.

Let's release this without any further commits. What typically happens, however, is that the release-1.0 branch may have other commits added to it as the release is finalized. Each of these commits would need to be merged back to develop. You won't study that case. Instead, type the following to release the code:

```
git checkout master
git merge --no-ff release-1.0
```

If Git's default editor appears, accept the autogenerated merge message by typing this:

```
:wq
```

Now type the following:

```
git tag -a V1.0 -m "Release 1.0"
```

In this command, `git tag`, you specify a message with the `-m` switch. If you omitted this switch, Git's default editor would open so you could enter a message.

Your repository looks like figure 17.6 now.

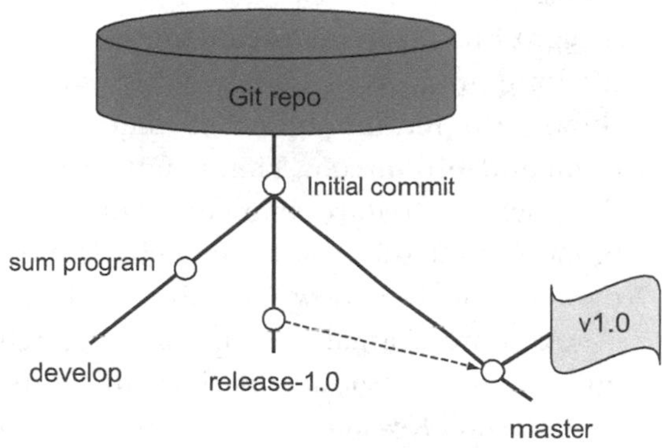

Figure 17.6 Releasing code with git-flow

Don't forget, per the workflow, you must merge the just-released code back into the develop branch, so that anyone who wants to implement a new feature will have the release-1.0 code as the starting point. Type the following:

```
git checkout develop
git merge --no-ff release-1.0
```

If Git's default editor appears, type the following to accept the autogenerated merge message:

```
:wq
```

Finally, you can delete the release branch. It's no longer needed. Type this:

```
git branch -d release-1.0
```

Your repository finally looks like figure 17.7.

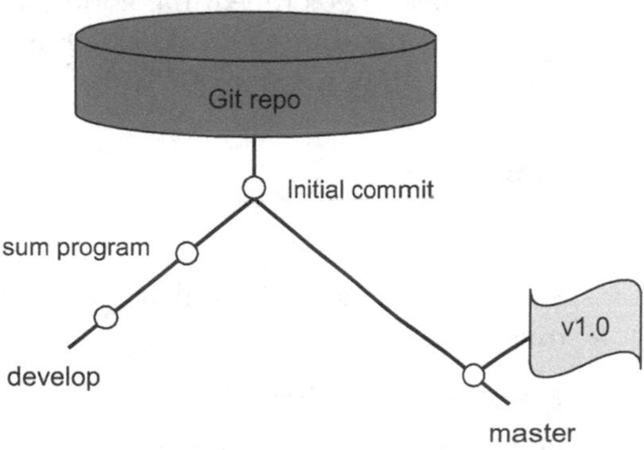

Figure 17.7 Merging the release back to develop

The sessions you did with git-flow aren't the entirety of this workflow. A workflow exists for bug fixing, and it looks roughly the same as the flow from feature to development or from development to production. When performing a bug fix, you must branch off master (which represents the production code), and after you're done with the bug fix, merge that fix back to develop and master. The workflow has a convention for the bug fix branch name: hotfix-*.

17.4 GitHub's flow

Leading Git evangelist and author Scott Chacon described the workflow used at GitHub, named GitHub Flow. His post is at http://scottchacon.com/2011/08/31/github-flow.html.

Figure 17.8 depicts an imaginary repository using GitHub Flow. The straight line illustrates the master branch. It's long-lived, as in git-flow. This figure shows two feature branches that have branched off master. This is unlike git-flow, where feature branches must be branched off the develop branch. These feature branches are later merged back to master. Unlike git-flow, GitHub Flow doesn't delete these feature branches.

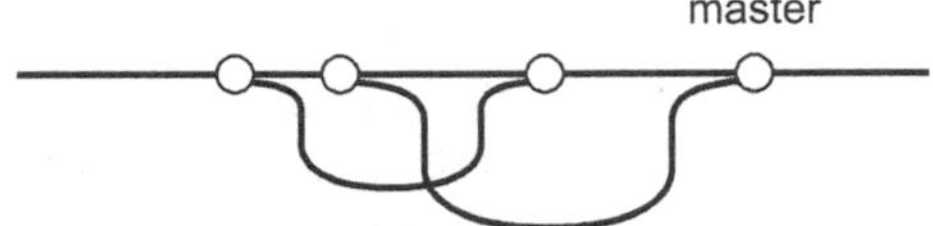

Figure 17.8 A depiction of GitHub Flow, with two feature branches. Only master is long-lived.

A more formal description of GitHub Flow can be found at https://guides.github.com/introduction/flow/index.html.

GitHub Flow and git-flow both designate master as the branch that represents the released code. But GitHub Flow doesn't describe any use of tags and the long-lived develop branch. Instead, a new developer who wants to add code to the code base must create only a descriptively named branch. After the branch is finished and signed off by a counterpart, it can be merged into master. Because master contains code that can be released to production, the developer is encouraged to deploy the code as well.

> **TRY IT NOW** Let's repeat the code that you produced in the previous section by using this simpler workflow. First, you'll make a new repository in the directory gh-flow (*gh* for GitHub):
>
> ```
> cd $HOME
> mkdir gh-flow
> cd gh-flow
> git init
> ```
>
> As with git-flow, you'll initially create an empty commit. This step creates the master branch:
>
> ```
> git commit --allow-empty -m "Initial commit"
> ```
>
> Per the workflow, you'll now create a descriptively named branch. This is similar to git-flow. Type the following:
>
> ```
> git checkout -b sum_program
> ```

Remember that this variation of the `git check-out` command was introduced in chapter 9. This command creates the situation shown in figure 17.9.

Your working directory is now the sum_program branch. This variation of the `git checkout` command normally takes a branch as its starting point, but because this version omits the branch, `git checkout` defaults to the current branch. Confirm this by typing the following:

```
git branch
```

You should see sum_program as the selected branch. Now let's create the sum.sh file as in the previous section. In your favorite editor, create a file called `sum.sh` and add these lines into it:

Figure 17.9 Making a feature branch off the master branch

```
# Add a and b
a=1
b=1
let c=$a+$b
printf "This is the answer: %d\n" $c
```

After you save this file, you can run it by typing this:

```
bash sum.sh
```

Now add and commit this to the branch:

```
git add sum.sh
git commit -m "The sum program"
```

The feature is saved on the sum_program branch. To bring this into production using this workflow, merge this into master. Type the following:

```
git checkout master
git merge sum_program
```

The `git checkout` command puts you in the master branch. Then you merge in the sum_program branch.

After the preceding steps, your repository can be drawn like figure 17.10.

When you merged sum_program into master, the code can officially be considered part of production. Depending on your project, a further deployment step is usually performed at this point. If the repo represents a website, a deployment script would retrieve the contents of the master branch and push it up to the web servers,

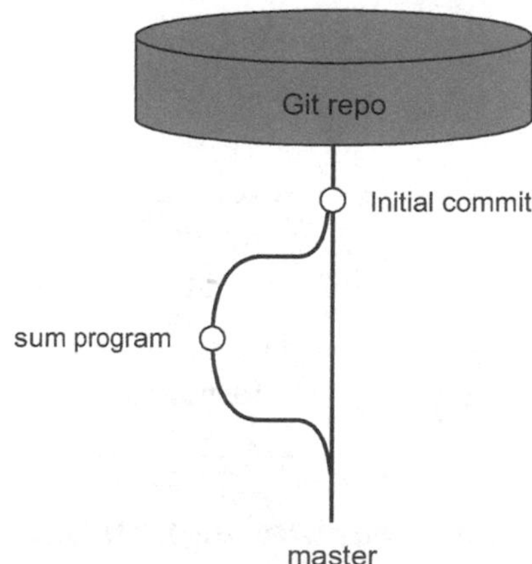

Figure 17.10 Your simple repository following the GitHub Flow process

for example. If the repo is server code, a continuous integration tool would produce a new build based on this update to master, for example. The emphasis on this workflow is fast development, rapid deployment, as well as fast collaboration (using GitHub features such as pull requests). You'll read about GitHub (and pull requests) in the next chapter.

17.5 Lab

This chapter didn't cover a lot of new Git mechanics. Instead, it focused on how Git should be used. Git's flexibility lends itself to multiple workflows. Steps 4 and 5 that follow are starting points for exploring more complicated workflows and conventions.

1. This chapter refers to external sources for official documentation of workflows. If you didn't peruse the URLs during your reading, please give them a read now.
2. Look at the repository produced by the GitHub workflow. Now delete this repository and repeat the steps, with one change to the last `git merge` command: add the `--no-ff switch`. Now look at the repository and confirm that you have one additional commit.
3. The book's website contains a zip file of simple scripts. Two of these scripts make the sample repositories you discussed in this chapter: make_nvie_repo.sh and make_gh_repo.sh. Download and examine these scripts if you haven't already.
4. The open source project Drupal (a content management system written in PHP that is similar to WordPress) has lengthy guidelines for using Git. Skim through it at www.drupal.org/documentation/git.

   One interesting aspect is that Drupal module developers shouldn't use the master branch. Instead, developers are asked to use a major-version branch (for example, 7.x–1.x), and are instructed to delete the master branch. See www.drupal.org/empty-git-master.
5. The Git project describes its branching process at https://code.google.com/p/git-core/source/browse/MaintNotes?name=todo.

   Like git-flow, Git relies on a few long-lived branches (master, maint, next, and proposed updates—abbreviated as pu).

   If you wanted to contribute to Git, you'd follow the steps here:

   https://code.google.com/p/git-core/source/browse/Documentation/SubmittingPatches?name=master

   The Git documentation discusses workflows as well. Type the following for more information:

   ```
   git help gitworkflows
   ```

17.6 Further exploration

The git-flow conventions have been encapsulated into command-line software by the gitflow project at GitHub. Visit https://github.com/nvie/gitflow.

The gitflow software can be installed on Windows, Mac, or Unix/Linux, and its documentation has detailed installation instructions for all three platforms. After you have it installed, you can use the command-line helper functions to move code between feature and develop branches, and between develop and release branches, and finally between release and master branches. These helper functions are all in the command line, and they take care of tagging and merging code as necessary.

The following listing shows an example session on the command line with gitflow.

Listing 17.3 An example session with the gitflow helper command

```
% git flow                                   <-- ❶ Type this to get help in gitflow
usage: git flow <subcommand>

Available subcommands are:
   init      Initialize a new git repo with support for the branching
  model.
   feature   Manage your feature branches.
   release   Manage your release branches.
   hotfix    Manage your hotfix branches.
   support   Manage your support branches.
   version   Shows version information.
% git flow feature start dumpstamp           <-- ❷ Starting a new feature
Switched to a new branch 'feature/dumpstamp'

Summary of actions:
- A new branch 'feature/dumpstamp' was created, based on 'develop'
- You are now on branch 'feature/dumpstamp'

Now, start committing on your feature. When done, use:

     git flow feature finish dumpstamp
% git branch                                 <-- ❸ Checking the branche
  develop
 *feature/dumpstamp
  master
% git flow feature finish dumpstamp          <-- ❹ Finishing the feature
Switched to branch 'develop'
Merge made by the 'recursive' strategy.
 README.txt        |  1 +
 dumpstamp.info    |  4 ++++
 dumpstamp.module  | 21 +++++++++++++++++++++
 3 files changed, 26 insertions(+)
 create mode 100644 README.txt
 create mode 100644 dumpstamp.info
 create mode 100644 dumpstamp.module
Deleted branch feature/dumpstamp (was 28e24e8).
```

In the session listing, first notice that the command presents its own help text ❶, when no subcommand is given. You create a new feature ❷ in one step with the command `git flow feature start dumpstamp` (dumpstamp is the name of the feature in the example). You can confirm that this creates the appropriate branches ❸. Finally, after you commit some new code into this branch, you can finish this feature ❹ with the `git flow feature finish` command. This takes care of merging the code to develop.

The git-flow workflow also has support in a third-party GUI tool called SourceTree, which you'll learn about in chapter 19.

17.7 Commands in this chapter

Table 17.2 Commands used in this chapter

Command	Description
`git commit --allow-empty -m "Initial commit"`	Create a commit without adding any files.
`git merge --no-ff BRANCH`	Merge BRANCH into the current branch, creating a merge commit even if it's a fast-forward commit.
`git flow`	A Git command that becomes available after installing gitflow.

18 Working with GitHub

GitHub (http://github.com) is the Git project-hosting website that has done much to increase Git's popularity. GitHub takes advantage of Git's distributed architecture, but offers additional project management and collaboration features. These features include wiki documentation, issue tracking, and basic collaboration management. GitHub's social coding features have made it a popular platform for hosting and sharing code.

In this chapter, you'll create a project on GitHub. I'll relate GitHub repos to our earlier `git clone` exercises. You'll then practice the two important GitHub features for collaboration: forks and pull requests. These two features take the standard `git diff` and `git push/pull` commands and add a bit more formality. Developers must understand forks and pull requests in order to collaborate on GitHub, as GitHub encourages this mode of collaboration.

18.1 Understanding GitHub basics

In chapter 11, you created a small set of Git repositories based on the math repository. The set of repos looks like figure 18.1.

In figure 18.1, the folder in the bottom row (math.clone, math.bob, math.carol) are cloned from the folder at the top (math.git). The folders on the bottom are clones of the main repository, math.git. All the clones can perform a `git push` to and a `git pull` from this main repository.

In chapter 11, you learned how to create this main repository by using the `git clone --bare` command to create a bare directory. Remember, the bare directory

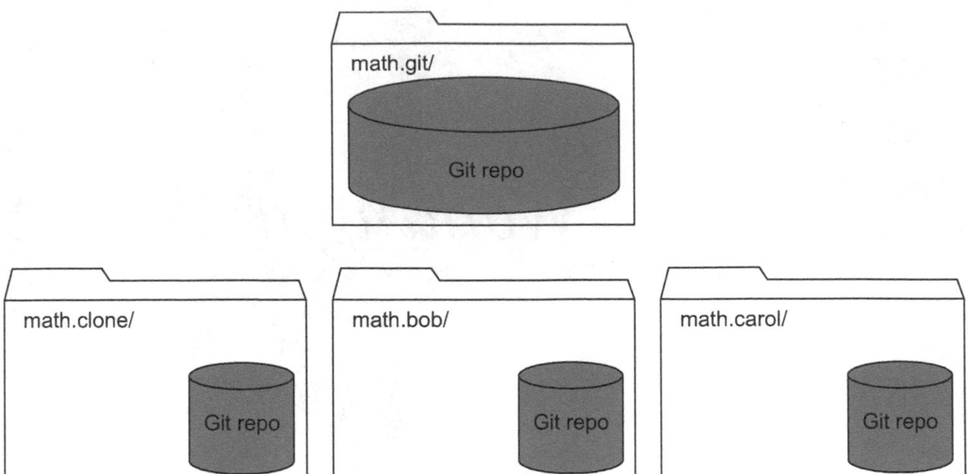

Figure 18.1 A set of Git repositories

is what is manipulated in collaboration. This is depicted in figure 18.2. To refresh your memory, visit section 11.2. That section introduced the convention of drawing the bare repository with a repo that fills up the entire folder. (This leaves no room for a working directory.)

In chapter 11, you used the bare directory as a way to simulate Git repository hosting on an external server. Your bare directory pretended to be a GitHub-hosted repo.

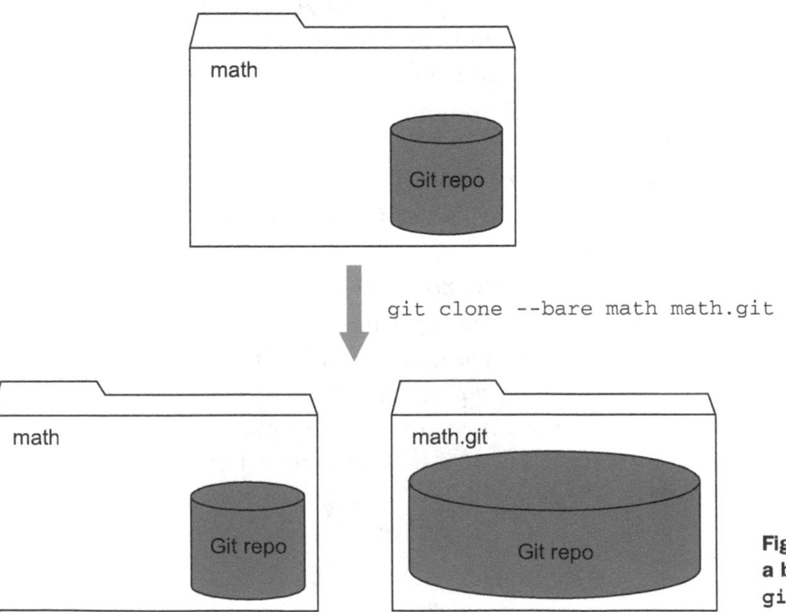

Figure 18.2 Creating a bare directory, using `git clone --bare`

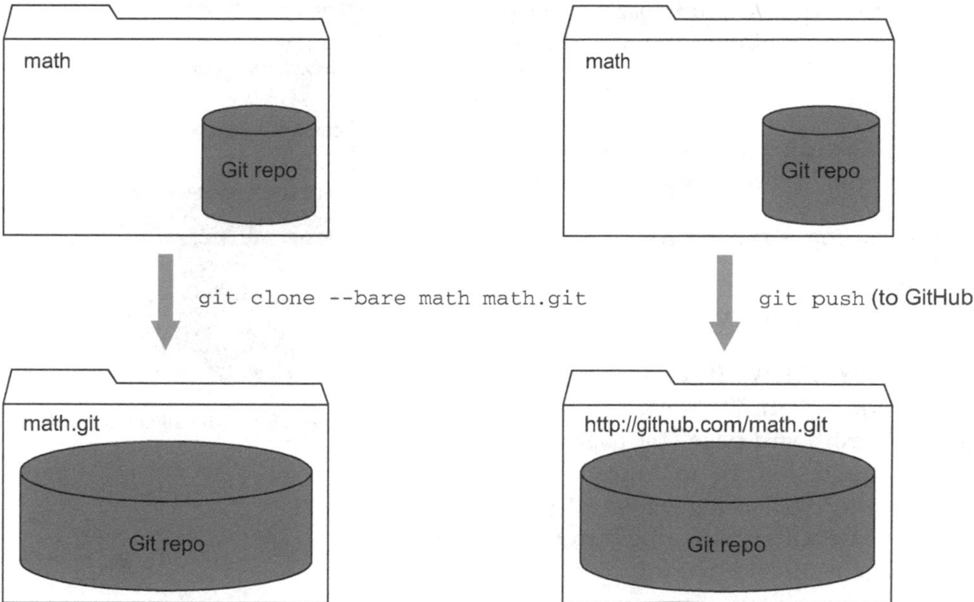

Figure 18.3 Relating your bare directory to GitHub

You created your repository on a local directory of your computer, and cloned it directly from that directory. Now you're ready to try the real thing.

In this chapter, you'll create a repository on GitHub and push your math repository to it. GitHub will replace your bare directory. As a Git hosting site, GitHub enables you to share your repository with any number of collaborators. Your project is also public for anyone else to find and clone.

In figure 18.3, the left side shows how you cloned your repository to a bare directory (via `git clone --bare`) that exists on your computer. The right side shows what you'll perform: pushing an existing repository into a bare directory that you'll create on GitHub.

18.1.1 Creating a GitHub account

Now is a good time to create a GitHub account. It's free, and more important, you'll need it to explore the other facilities. If you've already created an account, go to the next TRY IT NOW, where you'll create a new repository.

> **TRY IT NOW** In your browser, visit the GitHub website at http://github.com.
>
> On this first page, you can enter a username, a password, and an email address, as in figure 18.4. Enter your information and then click Sign Up for GitHub.

On the second page, Welcome to GitHub, choose the free plan and then click the Finish Sign Up button to continue.

The third page is your dashboard. Success! You're now a member of GitHub.

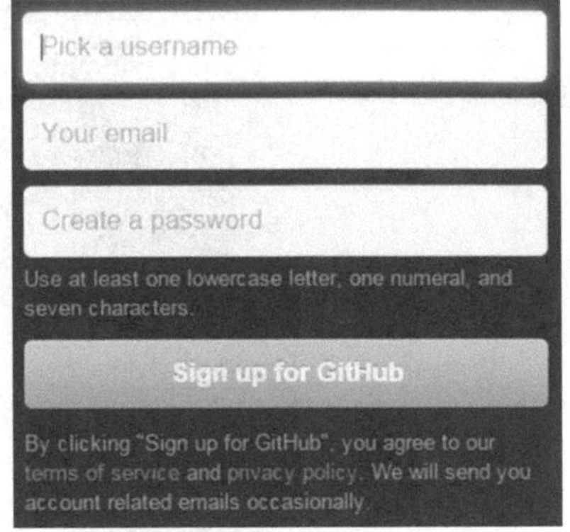

Figure 18.4 The signup for GitHub

18.1.2 Creating a repository

Next you'll create a repository on GitHub. This is also straightforward.

TRY IT NOW If you've followed the previous TRY IT NOW, you'll be on the dashboard page (the page that appears when you go to http://github.com, after you've logged in). If you already have an account, log into GitHub and go to the dashboard.

On this page (and on practically every page) is a header that contains your username. A plus sign is next to your username, as in figure 18.5.

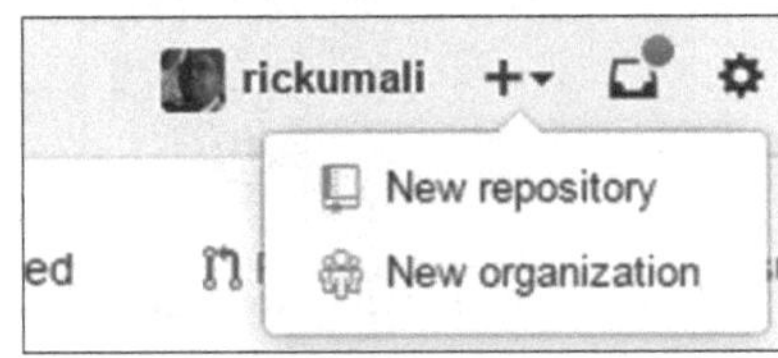

Figure 18.5 Creating a new repository

Click this plus sign, and in the pull-down menu that appears, click New Repository.

When you click this, the URL in your browser should be http://github.com/new, and your page should look like figure 18.6.

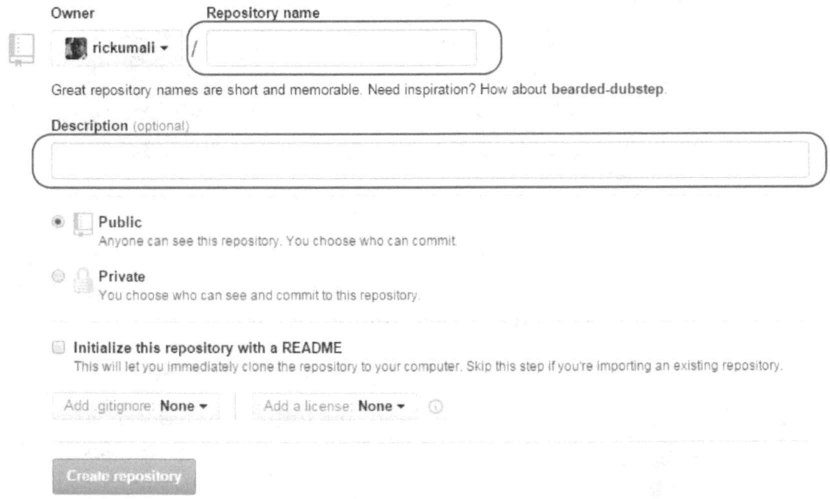

Figure 18.6 Creating a repository in GitHub

On this page, enter a repository name in the Repository Name text box; for now, enter the string `math`. In the Description text box, enter something that will remind you that this repository was created from this book. Next, click the green Create Repository button.

18.1.3 Interacting with the repository

The next page on the GitHub website contains information about how to interact with this repository you just created. It may be helpful to consider figure 18.7.

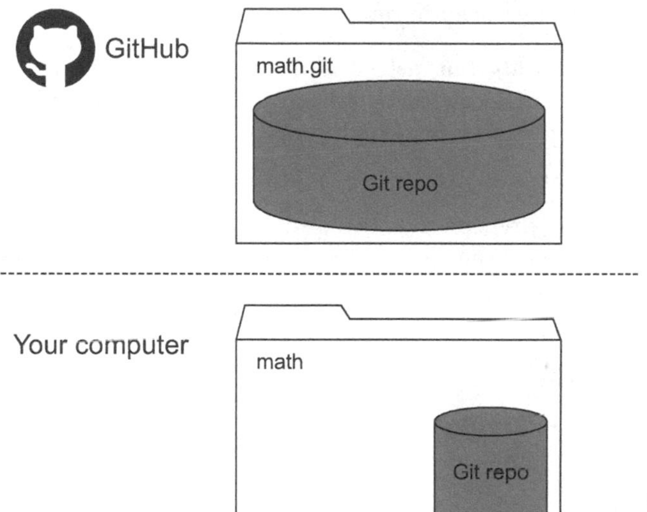

Figure 18.7 Creating a GitHub repository

You've created an empty repository on GitHub, giving it the name *math*. Behind the scenes, Git has created a bare directory for your code. GitHub then gives you instructions for interacting with your new empty repository on GitHub. Hopefully, all the instructions will look familiar, because you went through them in the collaboration chapters, starting with chapter 11.

TRY IT NOW The GitHub web page includes instructions that describe how to push an existing repository from the command line.

At the top of the GitHub web page, as in figure 18.8, you'll see a Quick Setup section. Click HTTPS, changing the URL to one that begins with *https* (shown in figure 18.8).

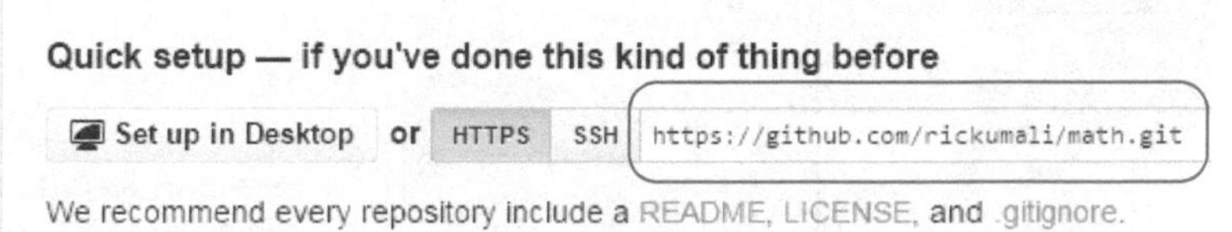

Figure 18.8 The clone URL in the Quick Setup question

The page contains a section titled Push an Existing Repository from the Command Line. This section contains the instructions shown in the following listing.

Listing 18.1 Instructions to push an existing repository to GitHub

```
git remote add origin https://github.com/yourname/math.git
git push -u origin master
```

As you may remember from chapter 13, these are the steps that you used to push your master branch of your math repo to math.git.

Let's type these two commands, but instead of using the remote name *origin*, you'll use `github`. In chapter 12, you learned that you can pick any name for the remote. Type the following command, which creates a remote named *github*:

```
git remote add github https://github.com/yourname/math.git
```

Replace *yourname* in the preceding command with your login name at GitHub. Also, if you named your repository something else besides math, replace *math* with that name.

Now that you have this remote, type the following:

```
git push -u github master
```

The `git push` command sends the master branch of your math repository to your math.git repository on GitHub.

GitHub has good documentation for how to create repositories and upload code to those repositories via `git push`.

Figure18.9 depicts the creation of the repository on GitHub. After you click Create Repository, you push code into the repository via the `git push` command.

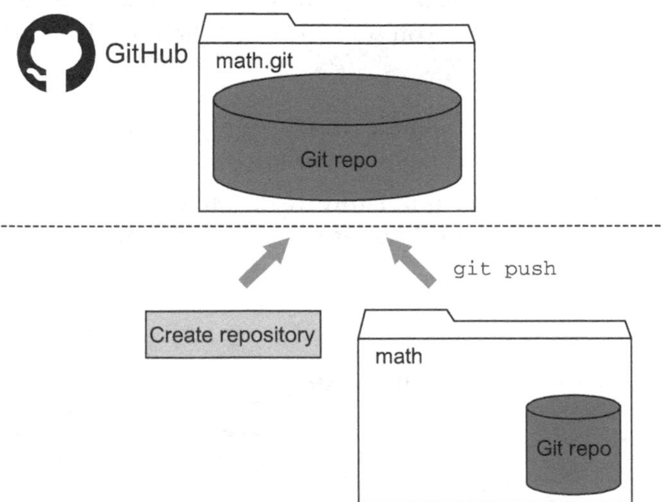

Figure 18.9 Creating your repository on GitHub and adding your files to it

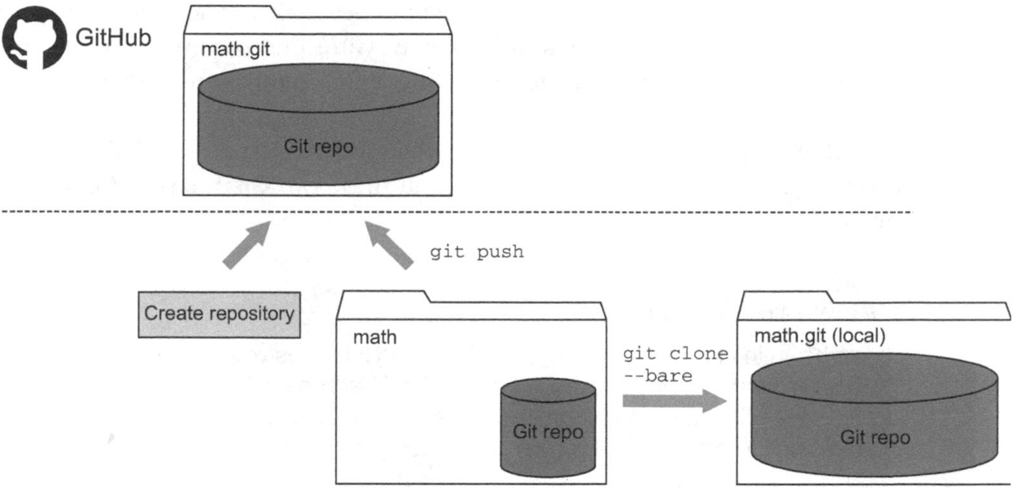

Figure 18.10 Making clones locally versus on GitHub

Figure 18.10 compares making clones locally (below the dashed line) to making a repository on GitHub and pushing to it.

The GitHub repository name is math.git, the same name you created for your bare directory (the right side of figure 18.10). Using *git* as the suffix of a directory is a Git convention indicating that this is a bare directory. GitHub follows this convention.

Everything you learned about collaborating, you can now do with your GitHub repository, provided you set the remote properly.

TRY IT NOW You'll now attempt to delete this repository. This will set you up for the next section.

In the GitHub web page for your math repository, click Settings. This button is on the right-hand side of the page, as in figure 18.11. (Note that it's not the little cog icon, which is also on the right-hand side of the page.)

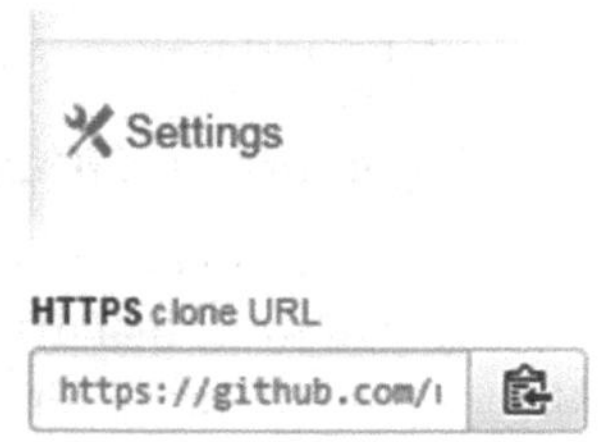

Figure 18.11 The Settings button

On the Settings page, scroll down to the Danger Zone section, marked in red. This is an appropriately named section, because it provides you the ability to delete this repository. Click it, and then follow the prompts to properly remove this repository.

One thing to notice about deleting repositories on GitHub: it has no effect on clones of this repository. The only way they'll know this repository is gone is if they try to pull from or push to it.

Everything you did in this section would be repeated if you accessed another Git hosting platform such as Atlassian's Bitbucket, or Gitorious, or GitLab. You need to create a login, create a project, and then use Git collaboration commands (`git remote`, `git clone`, `git push`, `git pull`).

GitHub repositories can be copied using forks, and you can collaborate with repositories by using pull requests. You'll look at these two GitHub-specific mechanisms in the next two sections.

Above and Beyond

As shown in figure 18.8 and figure 18.11, GitHub gives you two clone URLs with which you can clone a repository: HTTPS and SSH. If you select the HTTPS URL as your clone URL, the `git push` command prompts you for your username and password. Constantly entering your credentials can become tedious.

You can avoid constantly entering your username and password by selecting the SSH clone URL and setting up an SSH key. This is beyond the scope of this book, but GitHub has plenty of help for this configuration at the following site:

https://help.github.com/articles/generating-ssh-keys

If you start working with GitHub a lot, setting up an SSH key becomes an essential configuration and is worth exploring.

18.2 Working with forks

A *fork* is a copy of a GitHub repository that you have full rights to push to and pull from. Remember that with GitHub, practically any repository can be cloned, but by default not everyone can push changes back to it. To push changes back to a repository that you didn't create, you must be added as a collaborator. Forks get you past this limitation, enabling you to make changes to a repository that you ordinarily wouldn't be able to update.

18.2.1 Making a fork on GitHub

If you want to make changes to an existing GitHub project that you're not a collaborator for, you must first make a fork of this repository. In figure 18.12, you make a fork of the rickumali/math repository. Then, after you have this fork on GitHub, you clone that in order to make changes.

When you're ready to submit your changes back to the original repository, you initiate a pull request to the originating repository. You'll learn more about pull requests in the next section.

The number of forks made from a repository on GitHub is a measure of its popularity. It gives a measure of developers' interest in contributing to a code base, and GitHub has areas that track this measurement.

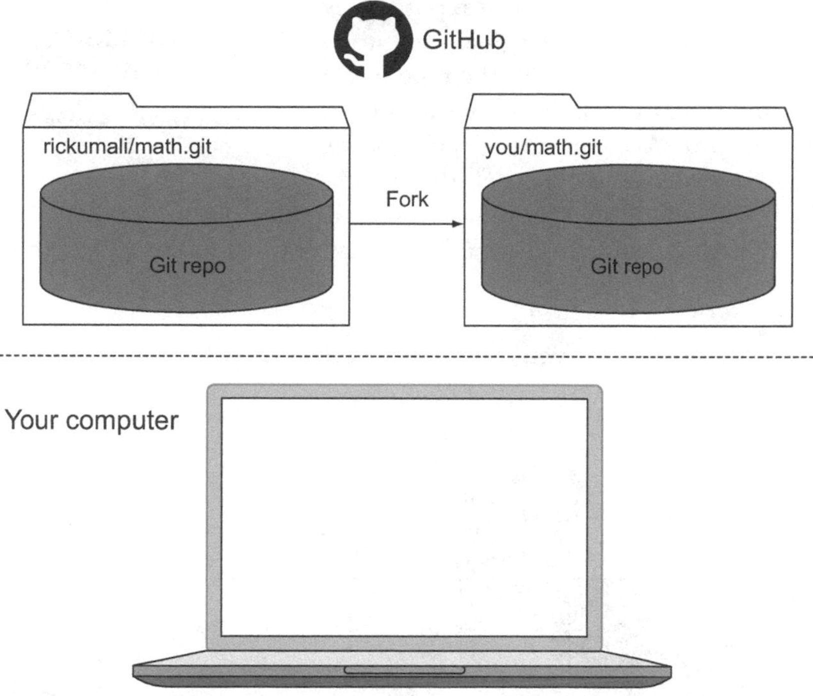

Figure 18.12 Making a fork of a repository

TRY IT NOW You'll now practice making a fork. You can make a fork of any repository, but in this exercise, you'll make a fork of my math repository. In your browser, visit this URL:

https://github.com/rickumali/math

This is the URL for my math repository. In the upper-right part of this page, you'll see a Fork button, shown in figure 18.13.

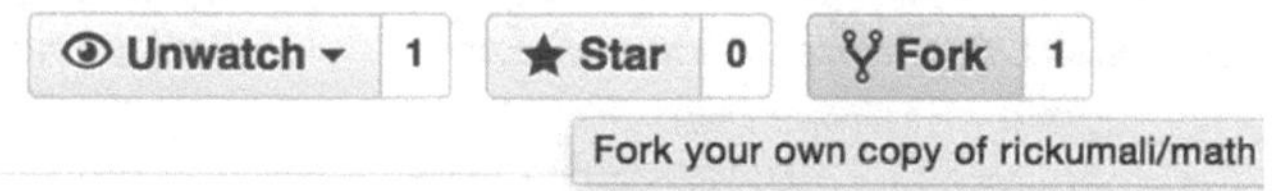

Figure 18.13 The Fork button on a project's home page

Click the Fork button. GitHub completes its steps to make a copy of the math repository into your account. When it's complete, the math repo will exist in your GitHub account.

Notice that the name of the forked repository is the same as the name of the original repository, and that immediately underneath the repo name is a small line of text that indicates it's a fork. In figure 18.14, your account name will appear where the name *rodrigoumali* appears (rodrigoumali is a second account I used to simulate the pull request flow). At this point, you've created the situation shown in the top half of figure 18.12, at the start of this section.

Figure 18.14 A forked repository

18.2.2 Cloning your fork

Now that you have this fork, you can make a clone of it to your local machine. Figure 18.15 depicts cloning the fork to your computer.

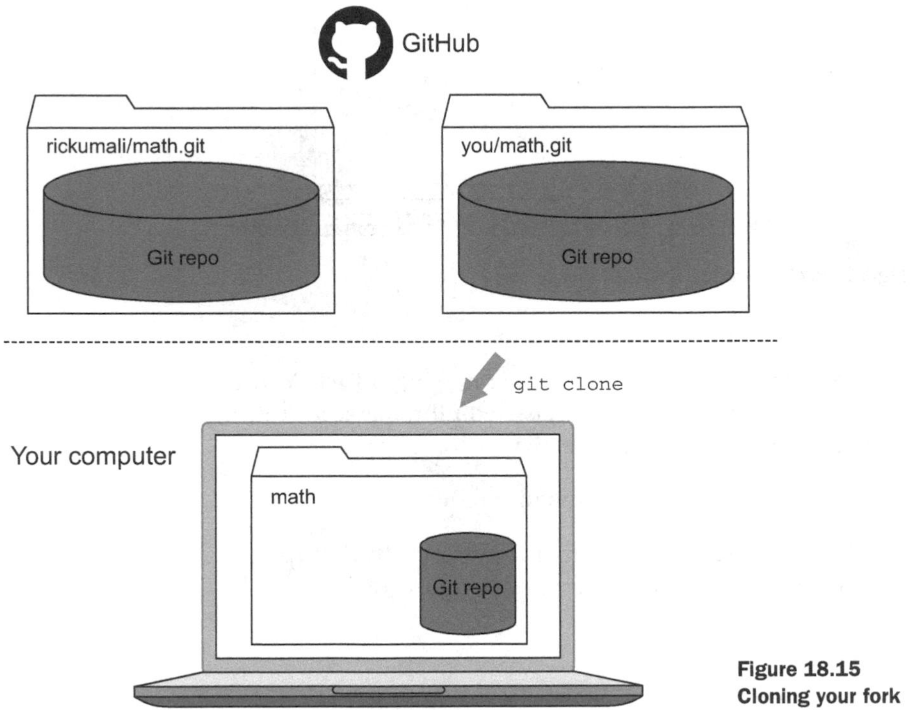

Figure 18.15 Cloning your fork

TRY IT NOW Let's make a clone of the forked repository. To do that, you'll need the clone URL. This can be obtained from the right-hand sidebar of your repo's main page (see figure 18.16). You looked at this way back in chapter 3.

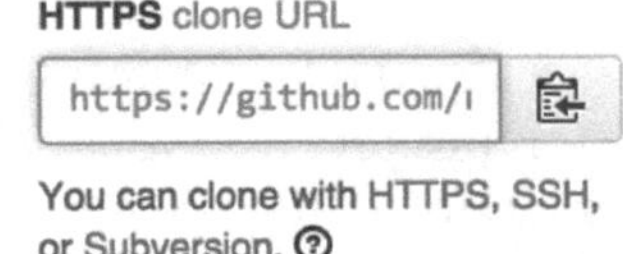

Figure 18.16 The clone URL, which is on the right-hand side of your project's main page

Make sure to use the HTTPS form of the clone URL, by clicking the HTTPS link shown in figure 18.16. Then, in your home directory, remove the existing math repository by typing this:

```
cd $HOME
rm -rf math
```

Type the following, replacing the string `yourname` with your GitHub username:

```
git clone https://github.com/yourname/math.git
```

Now you have a GitHub clone of your fork!

At this point, you have the situation shown in figure 18.15 at the start of this subsection. Working on a fork allows you to push your changes to your own copy on GitHub. The original repository isn't modified when you do this. But how would you contribute a change back to the original project? By performing a pull request with the fork you've just created.

18.3 Collaborating with pull requests

You learned in the preceding section that a fork is your personal copy of another repository. If you make changes to your personal copy, you can push it back up to GitHub, because the clone of your fork is yours and yours alone. But what if you want to push your change to the originating project (as in figure 18.17)?

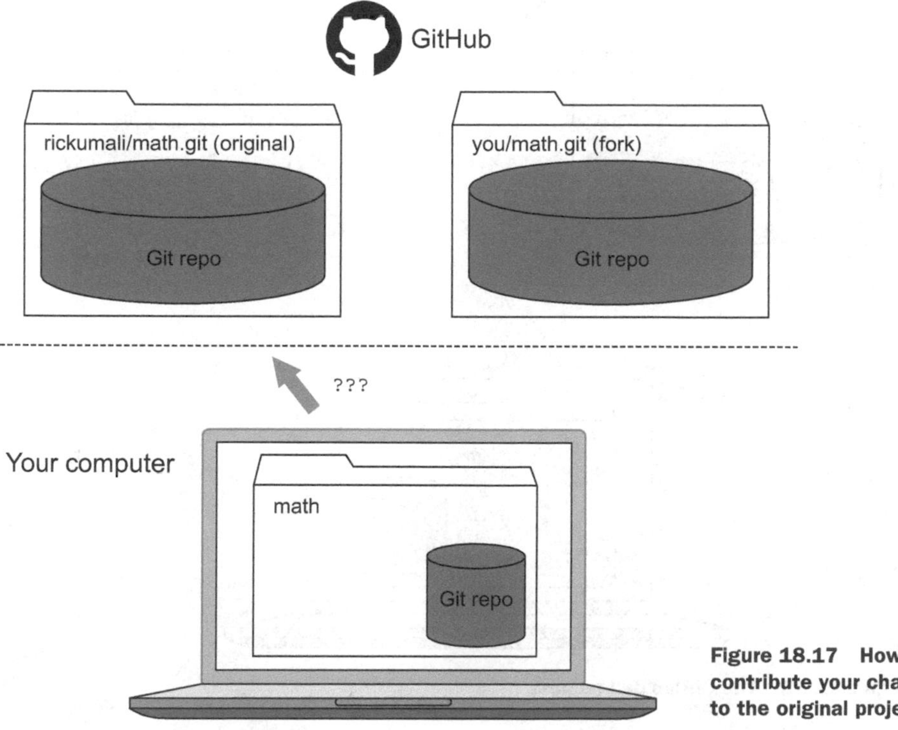

Figure 18.17 How do you contribute your change back to the original project?

To do this, you might think to notify the original repository's owner of your change. Instead of doing that, GitHub has a mechanism to propose changes back to the original repository: the pull request.

The GitHub pull request was introduced in 2008. Pull requests are like messages back to the original repository's owner, indicating a change request. As shown in figure 18.18, pull requests are always made from the fork (indicated by you/math.git) to the original repository.

This feature enables you to make a fork of a repository and test your contribution on your local clone. After you're satisfied with your change, you can use the pull request feature to present the change to the original repository.

18.3.1 Making a change to your fork

To try out a pull request, you must first make a change to your repository, and then push it back to your fork (as in figure 18.18).

TRY IT NOW You'll now use the clone of the fork that you made.

```
% cd $HOME
% cd math
% git branch
```

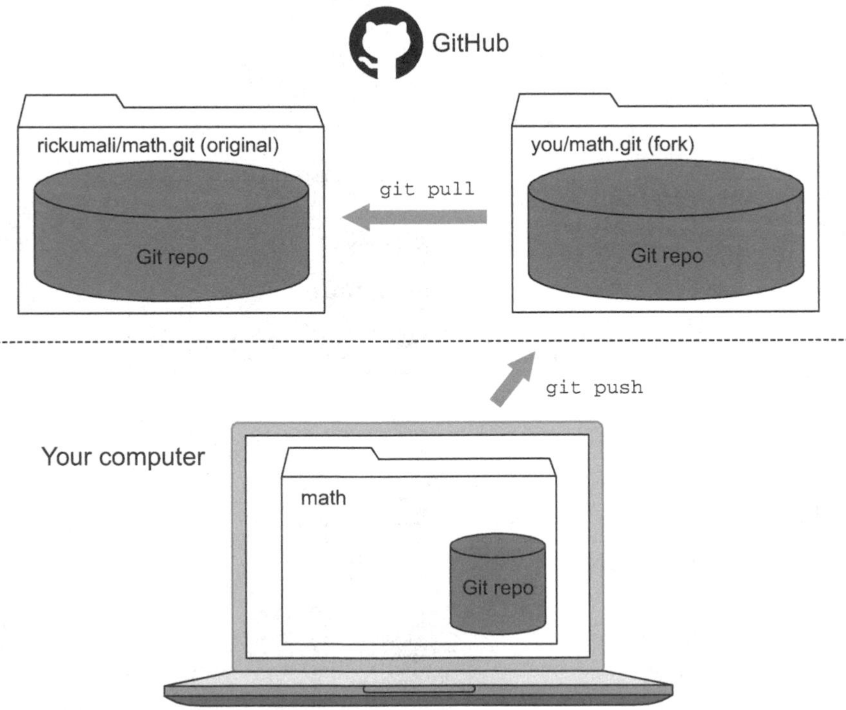

Figure 18.18 The GitHub pull request

Confirm that your branch is master. Then make a small change to the repository by typing the following:

```
% echo "Small change to fork"  >> readme.txt
```

Commit this change:

```
git commit -a -m "Small change to fork"
```

Now push this change. Because your clone is a clone of your fork, the change won't go to the rickumali/math.git repository. The change will instead go to your fork. Type this:

```
git push
```

You'll be prompted for your GitHub username and password. The output looks like the following listing.

Listing 18.2 Output from `git push` (back to your fork)

```
Username for 'https://github.com': yourusername
Password for 'https://rickumali@github.com':
Counting objects: 5, done.
Delta compression using up to 4 threads.
Compressing objects: 100% (3/3), done.
Writing objects: 100% (3/3), 308 bytes | 0 bytes/s, done.
Total 3 (delta 1), reused 0 (delta 0)
To https://github.com/yourusername/math.git
   2e044d8..5bc718c  master -> master
```

If you have problems pushing back to your repository from your computer, consult the help from GitHub or visit this book's website. Pushing code involves going over the network (as does `git clone` and `git pull`), but pushing code also requires obtaining authorization (your username and password). On GitHub, you can push code only to repositories that you're authorized on. You can push only to repositories you've created.

18.3.2 Making a pull request

At this point, if you visit your repository's page on GitHub, you'll see that it's ahead of its original repository, as shown in figure 18.19.

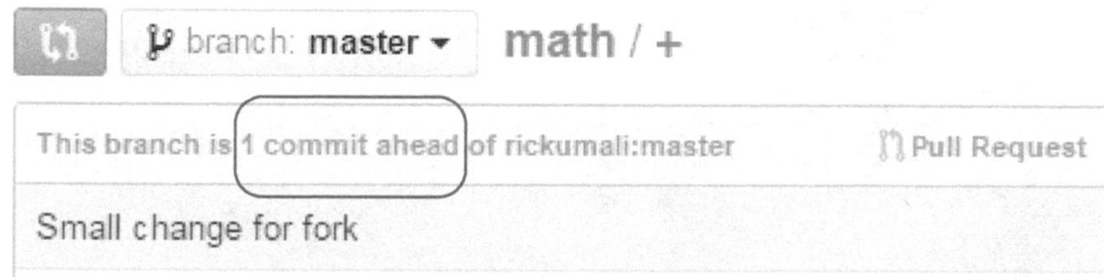

Figure 18.19 Your fork is now one ahead of the original repository (see text in square).

On GitHub, the title section of your fork's main page should look like figure 18.19. To the right of the announcement that the fork is one ahead of rickumali:master, you should see a Pull Request link. Let's use this.

TRY IT NOW On the GitHub page for your fork, click Pull Request. You'll be taken to a page that prepares your pull request (partially shown in figure 18.20). It isn't the pull request yet. On this pre-pull-request page, you're able to review the change you're going to send to the owner of the originating repository.

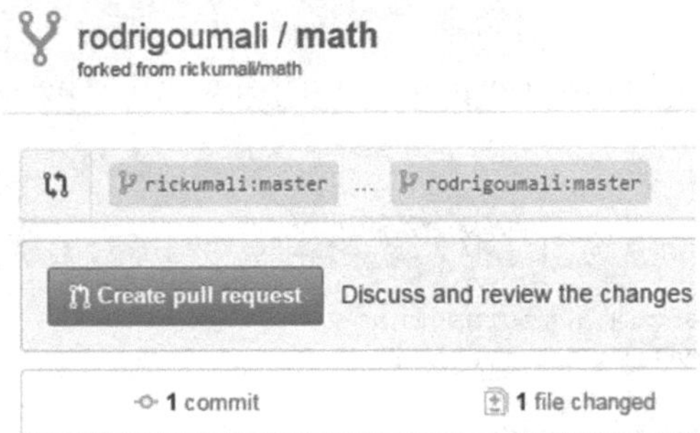

Figure 18.20 Preparing a pull request

In your pull request, you changed only one file on the master branch. The GitHub UI does show a diff window (figure 18.21) that should be familiar (it's a variation of the `git diff` output that you've used in the past).

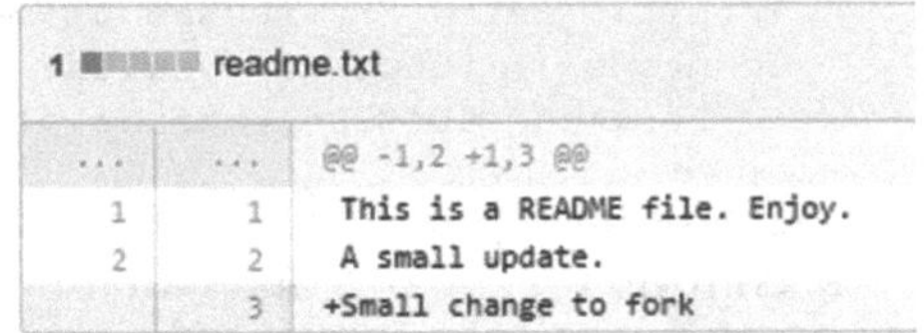

Figure 18.21 The difference viewer in a GitHub pull request

Click the green button on the web page. The next page your browser displays (figure 18.22) offers you a few fields for documenting the change you're proposing. This page also gives you a status indicating whether you can merge the change immediately.

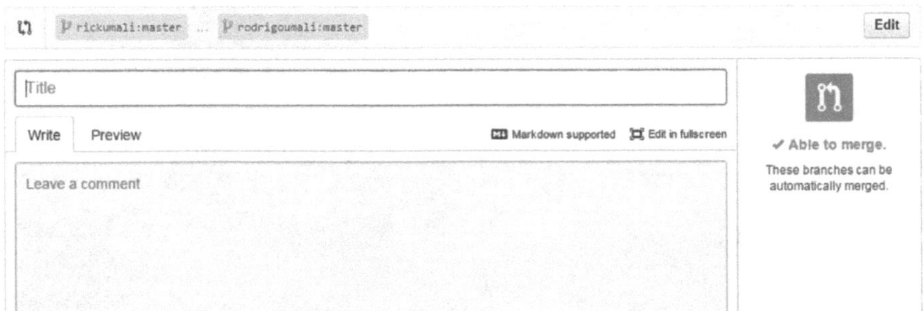

Figure 18.22 Documenting your pull request

Enter any title and any comment you want. Remember that the owner of the originating repository will see this.

At the bottom of the form is another green button, Create Pull Request. Click it, and the pull request will be given a number and sent to the originating repository.

18.3.3 *Closing the pull request*

After the pull request has been made, the owner of the original repository is notified. When logging into GitHub, the owner will see a web page like figure 18.23.

The owner will see the change that's proposed, who it's from, plus all the commits that led up to this change. The fork owner and the originating repository owner can now carry on a dialogue on the pull-request page. But notice that only the original repository owner can merge the code, per the message "Only those with write access to this repository can merge pull requests."

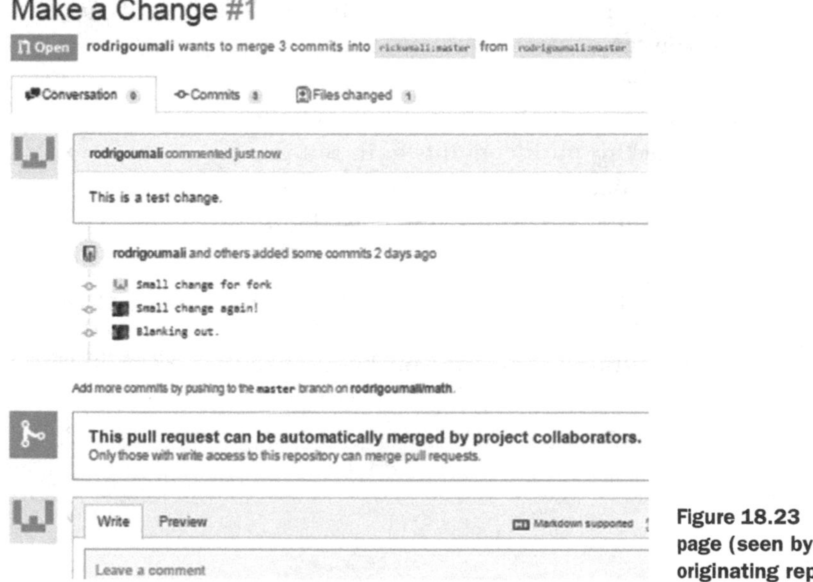

Figure 18.23 A basic pull-request page (seen by the owner of the originating repository)

In this TRY IT NOW, you won't be able to merge your change to the original repository, but you can close the pull request. This view is helpful because it's the same view that the owner of the originating repository sees.

TRY IT NOW On the pull-request page, click the Close Pull Request button. If you want to leave the page open, that's appropriate as well. I'll see your request (remember, GitHub sends an email to the originating repository owner) and I'll close it from my side.

On GitHub, if you explore any active repository, you'll see projects with any number of open pull requests. Like forks, pull requests are a measure of a project's activity and popularity.

18.4 Lab

Forks and pull requests are two features you'll use frequently if you collaborate on a project hosted on GitHub. Work through this lab to acquaint yourself with nifty features.

1. Walk through the Hello World GitHub guide at https://guides.github.com/activities/hello-world/.
2. Explore the many projects on GitHub by visiting the following site:

   https://github.com/explore

   Try to find a project of interest, mark it to be watched, and add a star to it (if you're so inclined). Visit a project's issue queue, labels, and wiki to see the variety of ways a project can be documented.
3. Follow some users, including myself! GitHub by default sends emails based on activity, and you can control this frequency.
4. Forks and pull requests are necessary because by default repositories on GitHub allow only the creator of the project to push commits to it. But you can enable other people to push to your project. Find your repository's Settings link (see figure 18.24).

   From the Settings page, you can add collaborators to your project. Click Collaborators, as in figure 18.25. Consider adding someone you know (like me!)

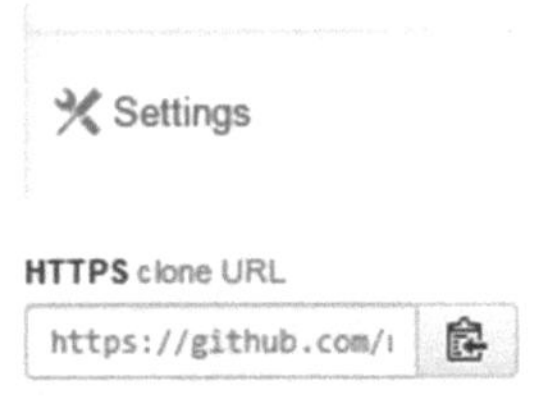

Figure 18.24 Your repository's Settings link (above the clone URL)

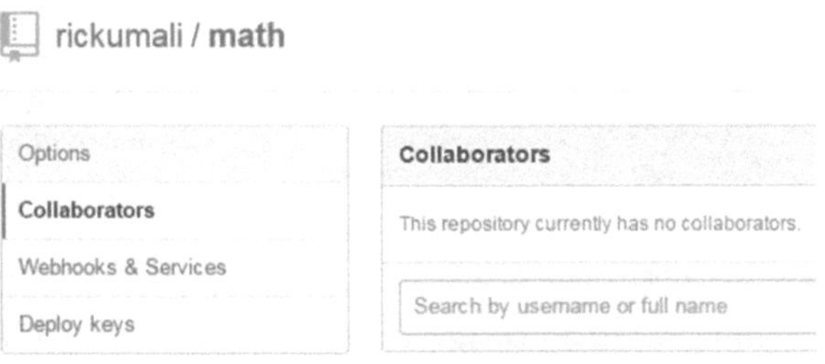

Figure 18.25 On the Settings page for a project, you can add other collaborators.

5. The pull request that you explored in the previous section was from the point of view of initiating a pull request. The easiest way to go through the process from the receiving end of a pull request is to create a second account on GitHub. This account could be considered your alter ego account, and like the

repositories you created for Carol and Bob in chapter 13, it's to simulate a second user.

Once you have created this second account, go through the process of forking a repository from your first account, making a change to it, and then initiating a pull request. From here, when you log back into GitHub using your first account, you'll see a notification for the pull request. You can visit the page, as shown in figure 18.23. Now, as the owner of this repository, you can merge this change.

The GitHub website has more resources on this workflow:

https://help.github.com/articles/merging-a-pull-request/

18.5 Further exploration

One of the best features of GitHub-hosted projects is its adoption of Markdown for its documentation format. *Markdown* is a syntax for text files that allows the author to mark sections, add links, make text bold or italicized, and more. It's similar to HTML, but somewhat easier. The following listing shows some text Markdown syntax.

Listing 18.3 Text utilizing Markdown

```
First Header
============

Second Header
-------------

This is text with *emphasis*.
```

A Markdown program will convert listing 18.3 into stylized text, as in figure 18.26.

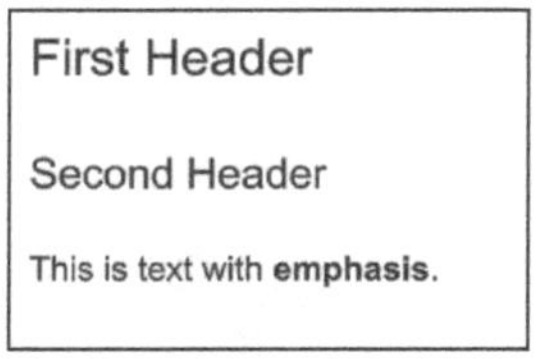

First Header

Second Header

This is text with **emphasis**.

Figure 18.26 Converted Markdown text

GitHub converts any text files in your repository named README into web-friendly documentation, if it's written using Markdown. Markdown is readable in the text file, but on the GitHub pages, this text becomes stylish and web-friendly.

Read these URLs for more information:

https://help.github.com/articles/markdown-basics/

https://help.github.com/articles/github-flavored-markdown/

As a way to explore this, rename the readme.txt file to `README.md` in the math repository that you pushed to GitHub in section 18.1. Edit this file to use Markdown syntax (discussed in the preceding URLs). Then push README to GitHub. Notice how the text is displayed compared with before.

18.6 Commands in this chapter

Table 18.1 Commands used in this chapter

Command	Description
`git remote add github https://github.com/yourname/math.git`	Add a rename named *github* that points to your math repo on GitHub. (Replace *yourname* with your GitHub username.)
`git push -u github master`	Push your master branch to the remote identified by GitHub, and set it to the upstream (review in chapter 13).
`git clone https://github.com/yourname/math.git`	Clone your GitHub repository named *math*. (Replace *yourname* with your GitHub username.)

19 Third-party tools and Git

The only Git tools that you've used so far have been the command-line and GUI tools that come with the standard distribution of Git. Let's consider these the Git native tools. In this chapter, you'll explore two third-party tools that can serve as supplements or even replacements for these native Git tools. These are Atlassian's SourceTree and the Git integration that comes with the Eclipse IDE.

I selected these two tools based on my familiarity but also their cost: both tools are free. SourceTree's singular focus on Git makes it similar to GitHub for Windows and Tower for Mac. Eclipse is popular, and its functionality is representative of other IDEs such as IntelliJ IDEA and NetBeans.

This chapter appears near the end of the book because it's important to form a good understanding of Git from the point of view of its stock installation and its native tools. Now that you've used the Git native tools over the past several lunches, you'll have a better understanding for the underlying Git functionality as supported by these third-party tools. But at this point, it's time to learn about the larger Git ecosystem.

One last note about this chapter: you'll be downloading software (in sections 19.1.1 and 19.2.1), which may take some time. Plan accordingly; this chapter may take longer than a typical lunch. Consider doing the downloads earlier!

19.1 SourceTree

A good number of third-party tools have been created to replace the gitk and Git GUI native tools. In this section, you'll install Atlassian's SourceTree, a powerful GUI that claims to eliminate the use of the command line. SourceTree is available for both Windows and Mac, but not other Unix/Linux platforms such as Ubuntu. (If you're on Unix/Linux, you can skip to section 19.2.)

19.1.1 Installing SourceTree

Installing SourceTree is a straightforward process, which you'll attempt now.

> **TRY IT NOW** Let's first install SourceTree on your computer. Visit www.sourcetreeapp.com. Then click the Download button. The web page detects whether you're running Windows or Mac, and downloads the appropriate binary.
>
> Next, locate the downloaded file in your browser's designated Downloads directory. For Windows, you'll look for an EXE file with *SourceTreeSetup* in its name (figure 19.1). For Mac, the download will result in a disk icon on your desktop. Double-click that.

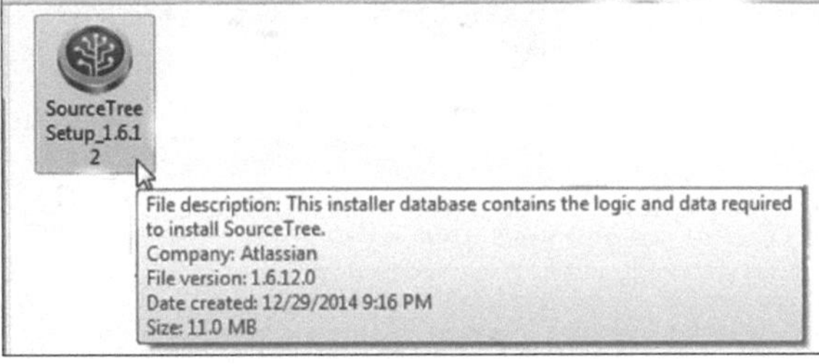

Figure 19.1 The SourceTree download for Windows, and its tooltip (when you hover the mouse over it)

> Now install the software by following your platform's standard installation process. The installation downloads the version control systems. Then the initial welcome screen appears (figure 19.2).
>
> Accept the license agreement (this is free software) and click Continue. The program then prompts you for other configurations. One such configuration is the setup of the global ignore file, as shown in figure 19.3. Accept this with Yes. I cover this configuration in the next chapter, and as the dialog box states, you can always change this later.

Figure 19.2 The initial welcome screen of Atlassian's SourceTree on Windows

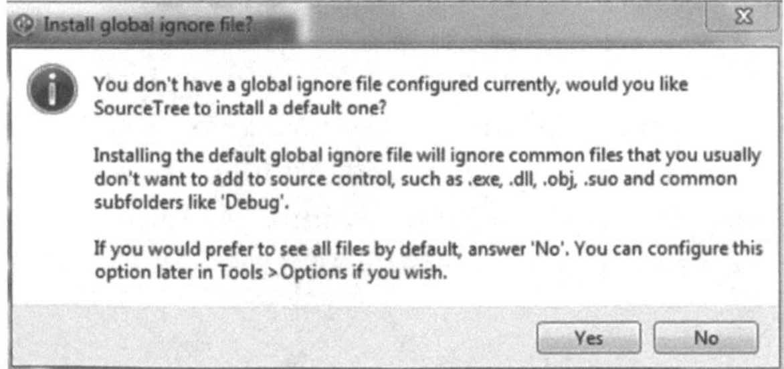

Figure 19.3 Global ignore file configuration

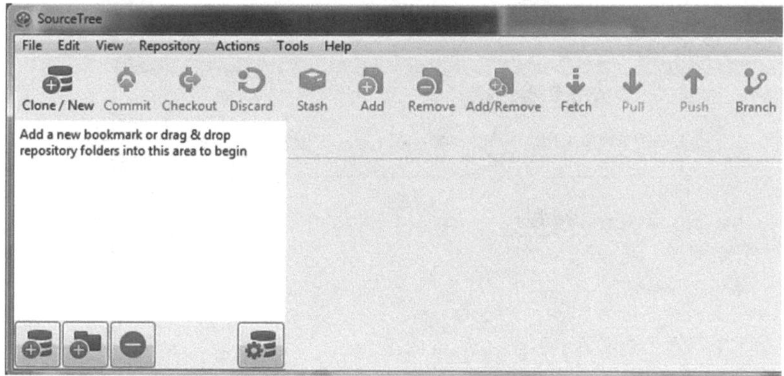

Figure 19.4 The default SourceTree window (in Windows)

You may be prompted for a username or password, but you can skip this step for now. Eventually, you'll see the main SourceTree window, which looks like figure 19.4 for Windows.

On the Mac, the initial screen looks like figure 19.5.

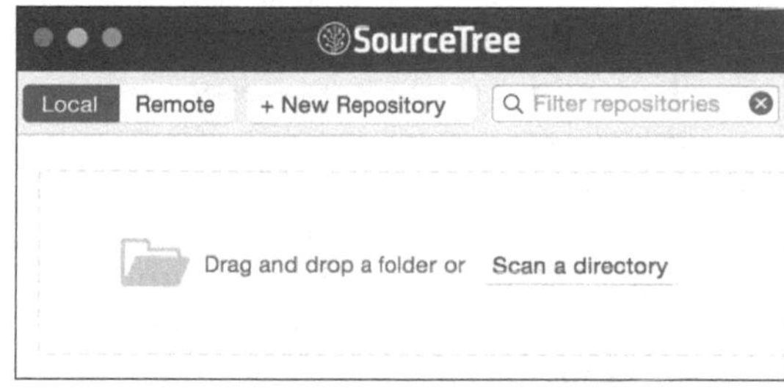

Figure 19.5 The default SourceTree window (in Mac)

What should be obvious right away is that SourceTree has a Windows- or Mac-specific look and feel, compared with the gitk and Git GUI tools. Next you'll need to add your existing math repository that you've been using throughout the book into SourceTree.

19.1.2 *Adding a repository into SourceTree*

You don't have to create repositories in SourceTree in order to use them in SourceTree. In this TRY IT NOW, you'll add an existing repository into SourceTree.

TRY IT NOW This exercise relies on the math directory that you've been using throughout the book. To follow along, start by downloading the make_math_repo.sh script from the book's website. The script will re-create the math directory, and you can then point SourceTree to it.

The make_math_repo.sh script does have an uncommitted change in the branch named another_fix_branch.

After you've re-created the math directory, continue with the appropriate section that follows for your platform.

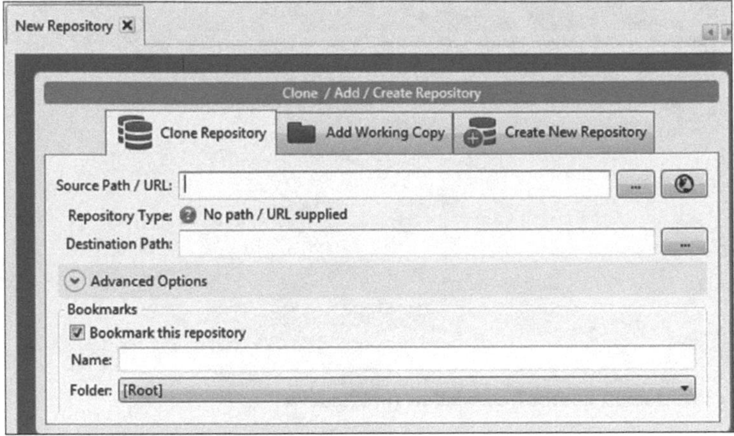

Figure 19.6 The New Repository tab

WINDOWS

Click the Clone/New button (in the upper-left window below the menu bar). A New Repository tab appears, as in figure 19.6.

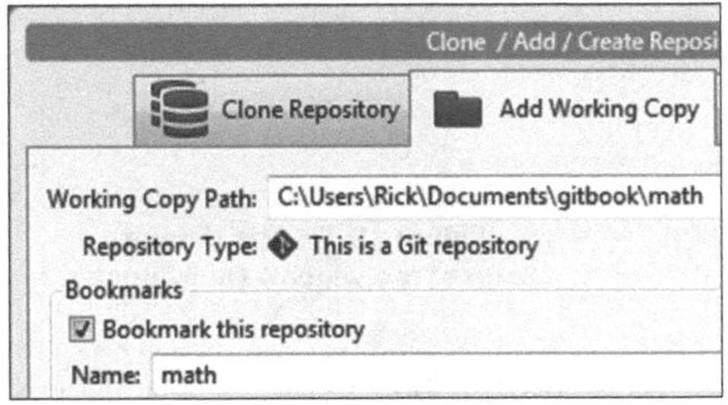

Figure 19.7 Choosing the math repository

To add your repository to SourceTree, click the Add Working Copy tab, and then browse to the math directory, as shown in figure 19.7. Remember that the math directory is your working directory, and that the Git repository is inside it.

Now click Add. Your math repository is shown in its current state in the SourceTree window. It should look roughly like figure 19.8. Notice the tab labeled *math* and its contents. If you open another repository, its content would appear inside its own

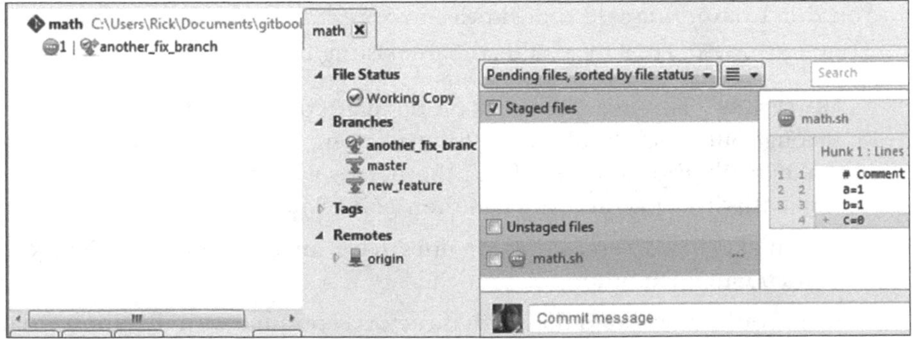

Figure 19.8 The standard SourceTree window, open to a repository

tab. If you're on Windows, you may see the leftmost pane showing any repositories that you've opened. This is the bookmarks pane, which is available only on the Windows version of SourceTree. In figure 19.8, only one repository is in this bookmarks pane.

MAC

Click the New Repository > Add Existing Local Repository option, shown in figure 19.9. (Alternatively, you could scan your home directory, and SourceTree will attempt to locate Git repositories from a specific directory.)

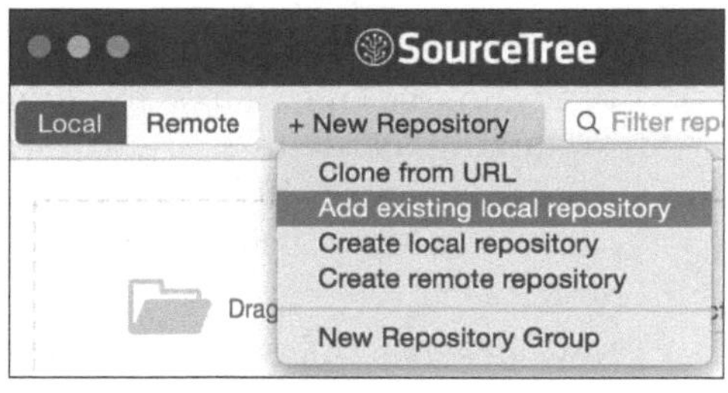

Figure 19.9 Adding a repository in the Mac version of SourceTree

From here, select the math repository. It then appears in your SourceTree window, as shown in figure 19.10. Double-click the repository to access the default repository view.

Figure 19.10 Your local repository in the Mac version of SourceTree

SourceTree detects the current state of your repository. If you followed the steps exactly, it recognizes that the repository is in the branch named another_fix_branch (shown previously in figure 19.8), and that the repository contains an unstaged file, math.sh.

19.1.3 Staging a file

The current SourceTree view is the File Status view, as indicated by the tabs at the bottom of SourceTree (figure 19.11).

Figure 19.11 The File Status view in the Windows version of SourceTree

On the Mac, the view is indicated by small icons at the top left of the main view (figure 19.12).

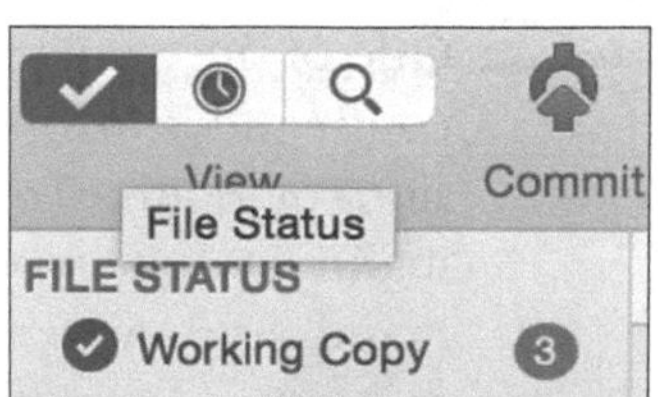

Figure 19.12 The File Status view in the Mac version of SourceTree

The icon next to math.sh is an indicator that the file has been modified (hover your mouse over it). As you learned in chapter 6, anytime Git detects that a file has been updated, it will be flagged as needing to be staged. (In this case, the last step of the make_math_repo.sh script is to modify a file.) You'll stage this file by performing the UI equivalent of `git add`.

TRY IT NOW In this exercise, you'll examine the contents of the math tab that shows the status of the files in your repository. On Windows, click View Show/Hide Bookmarks, which removes the bookmarks display to the left of the math tab.

On the math tab (shown previously in figure 19.8) SourceTree shows an information pane containing file statuses, branches, and any tags or remotes. Figure 19.13 shows this information pane on a Mac, against a newly re-created math directory.

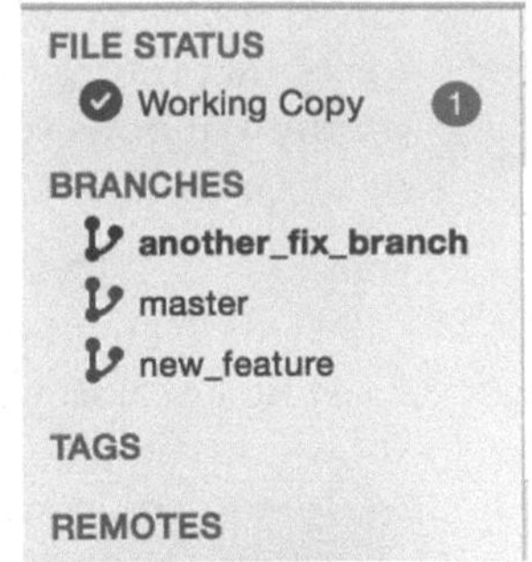

Figure 19.13 The SourceTree information pane

Now observe that the file view (figure 19.14) shows math.sh in the Unstaged Files section. To move this file to the Staged Files section, click the check box next to the filename to select it. This is the equivalent of the `git add` command.

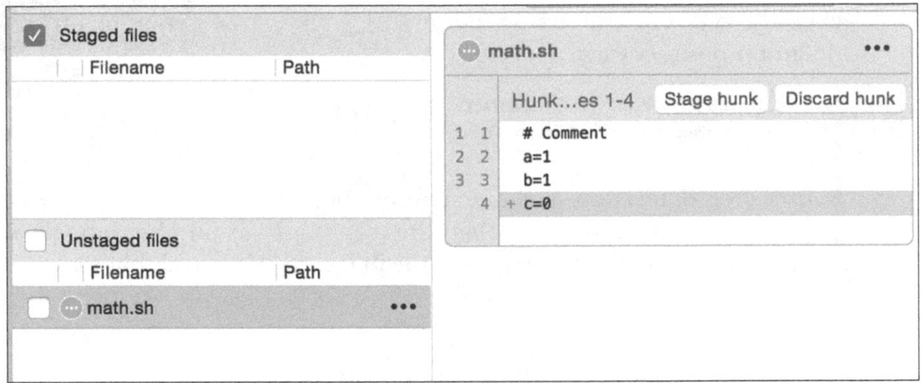

Figure 19.14 The file view

If you decide that you want to make further changes to the file, you can remove it from the staging area. To remove it from staging, click the check box next to the filename (to deselect it) when it's in the Staged Files section.

19.1.4 Tracking underlying Git commands in SourceTree

One of the challenges of working with any third-party tool is understanding which Git command is invoked with mouse-button presses and menu selections. In the Mac version of SourceTree, you can enable Show Command Output mode, which lets you see Git commands as they're invoked (see figure 19.15). This is one way to learn the tooling.

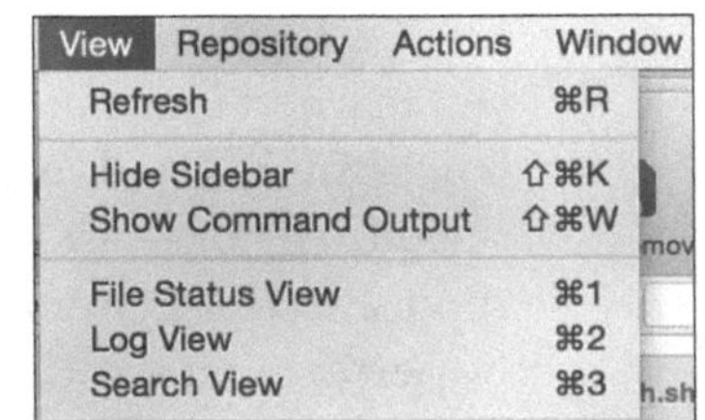

Figure 19.15 On the Mac, enabling Show Command Output will display Git commands as you operate.

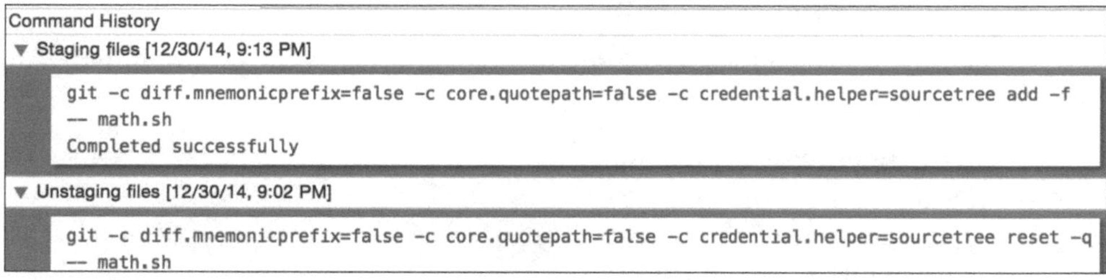

Figure 19.16 The Command History mode (on SourceTree for Mac)

Unfortunately, this mode doesn't exist in Windows as of this writing.

On the Mac, if you performed the preceding steps of moving a file into and out of the staging area, and you've enabled Command History mode, you'll see the `git add` and `git reset` commands being performed under the covers, as in figure 19.16.

TRY IT NOW Let's make sure the math.sh file is back in the staging area. If it isn't, click the check box next to math.sh to select this file in the Unstaged Files area. As in figure 19.17, you should see that the file appears in the staging area of the UI. The Unstaged Files section is empty.

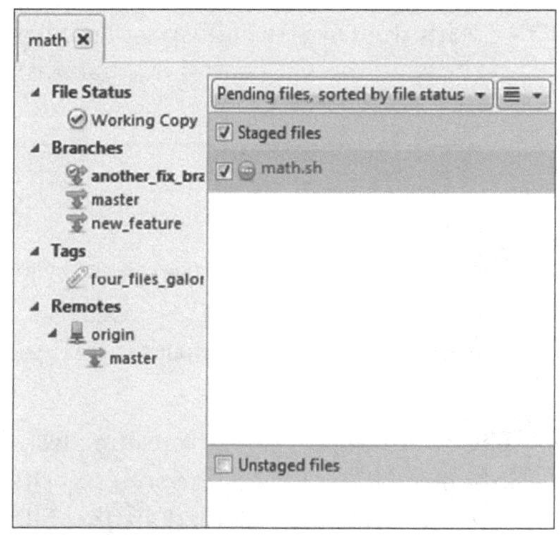

Figure 19.17 The staging area depicted in SourceTree

One other tracking option available for both Mac and Windows is displaying the console output. To enable this, open the Options dialog box. On Windows, you access this from the Tools menu, and on the Mac you use the standard Preferences menu. Select the General tab. Now enable the setting Always Display Full Console Output (figure 19.18). This displays the console output of most Git commands after they're run by SourceTree, which you'll see after you complete the next step.

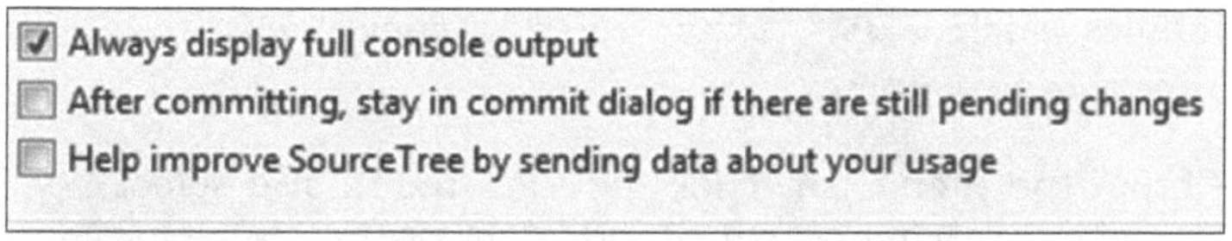

☑ Always display full console output
☐ After committing, stay in commit dialog if there are still pending changes
☐ Help improve SourceTree by sending data about your usage

Figure 19.18 Enable the Always Display Full Console Output option

19.1.5 Committing a file in SourceTree

Now that you've added a file to the staging area, you can commit that file to the repository.

TRY IT NOW Enter a commit log message in the text area at the bottom of the Staged Files section, as in figure 19.19. This commit log message is always available, and the editor is native for your platform.

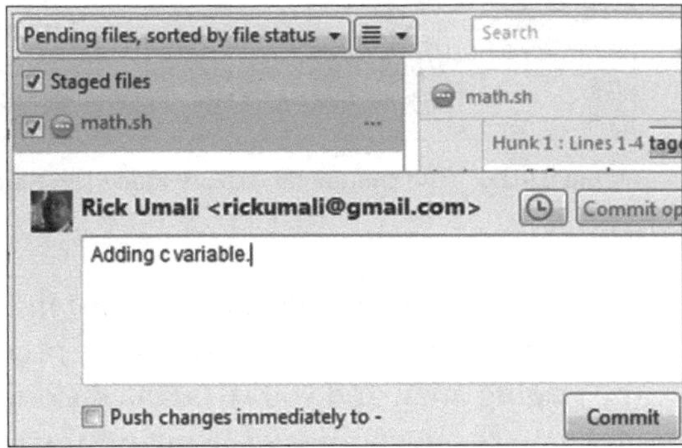

Figure 19.19 Entering a commit message in SourceTree

With the Display Full Console Output option enabled, you'll see a dialog box showing the Git command that was performed, as in figure 19.20.

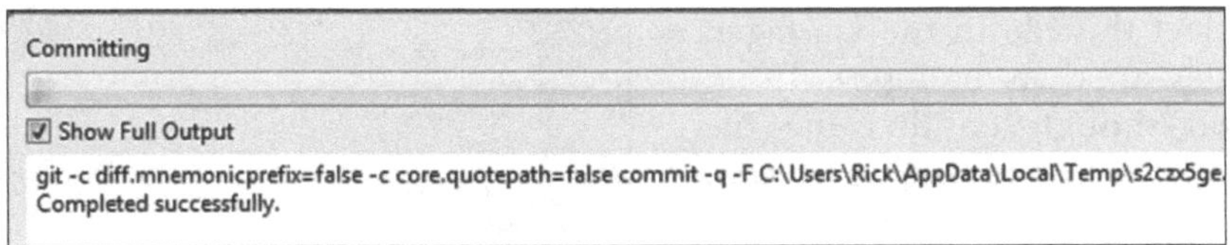

Figure 19.20 Console output for `git commit` in SourceTree

The common Git operations are always on display in the top ribbon of commands, as in figure 19.21. (The figure doesn't show all the options.) These commands will act on any files that are selected in the File Status view. Pay attention to the list of files selected before clicking any buttons!

Figure 19.21 The ribbon of Git commands

19.1.6 History view

Other Git capabilities (such as `git rebase` and `git reset`) are present in the history view. Let's examine this view now.

TRY IT NOW From the menu bar, click the View menu and select Log View. The main window pane switches to a view that shows the entire log history, as in figure 19.22.

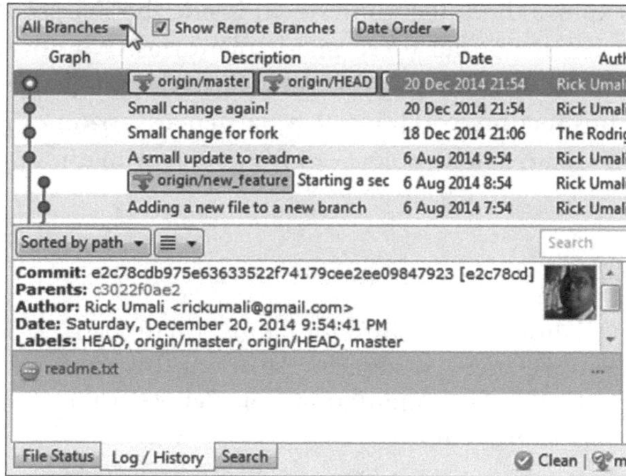

Figure 19.22 The log history view

This view is the equivalent of the `git log` output, which should be familiar. SourceTree shows all branches. If you access the context menu (via a right mouse-button click on Windows or Linux, or a two-finger click on the Mac), you'll see other Git operations, as in figure 19.23.

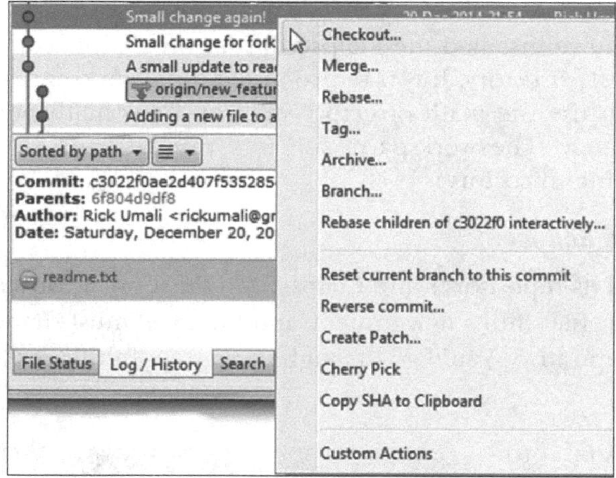

Figure 19.23 The context menu available from the history view

At this point, you could experiment on your own with the various capabilities such as tagging or checking out a particular SHA1 ID.

19.2 Git and the Eclipse IDE

An integrated development environment (IDE) is a natural place to incorporate Git. Some developers do all their work in an IDE, and having access to Git inside this tool makes sense.

In this section, you'll explore the integration of Git inside the Eclipse IDE. Eclipse is a popular IDE that can be run on all three major platforms: Windows, Mac, and Linux. Moreover, variants of the IDE are available for a variety of languages, including Java, PHP, and C/C++. In this section, you'll install a particular version of Eclipse into its own directory, and then create a new project pointing to your math repository.

19.2.1 Installing Eclipse

Eclipse is easy to install because all of its files go into one directory.

TRY IT NOW You'll install Eclipse C/C++ Development Tooling (CDT). Installing any variant of Eclipse is relatively straightforward, but this particular variant includes Git support. The documentation for Eclipse CDT is at the following site: https://eclipse.org/cdt/.

I'll defer to the documentation for details, but at a high level, the steps are as follows: download and install a current version of Java (Eclipse uses the Java runtime environment), and then download and install Eclipse CDT into its own directory. These two downloads may make this a long lunch.

Once you have Eclipse CDT installed, you can run the tool by double-clicking its icon or typing the following at the command line:

```
cd $HOME/eclipse-cdt
./eclipse
```

These steps assume you've installed the Eclipse files in the eclipse-cdt directory, inside your $HOME directory. If you're prompted for a workspace directory, make sure not to use the math directory yet. You'll explicitly add the math directory separately. The workspace directory can be your $HOME/workspace (or some other directory).

19.2.2 Adding a repository into Eclipse

To add your project (and its repository) into Eclipse, you must use a two-step process: you must first import your files into a new project, and then you must share your project, specifying your Git repository. You'll go through these steps in the next two TRY IT NOW sections.

TRY IT NOW The first step is to create a new project from the existing math directory. In Eclipse, files belong to projects. From the Eclipse menu, click the New Project item. As in figure 19.24, you'll see the New project wizard. Under the General folder, click Project.

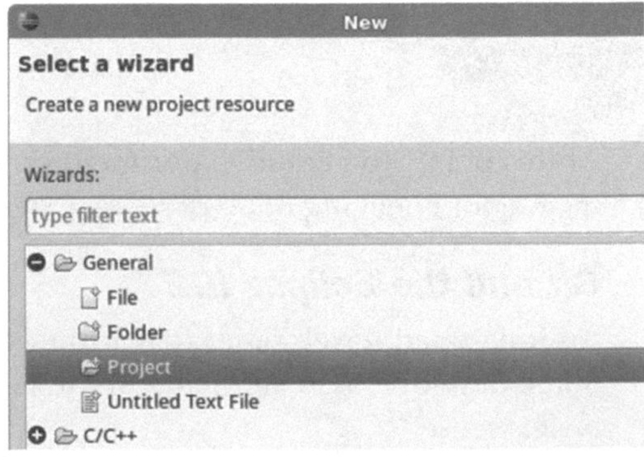

Figure 19.24 The New project wizard in Eclipse

Now click Next. The next dialog box that appears (figure 19.25) prompts you for the directory containing your project. Browse to the math directory.

Click Next. The math directory appears in the Project Explorer (the pane on the left side of Eclipse).

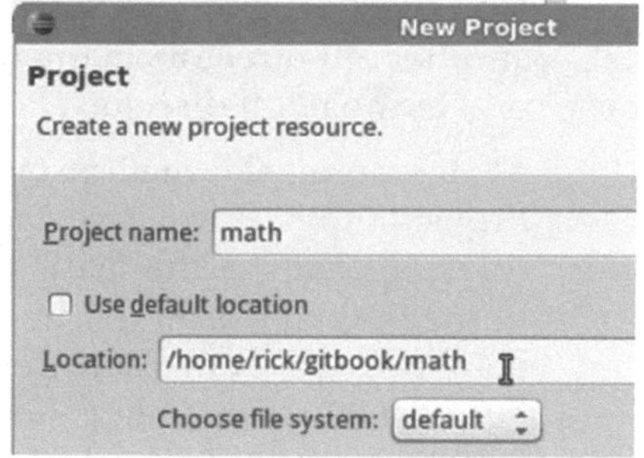

Figure 19.25 Associating your directory with the project

The project is now in Eclipse, as you can see in the Project Explorer. But in order to associate your math Git repository with this project, you have to share the project via the Team menu. This is the second step of the two-part process.

TRY IT NOW With your mouse over the math project in the Project Explorer, open the context menu (via a right mouse-button click on Windows or Linux, or a two-finger click on the Mac) and choose Team > Share Project (figure 19.26).

If you don't see the Share Project option, but instead the larger menu (shown later in this chapter, in figure 19.30), then you can jump ahead to figure 19.29 and confirm that your project looks like that image, and then move to the next section (section 19.2.3).

Figure 19.26 Sharing your project in Eclipse

This brings up another dialog box, prompting you for the version-control system. Select Git (figure 19.27). Eclipse's architecture is such that it can use other version-control systems, provided it's implemented as an Eclipse plugin, and that this plugin is installed.

On the next screen, you have to select the appropriate .git directory. Pick the one that is in the math directory. In figure 19.28, the dialog box displays two

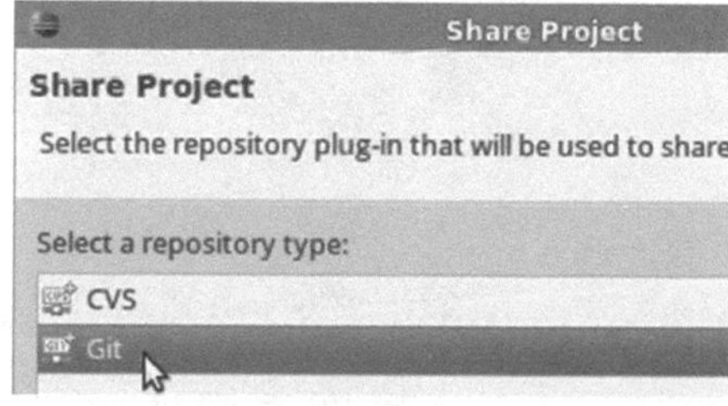

Figure 19.27 You can choose how to share your project in Eclipse.

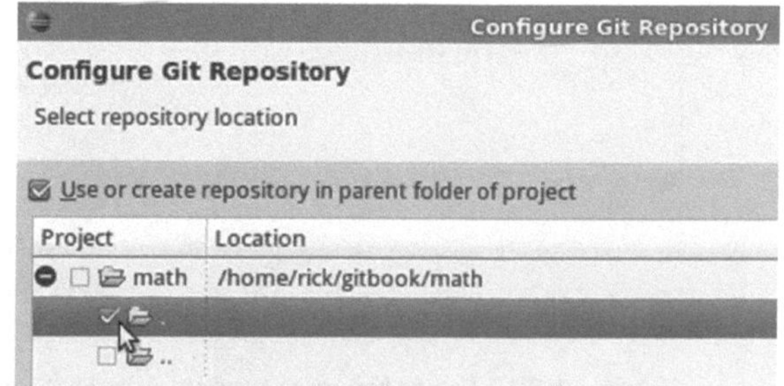

Figure 19.28 Picking the correct repository directory

.git directories, one in the same directory as math (indicated by the .), and another .git directory in the parent directory (indicated by the ..). Choose the one in the math directory.

At this point, the math project in the Project Explorer will have an indicator showing that it's the math repository in the branch master. (Your indicators may look different from those in figure 19.29, depending on which branch is currently checked out.)

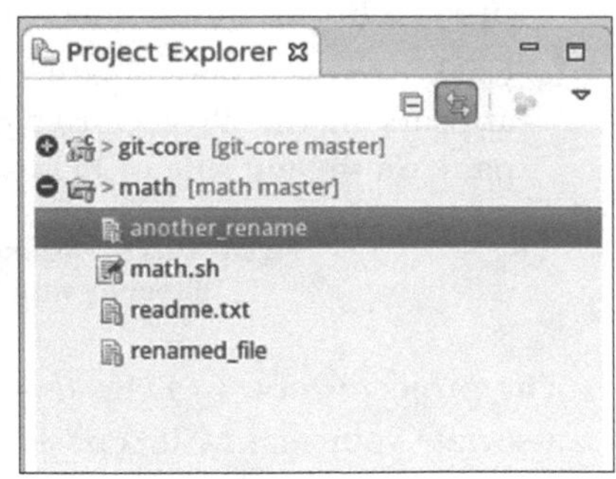

Figure 19.29 The project now has indicators showing that it's associated with the Git repository.

19.2.3 Staging and committing a file

Now that you've associated the repository to this project, you can perform the usual Git operations.

TRY IT NOW Let's modify a file and make a commit. You'll make these changes in the master branch, so let's switch branches inside Eclipse. Put your mouse on your project, and activate the context menu (via a right mouse-button click on Windows or Linux, or a two-finger click on the Mac). More menu items appear than before (compare figure 19.30 with figure 19.26).

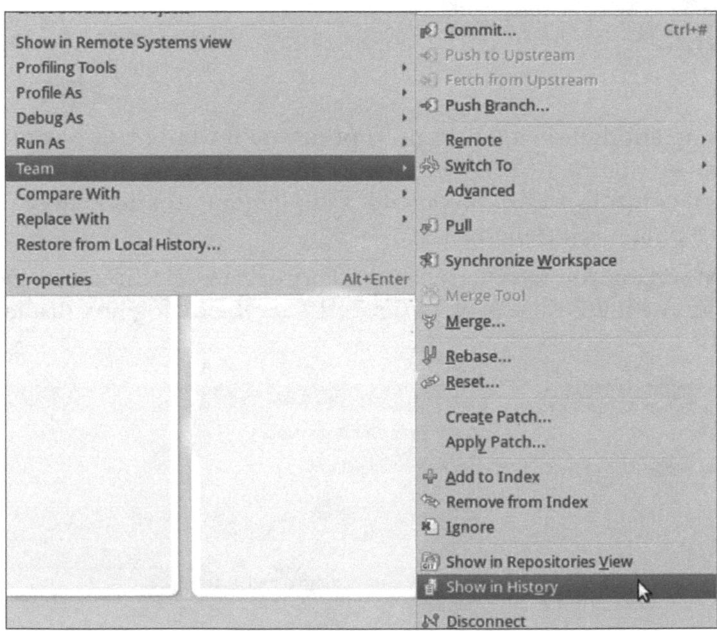

Figure 19.30 The new context menu, now that you've associated your project with the repository

Bring up the history view by clicking Team > Show in History, as shown in figure 19.30.

In the history view (indicated by the History tab), click the log entry message that contains the word *master*. Using the context menu of the history view (via a right mouse-button click on Windows or Linux, or a two-finger click on the Mac), select Checkout, as in figure 19.31. This is how to change Git branches in Eclipse.

If you don't see the branch master, you may need to turn on the Show All Branches and Tags toggle button, which is on the far-right side of the history view. This button is highlighted in figure 19.31.

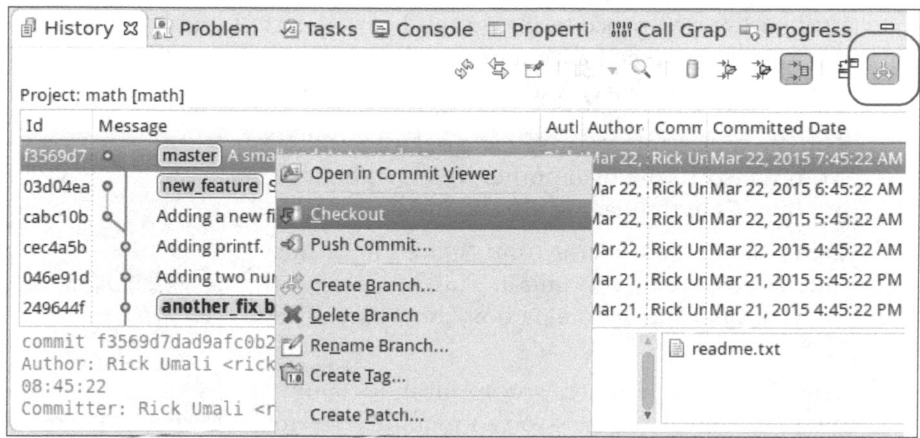

Figure 19.31 The context menu for the history view

The Project Explorer changes the text next to the math project to read *math master*. This indicates that you're in the master branch (figure 19.32). Now double-click the file named another_rename. The file opens in an editor window, with a tab containing its name (also shown in figure 19.32).

Enter some text in the editor. Notice that the file in the Project Explorer has a symbol next to its filename, indicating that it needs to be saved. Save the file now by clicking File > Save.

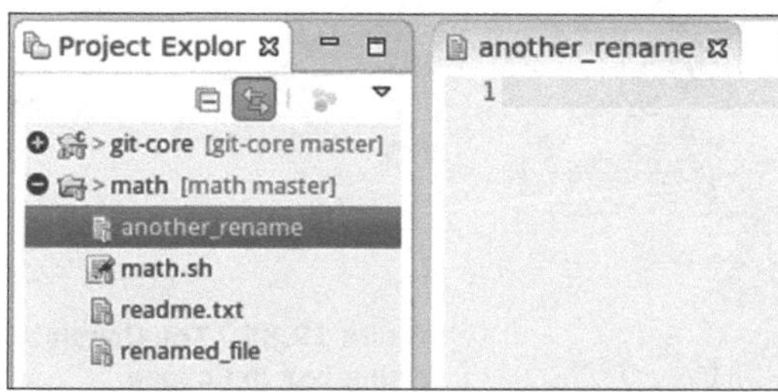

Figure 19.32 Your file inside the Eclipse editor (the filename is in the tab another_rename)

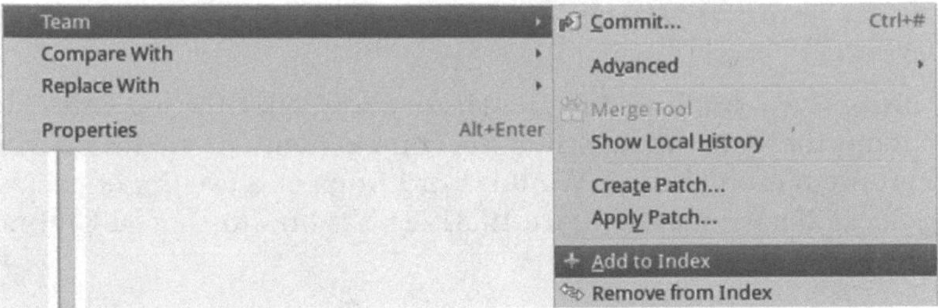

Figure 19.33 Adding a file to the staging area in the Eclipse IDE

Git recognizes that this saved file will need to be committed, using the by-now familiar two-step process: `git add` and `git commit`. To add the file to the staging area, use the context menu for the another_rename filename, and from the Team menu, select Add to Index (as in figure 19.33).

The Project Explorer window displays another symbol next to the file another_rename, indicating that it's in the staging area, as in figure 19.34.

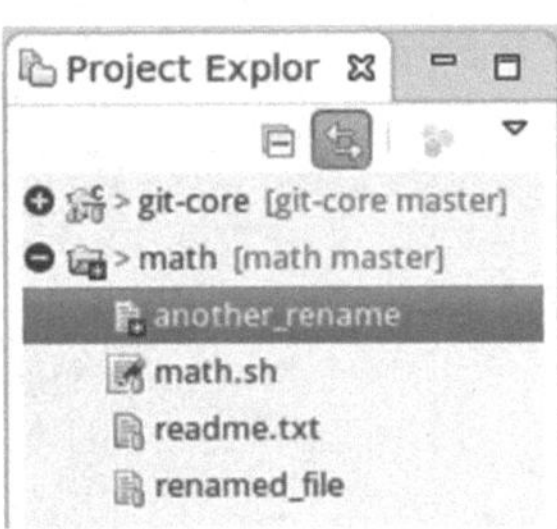

Figure 19.34 The icon next to another_rename indicates that it's in the staging area.

Using the context menu from figure 19.33, this time choose Team > Commit. This action opens the Commit Changes dialog box, shown in figure 19.35.

This dialog box requires you to fill in a commit message, and it requires you to pick the files to commit. Make sure that the filename another_rename has a check mark next to it. If

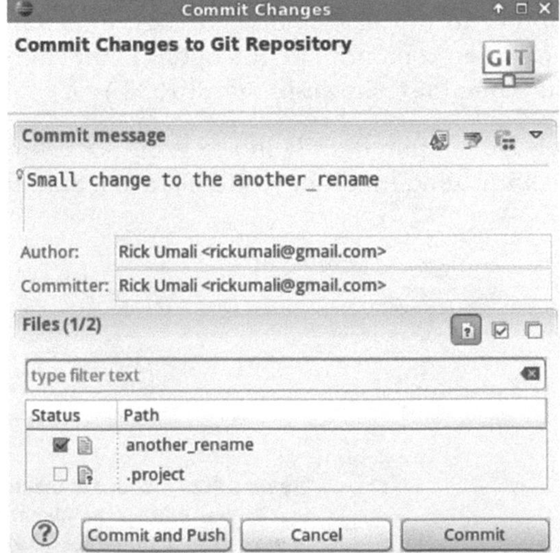

Figure 19.35 The Commit Changes dialog box in Eclipse

you see a filename .project, you can leave that unchecked. Click the Commit button to commit the change to the repository.

19.2.4 History view

Now that you've made changes to the repository, let's look at how the commit appears in the history view. This view is the equivalent of the `git log` command.

TRY IT NOW With the mouse over the math project in the Project Explorer, open the context menu (via a right mouse-button click on Windows or Linux, or a two-finger click on the Mac).

From the context menu, choose Team > Show in History. You'll see your project's commit log in the History tab of Eclipse, as in figure 19.36.

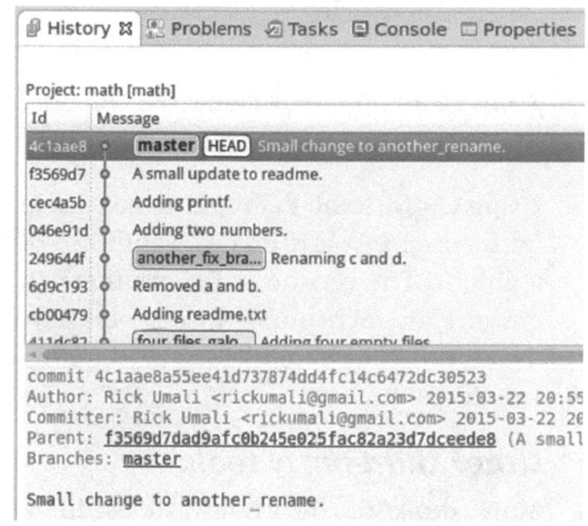

Figure 19.36 The history tab

If you click any of the commit messages in the History tab, and access the context menu Open in Commit Viewer, you'll see a dialog box that has two tabs of commit information: the commit message and the diff between the current version and the parent. Figure 19.37 and figure 19.38 show the commit message dialog box after selecting the second commit (for SHA1 ID `f3569d7d`).

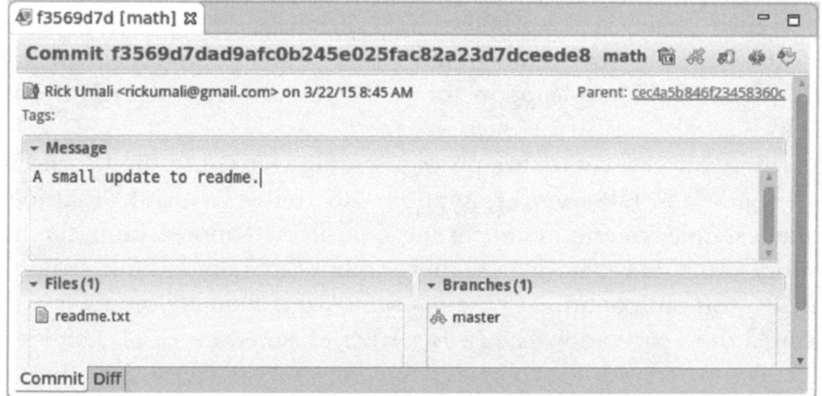

Figure 19.37 The commit message tab of the Commit detail window

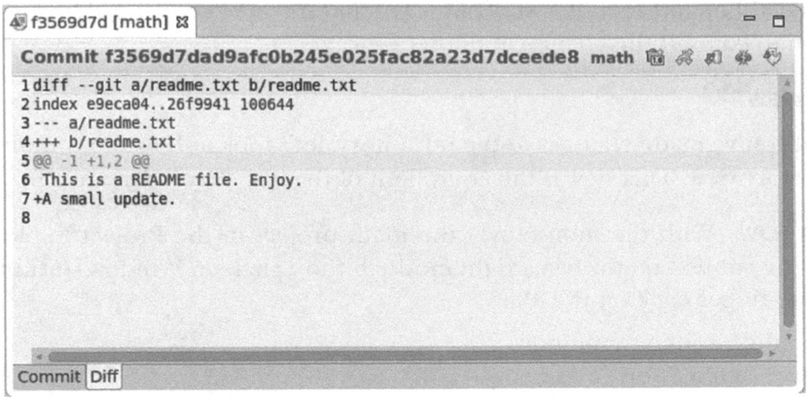

f3569d7d [math]

Commit f3569d7dad9afc0b245e025fac82a23d7dceede8 math

```
diff --git a/readme.txt b/readme.txt
index e9eca04..26f9941 100644
--- a/readme.txt
+++ b/readme.txt
@@ -1 +1,2 @@
 This is a README file. Enjoy.
+A small update.
```

Commit Diff

Figure 19.38 The diff of the Commit detail window

Using Git with the Eclipse IDE makes a lot of sense if your organization is standardized on Eclipse. If you use a different form of Eclipse, you can add Git into your Eclipse as a plugin. The version of Eclipse that you downloaded for this section has this Git plugin already incorporated, so if you have another version of Eclipse, explore how to install the EGit plugin (formally known as Team Provider). A similar plugin exists for IntelliJ IDEA.

19.3 Other third-party tools

Many more third-party tools exist for Git. For an up-to-date list, visit the following URL:

http://git-scm.com/downloads/guis

SourceTree and Eclipse are popular choices, for the reasons discussed in the opening section of this chapter. SourceTree might be easier to grasp initially than Eclipse, because it's closer in spirit to Git GUI and gitk. Eclipse, on the other hand, is first and foremost an IDE.

Your environment and workflow might dictate another choice. Microsoft's Visual Studio 2013 (an IDE for Windows) and Visual Studio Online (a cloud-based development environment) have support for Git. GitHub offers tooling for both Windows and Mac.

Learning any new Git tool requires orienting yourself to the UI, and relating it to the Git operations. Like your exploration with SourceTree and Git inside Eclipse, try to make a simple commit in an existing repository. Then examine the history of the repository. From there, move on to more complicated operations.

Finally, you can continue to use the standard command-line tools even if you opt for using a third-party tool. The GUIs are better suited for displaying log and file differences compared with the command line, but you can continue adding and committing with the command-line tools.

19.4 Lab

In this lab, you'll continue the exploration of SourceTree and Eclipse (with Git).

1 In SourceTree, create a new repository from scratch and add a file to that.
2 In SourceTree or Eclipse, switch between branches. Confirm that you're on the different branch by examining the `git branch` output on the command line of the math directory.

   It's important to know that you can use Git native tools in addition to using SourceTree or Eclipse.
3 SourceTree has built-in support for git-flow, which you learned about in chapter 17.

   To try out this support, click the Git Flow icon. The program prompts you for branch names and prefixes, and then converts the selected repository into one that uses git-flow. After the repository has been converted, click the Git Flow icon again, and you'll see the git-flow options to start a new feature, a new release, or a new hotfix. Try these, following along with what you did in section 17.3.

20 Sharpening your Git

At long last, you've reached the last chapter. This chapter covers the `git config` command. Its main benefit is to customize the Git commands and tools that you've learned over the past chapters. You've already used `git config` in earlier chapters to set or examine specific configurations such as what email address to use for commits (chapter 3) or how `git push` should behave (chapter 13). You've also used `git config` to create aliases (chapter 9). This chapter focuses on how to modify the behavior of other Git commands to suit your preferences.

Git is a tool, and keeping your Git sharp falls under the habit of always keeping your tools sharp. When your tools are sharp, they'll always be ready to use when the time comes. In the case of computer tools such as Git, the sharpness of the tool depends a lot on the knowledge and ability of the person using the tool.

20.1 Introducing the git config command

The `git config` command is the key command for advanced techniques with Git. This command lets you create aliases, modify the behavior of certain Git commands, and extend its capabilities. A great many Git features are controlled by Git configuration variables, which `git config` manipulates, so it's important to understand what these are, where they're stored, and how to change them.

20.1.1 Using Git configuration variables

Git configuration variables control how Git behaves. Some of these settings are cosmetic—for example, the color to display branches in the `git log` command. Some control Git's behavior, such as `push.default` (which you saw in chapter 13).

Let's use Git's help to see other configuration variables.

TRY IT NOW From your command line, type the following:

```
git config --help
```

Alternatively, to view the long list from the browser, visit this URL:

https://git-htmldocs.googlecode.com/git/git-config.html

You can see a wide variety of variables, from cosmetic configurations to behavioral ones. Do note that other Git configurations are described in their respective Git documentation. Before changing any settings, you should understand order of precedence for Git configurations.

20.1.2 Understanding Git configuration order of precedence

Git configuration has three levels: local, global, and system. Configurations that are set at the local level have the highest precedence, followed by global, followed by system. In the context of Git configuration, *local* means the current repository (repository-specific), *global* means global across all the repositories you have control over, and *system* means server-wide.

In figure 20.1, you have two users, mary and bob, on a server (the outermost box).

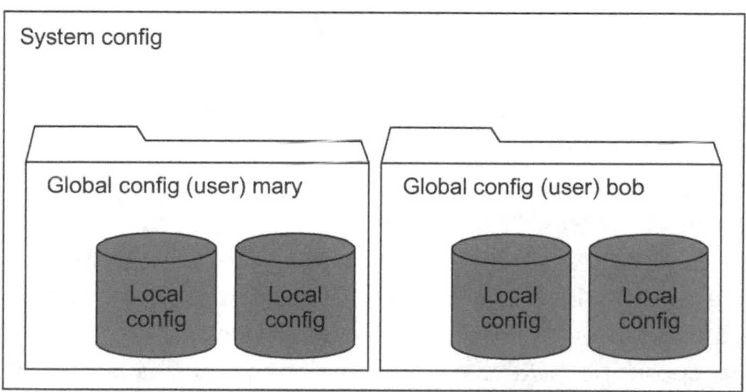

Figure 20.1 The levels of Git configuration

Both users (mary and bob) have their own global configuration (labeled *global config*) in figure 20.1. Any configuration at this level can be applied globally to the two repositories that each owns. Global configurations are the equivalent of user-specific configurations, but the word *global* has stuck from the earliest Git implementations.

Let's take a closer look at these configurations now.

TRY IT NOW Go into the math directory, and list the configurations in each of these areas by typing the following:

```
cd $HOME/math
git config --local --list
git config --global --list
git config --system --list
```

You should see output from at least two of these `git config` commands (the system configuration is often empty). For my computer, I see the session in the following listing.

Listing 20.1 Output from `git config --list` commands

```
> git config --local --list
core.repositoryformatversion=0
core.filemode=true
core.bare=false
core.logallrefupdates=true
core.ignorecase=true
core.precomposeunicode=true
> git config --global --list
user.name=Rick Umali
user.email=rumali@firstfuel.com
core.excludesfile=/Users/rumali/.gitignore_global
core.editor=vim
alias.lol=log --graph --oneline --all --decorate --simplify-by-decoration
color.ui=auto
color.diff=true
push.default=simple
gui.recentrepo=/Users/rumali/git-test
> git config --system --list
fatal: unable to read config file '/usr/local/etc/gitconfig': No such file
  or directory
```

Don't worry if your output doesn't look like listing 20.1. Do observe that your name and email are set in the global configuration. Also, depending on the state of your math repository, you may have branch configuration.

The error that you see in listing 20.1 (`unable to read config file`) for `git config --system --list` is typical, and it means that there are no server-level configurations. These would typically be configured by a system administrator, but don't be surprised if there aren't any on the system you're working on.

Visit the other repositories you've made in the course of this book and examine the local configuration. The following listing shows the `git config --local --list` output of my math.github repository (first created in chapter 12).

Listing 20.2 Output of `git config --local --list`

```
core.repositoryformatversion=0
core.filemode=true
core.bare=false
core.logallrefupdates=true
core.ignorecase=true
core.precomposeunicode=true
remote.origin.url=https://github.com/rickumali/math.git
remote.origin.fetch=+refs/heads/*:refs/remotes/origin/*
branch.master.remote=origin
branch.master.merge=refs/heads/master
```

Each configuration is specified as *name=value*. The name is typically separated by a period, as in `user.name=Rick Umali`. In Git, the string before the first period is considered the *section*, and the string after this period is the *key*. So in the configuration `user.name`, `user` is the section, and `name` is the key. See figure 20.2.

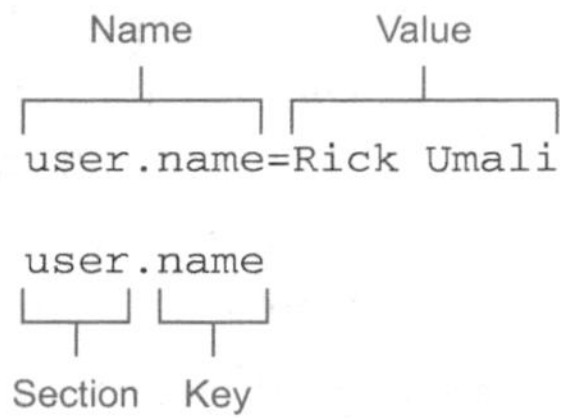

Figure 20.2 Visualizing the Git configuration name/value

When you have two periods (for example, `remote.origin.url` in listing 20.2), the period that separates the section and the key is the next-to-last period. So for `remote.origin.url`, the section is `remote.origin`, and the key is `url`. The string `origin`, in this case, is considered a subsection. The key is always the string after the last period (as in figure 20.3). This nomenclature of sections is necessary to understand Git config files, which you'll see later in this chapter.

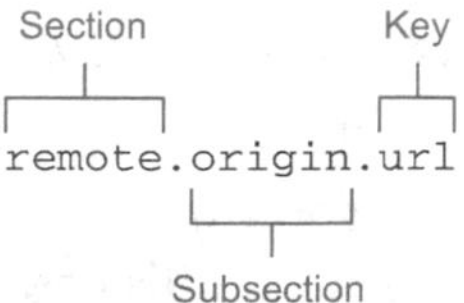

Figure 20.3 A Git configuration with a subsection

The `git config` help pages that you looked at in section 20.1.1 organize config variables into specific sections (for example, core, color, diff). Remember that the full name for a configuration variable is its section (or sections) and the key.

20.1.3 Setting Git configurations temporarily

In this section, you'll modify how the `git log` command displays the date and timestamp of each commit message. This is a cosmetic configuration, the easiest type of configuration to experiment with. Let's first set this configuration variable by using the `-c` Git command-line switch.

TRY IT NOW From the command line, type the following:

```
cd $HOME/math
git -c log.date=relative log -n 2
```

This should result in output like the following listing. (Remember, press Q to exit the pager.)

Listing 20.3 Output from `git -c log.date=relative log -n 2`

```
commit 7e8e188384ad0cd36af07f6035a1da7ac55b02cb
Author: Rick Umali <rickumali@gmail.com>
Date:   10 days ago

    Renaming c and d.

commit 2b12d3e602ea24d893b1c870d2f46f155d9dea11
Author: Rick Umali <rickumali@gmail.com>
Date:   10 days ago

    Removed a and b.
```

This output shows the date using the relative date format. The behavior of the `git log` command changed, because you applied a new configuration via the `-c` switch of the `git` command. The `-c` switch is a fast way to override a configuration quickly (see figure 20.4). Remember, `git -c` is not the `git config` command!

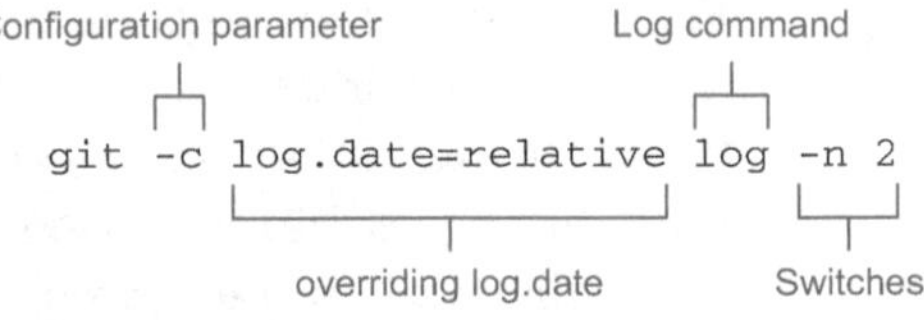

Figure 20.4 Changing the behavior of `git log` via the `-c` switch

Setting configurations in this way is a good technique for testing an unfamiliar configuration.

20.1.4 Setting Git configurations permanently

This configuration from the preceding section (`log.date=relative`) is temporary. If you run `git log` without the `-c` switch, you won't see dates in the relative format. To make this configuration change permanent, you must use the `git config` command to set the configuration. Let's do this now.

TRY IT NOW In the math repository, let's make the `log.date=relative` configuration permanent by typing the following:

```
git config --local log.date relative
```

After typing this, the configuration will be saved in the local Git configuration. Every time you access this repository, `git log` will use the relative date format. Type this:

```
git log -n 2
```

The output should look like listing 20.3. The `--local` switch to the `git config` command saves the configuration locally. Let's visit another repository—that math.github directory. If you don't have math.github in your $HOME directory, type the following to create it:

```
cd $HOME
git clone https://github.com/rickumali/math.git math.github
```

Now type this:

```
cd $HOME/math.github
git log -n 2
```

This output should show the dates in the default format, which is different from the other repository. Make sure to confirm this. *Local Git configuration* means local to the repository.

If you want to make this change applicable to any repository that you visit, use the `--global` switch instead of the `--local` switch. Remember, in Git parlance, `--global` means a configuration that can be applied globally to any repository you control. In practice, test switches by saving them locally first (the `--local` switch is the default, and it may be omitted), before promoting them to the global setting.

20.1.5 Resetting Git configurations

To go back to Git's default value for a particular configuration setting, you can unset the value. In the previous section, you set `log.date` to the value `relative`. Let's unset this, so you can view the dates in their original format.

TRY IT NOW In the math repository, let's reset the `log.date` configuration permanently by unsetting its value. Type the following:

```
git config --local --unset log.date
```

Now type this:

```
git log -n 2
```

The dates appear as they originally did.

20.2 Working with Git configuration files

Ultimately, Git configurations are stored in plain-text files. Table 20.1 lists the typical locations.

Table 20.1 Locations for Git configuration files

Location	Path	Notes
Local	$GIT_DIR/config	$GIT_DIR represents the working directory of your repository.
Global	$HOME/.gitconfig	This is in the home directory.
System	C:/Program Files (x86)/Git/etc/gitconfig (Windows) /Applications/Xcode.app/Contents/Developer/usr/etc/gitconfig (Mac) /etc/gitconfig (Unix/Linux)	This may vary, depending on how Git was installed.

If you stick with using the standard `git config` command, and use the `--local`, `--global`, and `--system` switches, you won't have to worry about knowing where these files are. But you might feel comfortable examining and even editing the config file.

20.2.1 Editing Git configuration files

Now that you know the location of the Git config files, you could edit them directly by using your favorite editor. If you're making a lot of changes, editing the file may be a good option. Some Git commands document their configuration by using the config file syntax, so editing the file directly will make the documentation easier to follow. In this section, you'll use the command line to start an editing session.

TRY IT NOW You'll work inside the math.github working directory. With the next commands, you'll go inside this directory to edit the local configuration file:

```
cd $HOME/math.github
git config --local --edit
```

This immediately opens the config file in the vi editor. The editor may be configured to show you the filename, shown in the bottom line (sometimes known as a *ruler*) in figure 20.5.

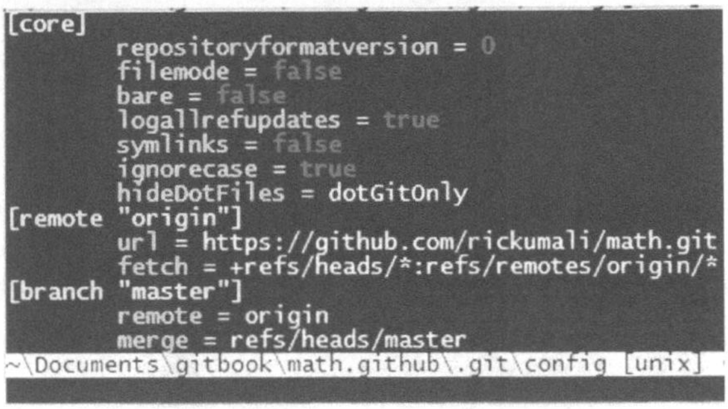

```
[core]
        repositoryformatversion = 0
        filemode = false
        bare = false
        logallrefupdates = true
        symlinks = false
        ignorecase = true
        hideDotFiles = dotGitOnly
[remote "origin"]
        url = https://github.com/rickumali/math.git
        fetch = +refs/heads/*:refs/remotes/origin/*
[branch "master"]
        remote = origin
        merge = refs/heads/master
~\Documents\gitbook\math.github\.git\config [unix]
```

Figure 20.5 Looking at the local configuration with `git config --edit`

You can exit the editor without saving anything by typing this:

```
:q!
```

This is a good practice if you're not sure you made any changes, and don't want to save what might be edited.

20.2.2 Using Git configuration file syntax

The Git config file has a particular syntax. The section is marked by brackets (for example, `[core]`), and some names have a subsection (for example, `[remote "origin"]`). The key and the value are then indented below their appropriate sections. The file format also supports comments, by placing a # or ; in front of any string on its own line, as in the following listing.

Listing 20.4 Comments in Git configuration files

```
#
# Comments
#

; User identity
[user]
        ; personal detail
        name = "Rick Umali"
        email = "rickumali@gmail.com"
```

TRY IT NOW Let's examine the other configurations. From the same math.github directory, type this:

```
git config --global --edit
```

For my Windows machine, I see figure 20.6.

```
[user]
        name = Rick Umali
        email = rickumali@gmail.com
[push]
        default = simple
[gui]
        recentrepo = C:/Users/Rick/Documents/gitbook
[help]
        autocorrect = -1
[core]
        autocrlf = true
        excludesfile = C:\\Users\\Rick\\Documents\\gitignore
~\.gitconfig [unix] (19:15 13/01/2015)
```

Figure 20.6 Looking at the global configuration inside with `git config --edit --global`

To exit, type the following:

```
:q!
```

Finally, let's look at the system configuration. On your machine, this command might not work, depending on whether you have a system Git configuration file. Type the following:

```
git config --system --edit
```

On my Windows machine, I see figure 20.7. (On my Mac and Ubuntu machines, I received an error because the system configuration file didn't exist. Your machine may behave differently!)

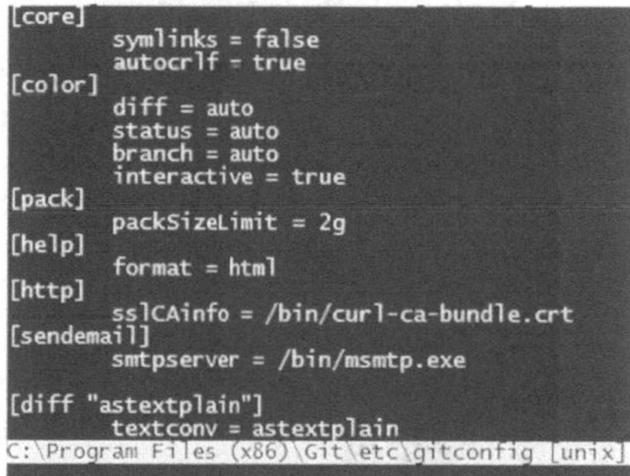

Figure 20.7 Looking at the system configuration inside with `git config --edit --system` (on Windows)

Clearly, a lot of configurations exist. You can look up all of these in the `git config` help file. On my Ubuntu machine, the file that the command opens is empty, as shown in figure 20.8. You can also see from the editor bar that it's a new file.

Make sure to exit the file:

```
:q!
```

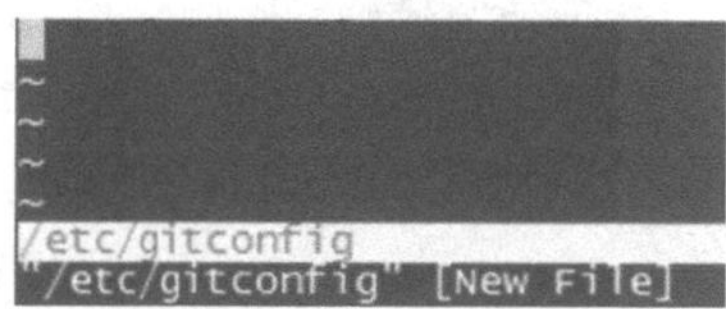

Figure 20.8 An empty Git system configuration file (on Ubuntu)

You won't edit the global or system configuration. Instead, let's examine one more big change: how to configure Git's default editor.

20.3 Configuring Git's default editor

During this entire book, you've been using the default editor that Git is configured with. But now that you've learned `git config`, you can override the default editor for a particular repository or globally (or even for the entire system) by configuring the `core.editor` configuration setting.

The `core.editor` configuration value is the name of an editor. It's typically already in your PATH, meaning that you can type its name, and it will start immediately. The value of the `core.editor` setting is substituted in the command line, as in the following listing.

Listing 20.5 How `core.editor` is used to edit a file

```
$(core.editor) temp_file
```

In the listing, the temp_file is the target file. This is usually a file containing a commit message, or one of the Git configuration files. Generally, if Git wants you to edit something, it will create a temporary file containing a template of what to edit, and then call the `core.editor` on that temporary file.

> **TRY IT NOW** From the math or math.github directory, type the following:
>
> ```
> cd $HOME/math
> git -c core.editor=echo config --local --edit
> ```
>
> If all goes well, you'll see a path returned to the screen, as in the following listing.

Listing 20.6 Output from fancy `git config` command

```
c:/Users/Rick/Documents/gitbook/math/.git/config
```

What just happened here? The breakdown of the command is in figure 20.9.

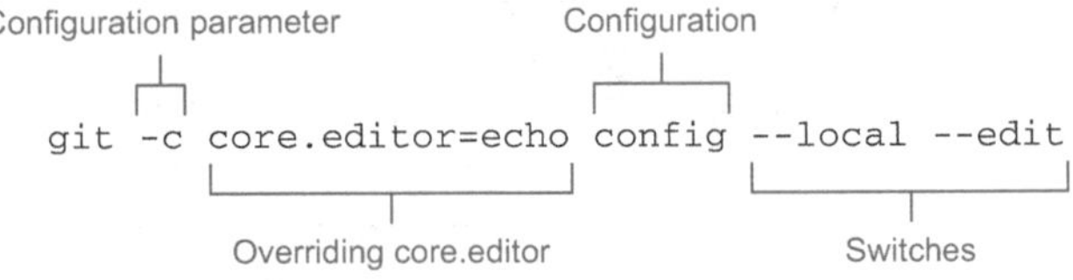

Figure 20.9 Overriding the `core.editor` via `git config`

You use the `-c` switch and an argument to that switch (`core.editor=echo`). In Git, the `-c` switch causes Git to override `core.editor` with the supplied value from the command line. In the case of the `git config` command, the name of the configuration file is passed to the `core.editor`, and since `core.editor` is set to the `echo` command, the resulting `git config` command prints the name of the file.

> **TRY IT NOW** Type the following to learn the location of the configuration files on your computer:
>
> ```
> git -c core.editor=echo config --global --edit
> git -c core.editor=echo config --system --edit
> ```

At this point, you have everything you need to change your `core.editor`. Two TRY IT NOW sections follow. One is for Windows users, and the second is for Mac and Unix/Linux users. Both sections do the same thing: change the `core.editor` into something else for your local repository.

TRY IT NOW (WINDOWS) If you're on a Windows machine, install Notepad++ from this URL:

http://notepad-plus-plus.org/download/v6.7.4.html

Once the executable is available, make sure that it starts properly. Now edit (perhaps using Notepad++) the command line's initialization file. This is either .bash_profile or .bashrc, in the $HOME directory. All the lines of the initialization file are executed every time you start Git BASH. Add this line at the bottom of the file:

```
PATH="$PATH:/c/Program Files (x86)/Notepad++"
```

Then save the file. The line appends the directory of Notepad++ to the end of the existing PATH. This is a standard technique. To make this change take effect, exit and then restart Git BASH. Now type this:

```
which notepad++
```

The response should look like the following listing.

Listing 20.7 Output of `which notepad++`

```
/c/Program Files (x86)/Notepad++/notepad++
```

Now you can take the intermediate step of setting the editor with the `-c` switch:

```
cd $HOME/math.github
git -c core.editor=notepad++ config --local --edit
```

You should see the editor appear (it's a GUI) with the local repository's configuration, as in figure 20.10.

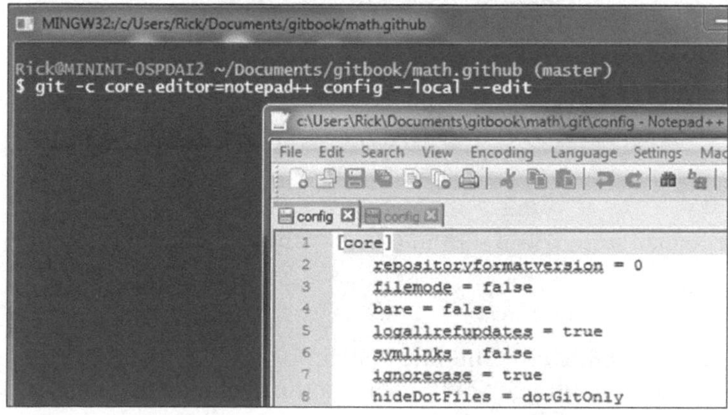

Figure 20.10 Using Notepad++ as the `git config` editor

Exit the editor (choose File > Exit). Now type the following to make this permanent for the current repository:

```
git config --local core.editor notepad++
```

The next time you're in this repository and you're making an edit that requires the `core.editor`, it will use Notepad++ instead of vi.

If you had difficulties with this, try putting the full path value in the `core.editor` setting. This skips the setting in the initialization file, reading the full path to Notepad++ from the config value. The Internet has resources on these kinds of configuration issues, but you can also visit the book's forum (from the book's website).

TRY IT NOW (MAC AND UNIX/LINUX) On the Mac or a Unix/Linux machine (such as Ubuntu), you'll change the `core.editor` from vi to nano. Nano is another text editor similar to vi, but perhaps slightly easier to learn.

A key assumption I'm making is that nano is already in your path. Type the following:

```
nano
```

You should see an editor window appear, similar to figure 20.11. The rows at the bottom of the window show how to operate the editor. (The ^ represents the Ctrl key.)

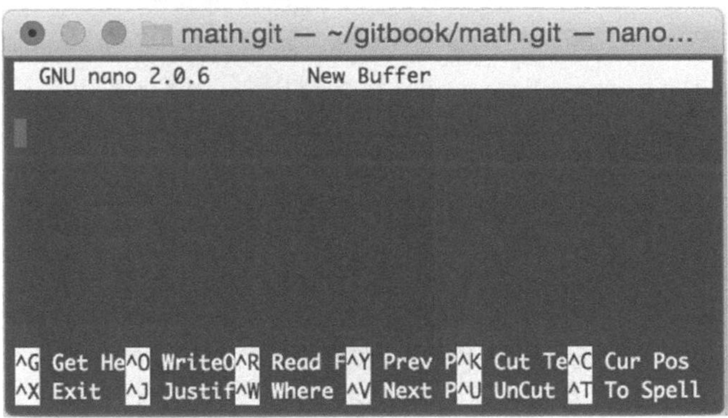

Figure 20.11 The nano editor

To exit, press Ctrl-X.

As an intermediate step, you'll set the editor with the `-c` switch:

```
cd $HOME/math.github
git -c core.editor=nano config --local --edit
```

You're using the `-c` Git command-line switch, for the `git config` command, as explained in section 20.1.3. You'll see something similar to figure 20.12.

```
math.github — ~/gitbook/math.github...
 GNU nano 2.0.6 File: ...h.github/.git/config

[core]
        repositoryformatversion = 0
        filemode = true
        bare = false
        logallrefupdates = true
        ignorecase = true
        precomposeunicode = true
        editor = nano
[remote "origin"]
        url = https://github.com/rickumali/math.git
        fetch = +refs/heads/*:refs/remotes/origin/*
[branch "master"]

^G Get He^O WriteO^R Read F^Y Prev P^K Cut Te^C Cur Pos
^X Exit   ^J Justif^W Where ^V Next P^U UnCut ^T To Spell
```

Figure 20.12 Editing files with nano

Exit the editor once again by pressing Ctrl-X.

Next, to make the nano editor the default editor for the current repository, type this:

```
git config --local core.editor nano
```

The next time you're in this repository and you're making an edit that requires the `core.editor`, it will use nano instead of vi.

20.4 Configuring files to ignore

Another common configuration is to enumerate the list of files that Git should ignore. These tend to be files that are generated (object files, output from scripts, and so forth) in your working directory, but that you don't want to store in the repository. To do this, you'll have to manipulate a Git ignore file.

In the previous chapter, for the installation of Atlassian SourceTree, you may have noticed the prompt about the global ignore file (figure 20.13).

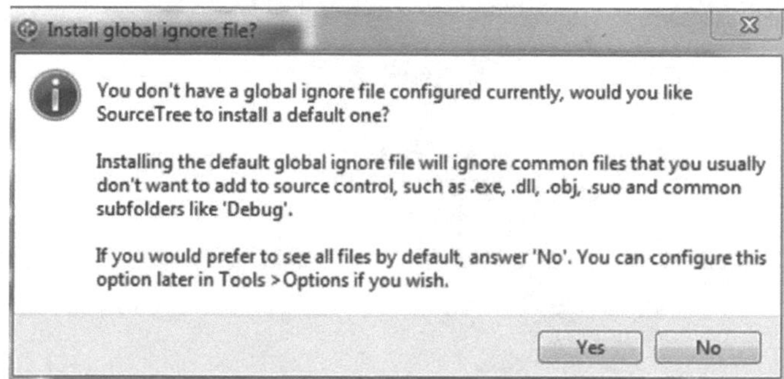

Figure 20.13 Global ignore file prompt, during Atlassian SourceTree installation

This configuration turns out to be a core configuration setting named `core.excludesfile`. This setting specifies a file containing a list of files to omit from source control. (This is a common requirement for projects that use languages that compile source files into object files. Typically, you don't want to store object files in version control.)

TRY IT NOW First, read the description of `core.excludesfile` in the `git config` help page, by typing this:

```
git config --help
```

The `core.excludesfile` entry on this help page reveals that the value of the `core.excludesfile` is used as a fallback file to the file .gitignore, which resides in your repository.

From the math repository, type the following to get the value of this configuration:

```
git config core.excludesfile
```

This command may show you a filename, or it may return nothing. Don't worry if you don't receive output. Continue reading to see what you might want to configure. If the preceding command does return a filename, type the following to display the contents of this file:

```
cat 'git config core.excludesfile'
```

This command-line technique retrieves the filename via the command in the backticks, and then displays the contents of that filename. Using this one-line command in Windows may show you the contents in the following listing.

Listing 20.8 Listing of `core.excludesfile` (on Windows)

```
#ignore thumbnails created by windows
Thumbs.db
#Ignore files build by Visual Studio
*.obj
*.exe
*.pdb
*.user
*.aps
*.pch
*.vspscc
*_i.c
*_p.c
*.ncb
*.suo
*.tlb
*.tlh
*.bak
*.cache
*.ilk
*.log
*.dll
*.lib
*.sbr
```

These lines are files that Git will ignore.

The page refers to the gitignore help page, which you can get to by typing this:

```
git help gitignore
```

The gitignore help page gives great guidance on which file you might use over another, so let's instead focus on the technique that .gitignore provides. The following TRY IT NOW will have you creating a .gitignore file in your math repository, and putting in a pattern that your repository will ignore.

TRY IT NOW Let's go into the math directory and create an empty file with an .obj extension. This file will represent your object file, which you don't want in the repository:

```
cd $HOME/math
touch file.obj
git status
```

The output of `git status` announces that a new file exists named file.obj. To make Git ignore this file, you'll create a .gitignore file in your directory. Type the following (being careful to put a period at the beginning of .gitignore):

```
echo "*.obj" > .gitignore
git status
```

Now `git status` will not report the file.obj file. But it does detect that the .gitignore file is a new file. At this point, you should commit the .gitignore file into the repository, so anyone else who clones the repository will ignore the same set of files.

20.5 Continually learning Git

You're at the end of the book, but it's safe to say that you're not finished learning about Git. "Manuals are most useful after you've used a product for a bit," is how author Larry Ullman sums it up. I'd add that most computer books fall under this guidance as well.

This book aims to go over all the techniques a beginner needs to know in order to competently use Git. But as you work with more varied repositories, or make more complicated changes, you'll need to do more than what is covered here. What follows are some techniques to help you keep on top of Git.

20.5.1 Work on a clone

Chapter 11 covered `git clone` extensively. If you find yourself in a confusing situation and aren't sure what Git operation to perform, make a clone of your repository and do your work on that clone first. This way, you can experiment without causing any damage to your working repository.

20.5.2 Work with the help

Git's documentation is still the authoritative help guide for Git. When you see techniques on the Internet that suggest using Git switches or commands that you're not sure of, look them up in the official Git help documentation first. Building up a good vocabulary obliges you to read with a dictionary by your side, and by extension, building up your Git knowledge requires you to access the `git help` command, so you can understand what you're typing. This technique, plus working on a clone of your work, is a good way to speed up your learning with Git.

20.5.3 Commit often

The more often you commit, the more confidence you'll have in being able to recover from any version-control issues. When files are committed into the repository, you'll always be able to get at them. And with `git rebase`, you can clear out these smaller intermediate commits before you publish your changes.

20.5.4 Collaborate

Be sure to collaborate with your fellow repository contributors. Because each repository comes with its own conventions, make sure you understand those conventions before you make contributions. Interacting with your co-contributors is key in this suggestion. They may have developed Git aliases or practices that are specific to their workflow.

To collaborate with other Git users outside a project, consider joining one of these Git groups:

- Git for Human Beings (Google Groups)
  https://groups.google.com/group/git-users
- Git Mailing List
  git@vger.kernel.org
  Details for joining this list are available at
  https://git.wiki.kernel.org/index.php/GitCommunity

Finally, I plan to participate on the book's website:
https://forums.manning.com/forums/learn-git-in-a-month-of-lunches

20.6 Lab

This is our last lunch together, so let's make this lab an easy one! Following are some questions to guide you along with `git config`:

1. In figure 20.1 (the bob and mary picture), how many Git configuration files are possible?
2. Using your favorite editor, add a fake section in your repository's configuration file, and add a few keys underneath it. Follow these steps:

```
% cd $HOME/math
% git config --add rick.set1 value1
% git config --add rick.set2 value2
```

   Using these commands, you've defined a section named `rick`. What `git config` switch will rename this section?

What `git config` switch will remove this section in one command?

3 In figure 20.6, one of the configurations is `help.autocorrect`. Look up what this does, and why this setting may be helpful. To try it in your local repository, type this:

```
% git config --local --add help.autocorrect -1
% git statsu
```

The misspelling in the preceding command is intentional!

4 If you installed Atlassian SourceTree and/or Eclipse, try examining their configuration-setting capabilities.

In SourceTree, choose Repository > Repository Settings. In the resulting dialog box, click Edit Config File (figure 20.14). This prompts you for a tool to edit the configuration.

Figure 20.14 Editing configurations with SourceTree

Pick an editor you're comfortable with. After you make that selection, the Git configuration file is loaded into the editor.

In Eclipse, you access the Git configuration settings via Window > Preferences. In the Preferences dialog box that appears, scroll down to the Team section, then the Git subsection, and click Configuration. You'll see figure 20.15. This dialog box offers a way to edit configuration values without resorting to an outside editor.

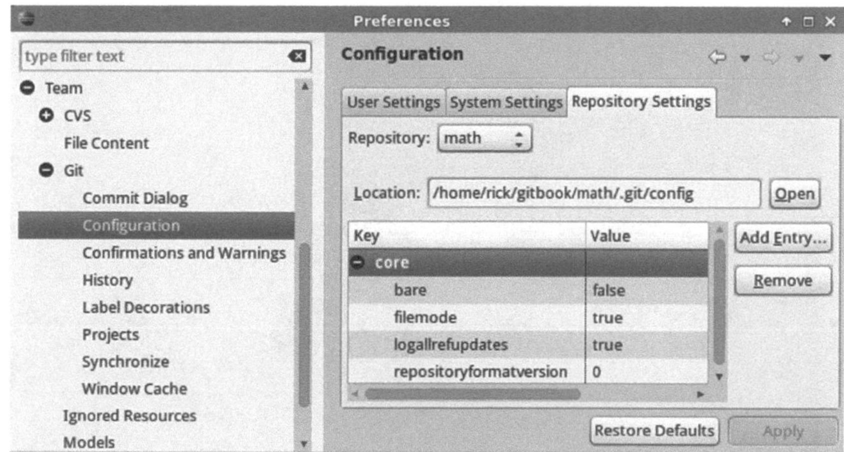

Figure 20.15 Editing configurations with Eclipse

20.7 Commands in this chapter

Table 20.2 Commands used in this chapter

Command	Description
`git config --local --list`	List the local (repository-specific) Git configuration.
`git config --global --list`	List the global (user-specific) Git configuration.
`git config --system --list`	List the system (server-specific) Git configuration.
`git -c log.date=relative log -n 2`	Show the last two commits using the relative date format.
`git config --local log.date relative`	Save the relative date format in the local Git configuration.
`git config --local --edit`	Edit the local (repository-specific) Git configuration.
`git config --global --edit`	Edit the global (user-specific) Git configuration.
`git config --system --edit`	Edit the system (server-specific) Git configuration.
`git -c core.editor=echo config --local --edit`	Print the name of the local Git configuration file.
`git -c core.editor=nano config --local --edit`	Edit the local Git configuration file using nano.
`git config core.excludesfile`	Print the value of the `core.excludesfile` Git configuration setting.

index

Git commands are indexed by their command name. To find "git clone," look for "clone command." Switches to the commands are indexed separately, followed by "switch." To find "- -bare," look for "- -bare switch."

Symbols

A

B

H

I

K

L

M

N

O

P

R

W

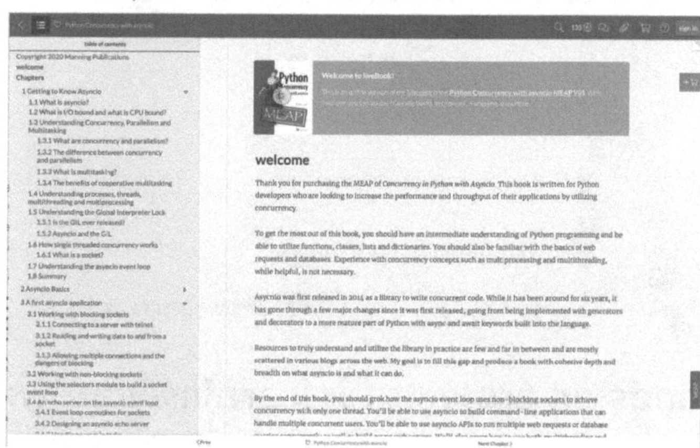

A new online reading experience

liveBook, our online reading platform, adds a new dimension to your Manning books, with features that make reading, learning, and sharing easier than ever. A liveBook version of your book is included FREE with every Manning book.

This next generation book platform is more than an online reader. It's packed with unique features to upgrade and enhance your learning experience.

- Add your own notes and bookmarks
- One-click code copy
- Learn from other readers in the discussion forum
- Audio recordings and interactive exercises
- Read all your purchased Manning content in any browser, anytime, anywhere

As an added bonus, you can search every Manning book and video in liveBook—even ones you don't yet own. Open any liveBook, and you'll be able to browse the content and read anything you like.*

Find out more at www.manning.com/livebook-program.

**Open reading is limited to 10 minutes per book daily*